THEODORE GRAY

THE Elements

A Visual Exploration of Every Known Atom in the Universe

Photographs by Theodore Gray and Nick Mann

BLACK DOG & LEVENTHAL PUBLISHERS NEW YORK

Distributed by
Workman Publishing Company
225 Varick Street
New York, NY 10014

Manufactured in China

Cover and interior design by Matthew Riley Cokeley.

Simulated atomic emission spectra by Nino Cutic based on data from NIST.

Other physical properties data from Wolfram *Mathematica*®; used with permission. All diagrams generated by *Mathematica*®.

ISBN-13: 978-1-57912-895-1

First paperback edition 2012

h g f e d c b a

Library of Congress Cataloging-in-Publication Data available on file.

ALL PHOTOGRAPHS BY NICK MANN AND THEODORE GRAY EXCEPT AS FOLLOWS:
Berkeley Seal © and TM 2001 UC Regents p. 222; Getty Images p. 15; courtesy of Hahn-Meitner Institute p. 230 (bottom right); istockphoto p. 155; courtesy of Lawrence Berkeley National Laboratory p.230 (top right and center right); courtesy of NASA, p.14; co[illegible] of NPL © Crown Copyright 2005; courtesy of Niels Bohr Archive, p.230 (bottom left); copyright © The Nobel Founda[illegible] 220, 226, 230 (top center), 232 (top center); Simon Fraser/Ph[illegible] Researchers, Inc. p. 149; courtesy of The University of Manchester p. 230 (center left); courtesy US Department of Energy p. 228; courtesy of the respective city or state 224, 230(center), 230 (bottom center), 232 (top left), courtesy Nicolaus Copernicus Museum, Frombork, Poland p. 232 (top right).

There is not anything which returns to nothing, but all things return dissolved into their elements.

–Lucretius, De Rerum Natura, 50 BC

THE PERIODIC TABLE is the universal catalog of everything you can drop on your foot. There are some things, such as light, love, logic, and time, that are not in the periodic table. But you can't drop any of those things on your foot.

The earth, this book, your foot—everything tangible—is made of elements. Your foot is made mostly of oxygen, with quite a bit of carbon joining it, giving structure to the organic molecules that define you as an example of carbon-based life. (And if you're not a carbon-based life-form: Welcome to our planet! If you have a foot, please don't drop this book on it.)

Oxygen is a clear, colorless gas, yet it makes up three-fifths of the weight of your body. How can that be?

Elements have two faces: their pure state, and the range of chemical compounds they form when they combine with other elements. Oxygen in pure form is indeed a gas, but when it reacts with silicon they become together the strong silicate minerals that compose the majority of the earth's crust. When oxygen combines with hydrogen and carbon, the result can be anything from water to carbon monoxide to sugar.

Oxygen atoms are still present in these compounds, no matter how unlike pure oxygen the substances may appear. And the oxygen atoms can always be extracted back out and returned to pure gaseous form.

But (short of nuclear disintegration) each oxygen atom can never itself be broken down or taken apart into something simpler. This property of indivisibility is what makes an element an element.

In this book I try to show you both faces of every element. First, you will see a great big photograph of the pure element (whenever that is physically possible). On the facing page you will see examples of the ways that element lives in the world—compounds and applications that are especially characteristic of it.

Before we get to the individual elements, it's worth looking at the periodic table as a whole to see how it is put together.

THE PERIODIC TABLE, this classic shape, is known the world over. As instantly recognizable as the Nike logo, the Taj Mahal, or Einstein's hair, the periodic table is one of our civilization's iconic images.

The basic structure of the periodic table is determined not by art or whim or chance, but by the fundamental and universal laws of quantum mechanics. A civilization of methane-breathing pod-beings might advertise their pod-shoes with a square logo, but their periodic table will have recognizably the same logical structure as ours.

Every element is defined by its atomic number, an integer from 1 to 118 (so far—more will no doubt be discovered in due time). An element's atomic number is the number of protons found in the nucleus of every atom of that element, which in turn determines how many electrons orbit around each of those nuclei. It's those electrons, particularly the outermost "shell" of them, that determine the chemical properties of the element. (Electron shells are described in more detail on page 12)

The periodic table lists the elements in order by atomic number. The sequence skips across gaps in ways that might seem quite arbitrary, but that of course are not. The gaps are there so that each vertical column contains elements with the same number of outer-shell electrons.

And that explains the most important fact about the periodic table: Elements in the same column tend to have similar chemical properties.

Let's look at the major groups in the periodic table, as defined by the arrangement of columns.

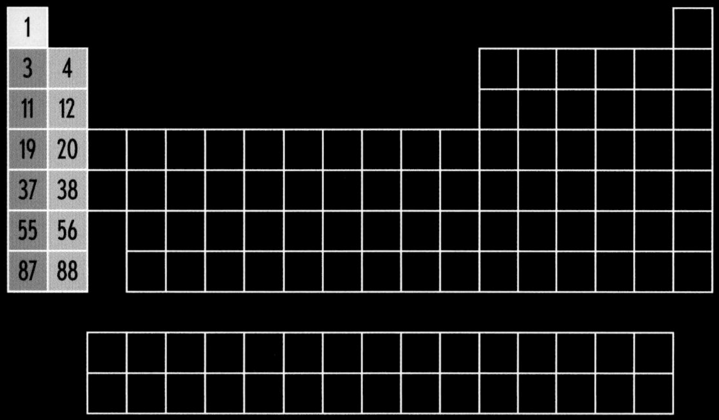

1	
3	4
11	12
19	20
37	38
55	56
87	88

THE VERY FIRST ELEMENT, hydrogen, is a bit of an anomaly. It's conventionally placed in the leftmost column, and it does share some chemical properties with the other elements in that column (principally the fact that in compounds, it normally loses one electron to form an H^+ ion, just as sodium, element 11, loses one electron to form Na^+). But hydrogen is a gas, while the other elements in the first column are soft metals. So some presentations of the periodic table isolate hydrogen in a category all its own.

The other elements of the first column, not counting hydrogen, are called the *alkali metals*, and they are all fun to throw into a lake. Alkali metals react with water to release hydrogen gas, which is highly flammable. When you throw a large enough lump of sodium into a lake, the result is a huge explosion a few seconds later. Depending on whether you took the right precautions, this is either a thrilling and beautiful experience or the end of your life as you have known it when molten sodium sprays into your eyes, permanently blinding you.

Chemistry is a bit like that: powerful enough to do great things in the world, but also dangerous enough to do terrible things just as easily. If you don't respect it, chemistry bites.

The elements of the second column are called the *alkali earth metals*. Like the alkali metals, these are relatively soft metals that react with water to liberate hydrogen gas. But where the alkali metals react explosively, the alkali earths are tamer—they react slowly enough that the hydrogen does not spontaneously ignite, allowing calcium (20), for example, to be used in portable hydrogen generators.

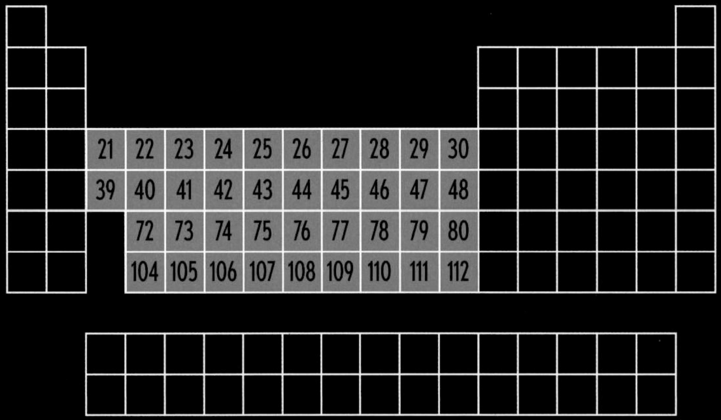

21	22	23	24	25	26	27	28	29	30
39	40	41	42	43	44	45	46	47	48
	72	73	74	75	76	77	78	79	80
	104	105	106	107	108	109	110	111	112

THE WIDE CENTRAL block of the periodic table is known as the *transition metals*. These are the workhorse metals of industry—the first row alone is a veritable who's who of common metals. All the transition metals except mercury (80) are fairly hard, structurally sound metals. (And so, in fact, is mercury, if you cool it enough. Mercury freezes into a metal remarkably like tin, element 50.) Even technetium (43), the lone radioactive element in this block, is a sturdy metal like its neighbors. It's just not one you'd want to make a fork out of—not because it wouldn't work, but because it would be very expensive and would slowly kill you with its radioactivity.

The transition metals as a whole are relatively stable in air, but some do oxidize slowly. The most notable example is of course iron (26), whose tendency to rust is by far our most destructive unwanted chemical reaction. Others, such as gold (79) and platinum (78), are prized for their extreme resistance to corrosion.

The two empty spots in the lower left corner are reserved for the lanthanide and actinide series of elements, highlighted on page 11. According to the logic of the periodic table, a fourteen-element-wide gap should appear between the second and third columns, with the elements of the lanthanide and actinide groups inserted in that gap. But because this would make the periodic table impractically wide, the convention is to close up that gap and display the rare earths in two rows at the bottom.

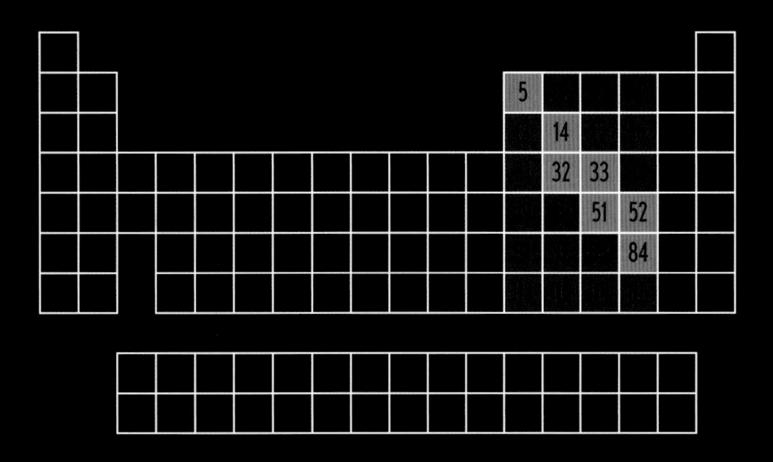

THE LOWER LEFT TRIANGLE here is known as the *ordinary metals*, though in reality most of the metals that people think of as ordinary are in fact transition metals in the previous group. (By now you may have noticed that the great majority of elements are metals of one sort or another.)

The upper right triangle is known as the *nonmetals*. (The next two groups, halogens and noble gases, are also not metals.) The nonmetals are electrical insulators, while all metals conduct electricity at least to some extent.

Between the metals and nonmetals is a diagonal line of fence-sitters known as the *metalloids*. These are, as you might expect from the name, somewhat like metal and somewhat not like metal. In particular they conduct electricity, but not very well. The metalloids include the semiconductors that have become so important to modern life.

The fact that this line is diagonal violates the general rule that elements in a given vertical column share common characteristics. Well, it's only a general rule—chemistry is too complicated for any rule to be absolutely hard and fast. In the case of the metal-to-nonmetal boundary, several factors compete with each other to determine whether an element falls into one camp or the other, and the balance drifts toward the right as you move down the table.

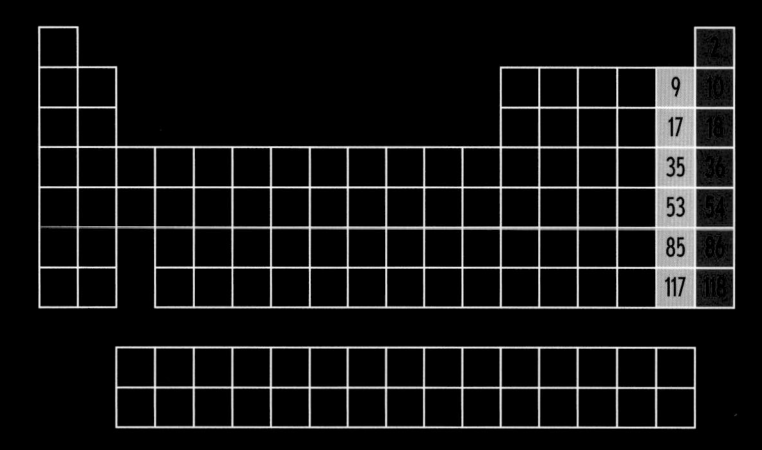

THE SEVENTEENTH (second-to-last) column is called the *halogens*, and its members are a pretty nasty lot in pure form. All the elements of this column are highly reactive, violently smelly substances. Pure fluorine (9) is legendary for its ability to attack nearly anything; chlorine (17) was used as a poison gas in World War I. But in the form of compounds such as fluoridated toothpaste and table salt (sodium chloride), the halogens are tamed for domestic use.

The very last column is the *noble gases*. Noble is used here in the sense of "above the business of the common riffraff." Noble gases almost never form compounds with each other or with any other elements. Because they are so inert, the noble gases are often used to shield reactive elements, since under a blanket of noble gas there's nothing for the reactive element to react with. If you buy sodium from a chemical supplier, it will come in a sealed container filled with argon (18).

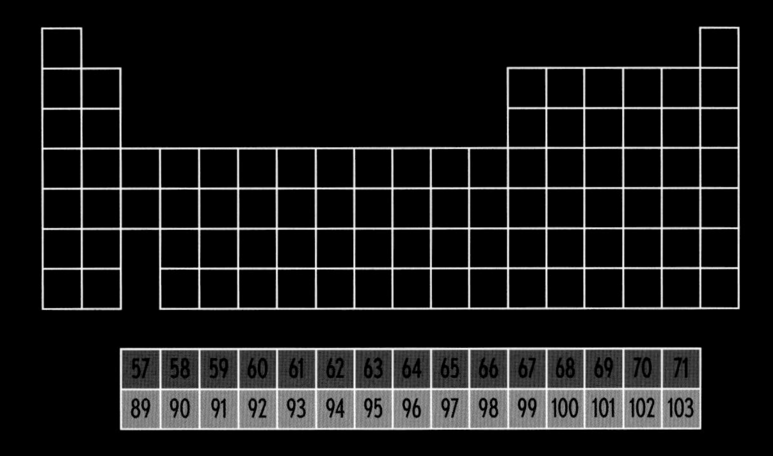

THESE TWO GROUPS are known collectively as the *rare earths*, despite the fact that some of them are not rare at all. The top row, starting with lanthanum (57), is known as the *lanthanides*; and you will not be surprised to learn that the bottom row, starting with actinium (89), is known as the *actinides*.

As you will read when you get to lutetium (71), the lanthanides are especially notorious for being chemically similar to each other. Some are so similar that people argued for years whether they were really separate elements at all.

All the actinides are radioactive, with uranium (92) and plutonium (94) being the most famous. Adding the actinides to the standard layout of the periodic table can be blamed on Glenn Seaborg, largely because he was responsible for discovering so many new elements in this range that a new row became necessary. (Although new elements have been discovered by many people, Seaborg is the only one forced to invent a row to display all of his discoveries.)

Now that we have seen the periodic table as a whole and in parts, we're ready to start our journey through the wild, beautiful, up-and-down, fun, and terrifying world of the elements.

This is all there is. From here to Timbuktu, and including Timbuktu, everything everywhere is made of one or more of these elements. The infinite variety of combinations and recombinations that we call chemistry starts and ends with this short and memorable list, the building blocks of the physical world.

Almost everything you see in this book is sitting somewhere in my office, except that one thing the FBI confiscated and a few historical objects. I had a great time collecting these examples of the vibrant diversity of the elements, and I hope you have as much fun reading about them.

See you at hydrogen!

How the Periodic Table Got Its Shape

HANG ON TIGHT, we're going to explain quantum mechanics in one page. (If you find this section too technical, feel free to skim it—there isn't going to be a quiz at the end.)

Every element is defined by its atomic number, the number of positively charged protons in the nucleus of every atom of that element. These protons are matched by an equal number of negatively charged electrons, found in "orbits" around the nucleus. I say "orbits" in quotes because the electrons are not actually moving around their orbits like planets around a star. In fact, you can't really speak of them as moving at all.

Instead, each electron exists as a probability cloud, more likely to be in one place than another, but not actually *in* any one place at any given time. The figures below show the various three-dimensional shapes of the probability clouds of electrons around a nucleus.

The first type, called an "s" orbital, is totally symmetrical—the electron is not any more likely to be in one direction than another. The second type, called a "p" orbital, has two lobes, meaning the electron is more likely to be found on one side or the other of the nucleus, and less likely to be found in any direction in between.

While there is only one "s"-type orbital, there are three "p" types, with lobes pointing in the three orthogonal directions (x, y, z) of space. Similarly there are five different types of "d" orbitals and seven different types of "f" orbitals, with increasing numbers of lobes. (You may think of these shapes as a bit like three-dimensional standing waves.)

Each shape of orbital can exist in multiple sizes, for example the 1s orbital is a small sphere, 2s is a larger sphere, 3s is larger still, and so forth. The energy required for an electron to be in any given orbital increases as the orbit becomes bigger. And all else being equal, electrons will always settle into the smallest, lowest-energy orbit.

So do all the electrons in an atom normally sit together in the lowest-energy 1s orbital? No, and here we come to one of the most fundamental discoveries in the early history of quantum mechanics: No two particles can ever exist in exactly the same quantum state. Because electrons have an internal state known as "spin," which can be either up or down, it turns out that exactly two electrons can reside in a given orbital—one with spin up and one with spin down.

Hydrogen has only one electron, so it sits in the 1s orbital. Helium has two, and they both fit into 1s, filling it to its capacity of two. Lithium has three, and since there is no room in 1s anymore, the third electron is forced to sit in the higher-energy 2s orbital. And so on— the orbitals are filled one at a time in order of increasing energy.

Look at the Electron Filling Order diagram on the right side of any element page in this book, and you'll see a graph of the possible orbitals from 1s to 7p, with a red bar indicating which ones are filled with electrons (7p is the orbital of highest energy occupied by electrons of any known element). The exact order in which orbitals are filled turns out to be surprisingly subtle and complex, but you can watch it happen as you flip through the pages of this book. Pay particular attention around gadolinium (64)—if you think you've got it figured out, your confidence might be shaken by what happens there.

It is this filling order that determines the shape of the periodic table. The first two columns represent electrons filling "s" orbitals. The next ten columns are electrons filling the five "d" orbitals. The final six columns are electrons filling the three "p" orbitals. And last but not least, the fourteen rare earths are electrons filling the seven "f" orbitals. (If you're asking yourself why helium, element 2, is not above beryllium, element 4, congratulations—you're thinking like a chemist rather than a physicist. Eric Scerri's book, referenced in the bibliography, is a good start toward answering such questions.)

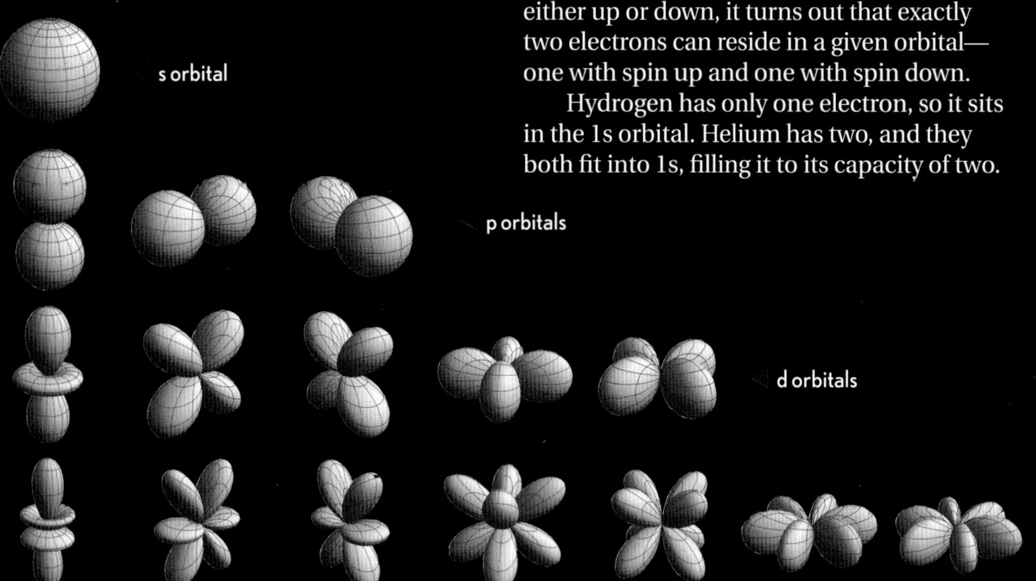

Elemental

Everything you need to know. Nothing you don't.

NAVIGATION TABLE
The mini table on every element page has one highlighted yellow square to show you where that element is located on the periodic table. The colors divide the table into the groups described on the preceding pages.

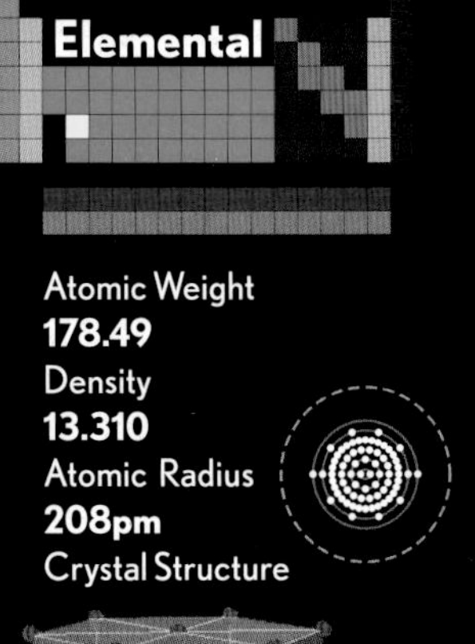

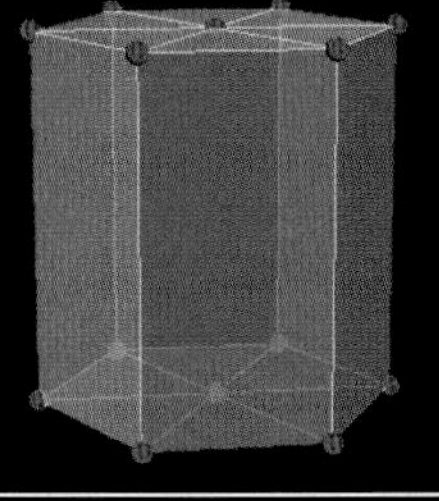

ATOMIC WEIGHT An element's atomic weight (not to be confused with its atomic number) is the average weight per atom in a typical sample of the element, expressed in "atomic mass units," or amu. The amu is defined as 1/12 the mass of a ^{12}C atom. Roughly speaking, one amu is the mass of one proton or one neutron, and thus an element's atomic weight is approximately equal to the total number of protons and neutrons in its nucleus.

However, you will notice that the atomic weights of some elements fall well between whole integers. When typical samples of an element contain two or more naturally occurring isotopes, the averaging of isotopic weights explains the fractional amu. (Isotopes are explained in more detail under protactinium, element 91; the basic idea is that an element's isotopes all have the same number of protons, and thus the same chemistry, but differ in the numbers of neutrons in their nuclei).

DENSITY The density of an element is defined as the idealized density of a hypothetical flawless single crystal of the absolutely pure element. This can never be realized exactly in practice, so the densities are generally calculated from a combination of the atomic weight and x-ray crystallographic measurements of the spacing of atoms in crystals. The density is given in units of grams per cubic centimeter.

ATOMIC RADIUS The density of a material depends on two things: how much each atom weighs, and how much space each atom takes up. The atomic radius shown for each element is the calculated average distance to the outermost electrons from the nucleus in picometers (trillionths of a meter). The diagrams are merely schematic—they represent all the electrons in their respective electron shells, with the overall size matching the size of the atom, but the position of individual electrons is not to scale, nor do electrons actually exist as sharp points spinning around the atom. The dashed blue reference circle shows the radius of the largest of all atoms, cesium (55).

CRYSTAL STRUCTURE The crystal structure diagram shows the arrangement of atoms (the unit cell that is repeated to form the whole crystal) when the element is in its most common pure crystalline form. For elements that are normally gas or liquid, this is the crystal form they take on when they are cooled enough to freeze solid.

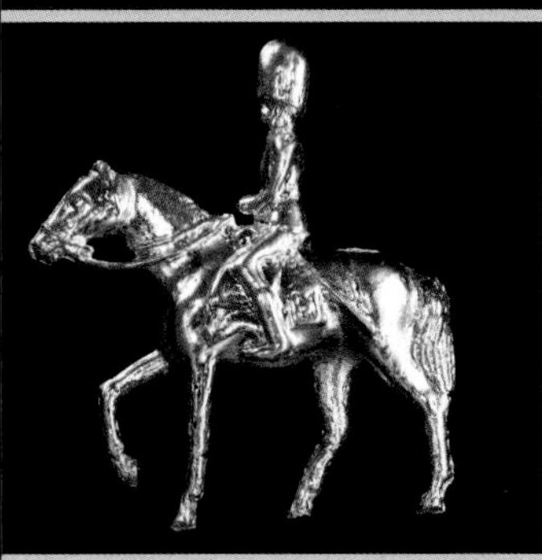

ELECTRON FILLING ORDER
This diagram shows the order in which electrons fill the available atomic orbitals, which are explained in detail on the preceding page.

ATOMIC EMISSION SPECTRUM
When atoms of a given element are heated to very high temperatures, they emit light of characteristic wavelengths, or colors, which correspond to the differences in energy levels between their electron orbitals. This diagram shows the colors of these lines, each one corresponding to a particular energy-level difference, arranged into a spectrum from the barely visible red at the top to the nearly ultraviolet at the bottom.

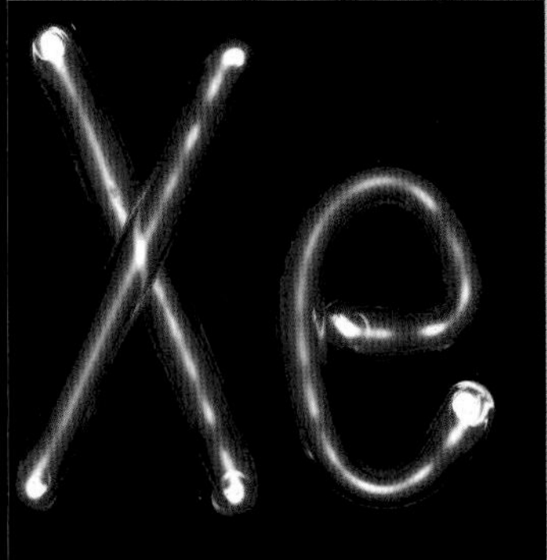

STATE OF MATTER
This temperature scale in degrees Celsius shows the range of temperatures over which the element is solid, liquid, or gas. The boundary between solid and liquid is the melting point, while the boundary between liquid and gas is the boiling point. Twist the pages of the book to spread the edges of the pages out, and you will see a graph of the melting and boiling points, which shows very pronounced trends across the periodic table.

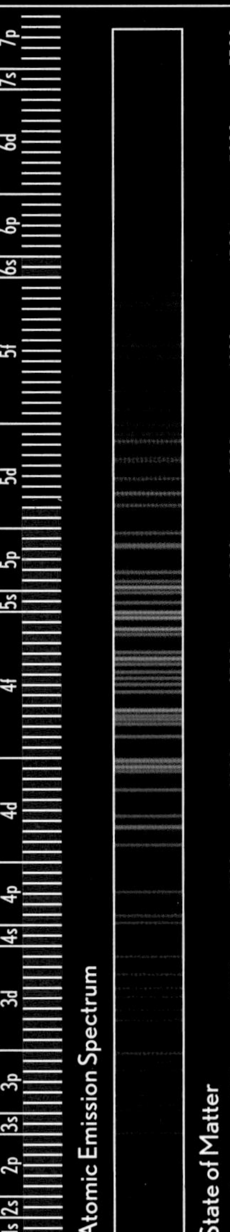

Hydrogen
H
1

Hydrogen

STARS SHINE BECAUSE they are transmuting vast amounts of hydrogen into helium. Our sun alone consumes six hundred million tons of hydrogen per second, converting it into five hundred and ninety-six million tons of helium. Think about it: Six hundred million tons *per second*. Even at *night*.

And where does the other four million tons per second go? It's converted into energy according to Einstein's famous formula, $E=mc^2$. About three-and-a-half-pounds-per-second's worth finds its way to the earth, where it forms the light of the dawn rising, the warmth of a summer afternoon, and the red glow of a dying day.

The sun's ferocious consumption of hydrogen sustains us all, but hydrogen's importance to life as we know it begins closer to home. Together with oxygen it forms the clouds, oceans, lakes, and rivers. Combined with carbon (6), nitrogen (7), and oxygen (8), it bonds together the blood and body of all living things.

Hydrogen is the lightest of all the gases—lighter even than helium—and much cheaper, which accounts for its ill-advised use in early airships such as the *Hindenburg*. You may have heard how well that went, though in fairness the people died because they fell, not because they were burned by the hydrogen, which in some ways is less dangerous to have in a vehicle than, say, gasoline.

Hydrogen is the most abundant element, the lightest, and the most beloved by physicists because, with only one proton and one electron, their lovely quantum mechanical formulas actually work exactly on it. Once you get to helium with two protons and two electrons, the physicists pretty much throw up their hands and let the chemists have it.

Elemental

Atomic Weight
1.00794
Density
0.0000899
Atomic Radius
53pm
Crystal Structure

Electron Filling Order
Atomic Emission Spectrum
State of Matter

The mineral scolecite, $CaAl_2Si_3O_{10}\cdot 3H_2O$, from Puna, Jalgaon, India.

The inside of a high-speed thyratron, a type of electronic switch filled with a small amount of hydrogen gas.

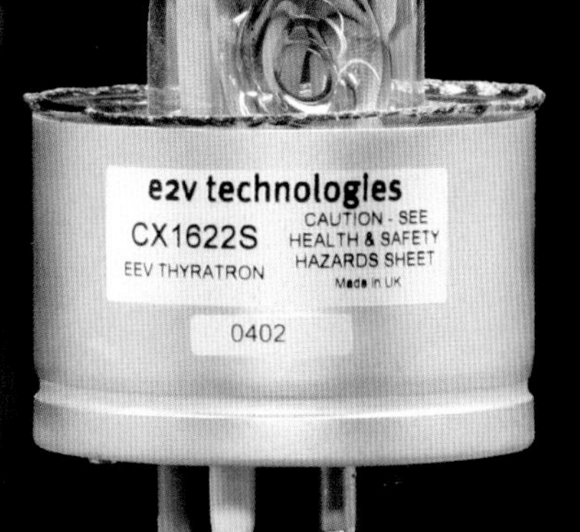

Tritium (^3H) luminous key chain, illegal in the U.S. because it is deemed a "frivolous" use of this strategic material.

The orange-red glow of an oxygen-hydrogen flame.

Tritium watches, on the other hand, are legal in the U.S.

The sun works by turning hydrogen into helium.

By weight, 75 percent of the visible universe is hydrogen. Ordinarily it is a colorless gas, but vast quantities of it in space absorb starlight, creating spectacular sights such as the Eagle Nebula, seen here by the Hubble Space Telescope.

Helium

2

Helium

HELIUM IS NAMED for the Greek god of the sun, Helios, because the first hints of its existence were dark lines in the spectrum of sunlight that could not be explained by the presence of any elements known at the time.

It might seem a paradox that an element common enough to fill party balloons with was the first element to be discovered in space. The reason is that helium is one of the noble gases, so named because they do not interact with the common riffraff of elements, remaining inert and aloof to nearly all chemical bonding. Because it does not interact, helium could not easily be detected by conventional wet chemical methods.

As a replacement for hydrogen in airships, helium, which is completely nonflammable, has much to recommend it. The main problem is that it's a lot more expensive, and provides somewhat less lift. Anyone want to go for a ride in the low bid model?

The helium we use today is extracted from natural gas as it comes out of the ground. But unlike all other stable elements, it was not deposited there when the earth was formed. Instead it was created over time by the radioactive decay of uranium (92) and thorium (90). These elements decay by alpha particle emission, and "alpha particle" is simply the physicist's name for the nucleus of a helium atom. So when you fill a party balloon, you're filling it with atoms that just a few tens or hundreds of millions of years ago were random protons and neutrons in the nuclei of large radioactive atoms. That, frankly, is weird. Though not as weird as the way lithium messes with your mind.

Helium-filled latex party balloons don't last long as this tiny atom escapes rapidly. Metalized Mylar balloons last days instead of hours.

Pure helium is an invisible gas, as in this antique sample ampoule.

Ordinarily a colorless, inert gas, helium glows creamy pale peach when an electric current runs through it.

A characteristic helium peach-colored glow is visible through the open side of this helium-neon laser. The laser light coming out the front is neon red.

Disposable helium tanks are available in party supply stores but often contain added oxygen to prevent suffocation if inhaled by children.

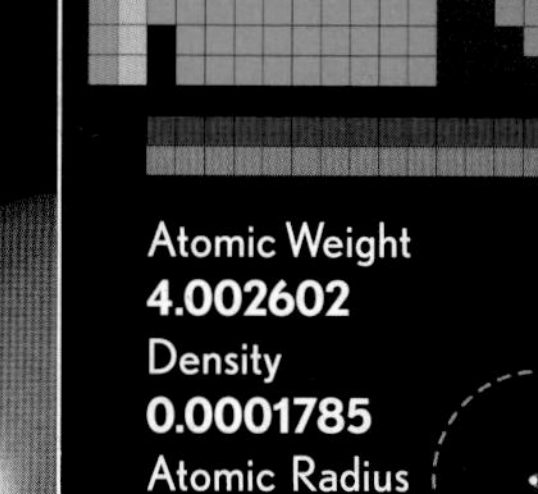

Atomic Weight
4.002602
Density
0.0001785
Atomic Radius
31pm
Crystal Structure

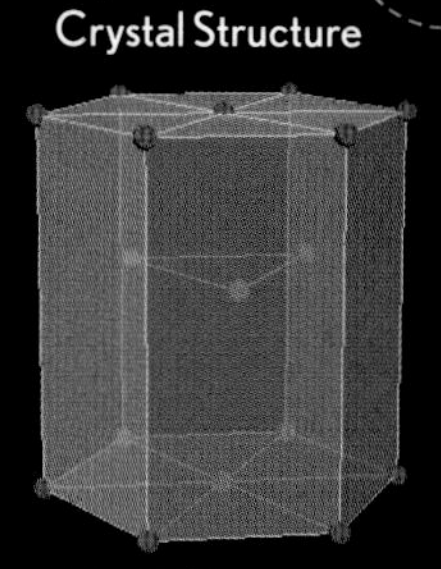

Electron Filling Order: 1s 2s 2p 3s 3p 3d 4s 4p 4d 4f 5s 5p 5d 5f 6s 6p 6d 7s 7p

Atomic Emission Spectrum

State of Matter: 0 500 1000 1500 2000 2500 3000 3500 4000 4500 5000 5500

Lithium
Li
3

Lithium

Atomic Weight
6.941
Density
0.535
Atomic Radius
167pm
Crystal Structure

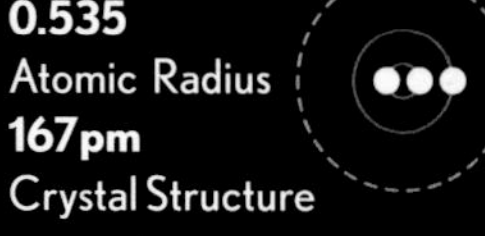

LITHIUM IS A VERY SOFT, very light metal. So light that it floats on water, a feat matched by only one other metal, sodium (11). While floating on water, lithium will react with that water, releasing hydrogen gas at a steady, moderate rate. (The real excitement in this department begins with sodium.)

Despite its reactive nature, lithium is widely used in consumer products. Lithium metal inside lithium-ion batteries powers countless electronic devices, from pacemakers to cars, including the laptop on which I am typing this text. Lithium-ion batteries pack tremendous power into not much weight, in part because of lithium's low density. Lithium stearate is also used in the popular lithium grease found on cars, trucks, and mechanics.

People who pay attention to these things have noticed an interesting fact: There's only one place in the world with a really large amount of easily recoverable lithium. If electric cars based on lithium-ion batteries ever become very widespread, you might want to keep an eye on Bolivia.

The lithium ion has another trick up its sleeve: It keeps some people on an even emotional keel. For reasons that are only vaguely understood, a steady dose of lithium carbonate (which dissolves into lithium ions in the body) smoothes out the highs and lows of bipolar disorder. That a simple element could have such a subtle effect on the mind is testimony to how even a phenomenon as complex as human emotion is at the mercy of basic chemistry.

Lithium is soft, reactive, and helps keep things in balance. Beryllium is, well, let's just say *different*.

Lithium batteries can be exotic, like the pacemaker battery above, or common, like this standard AA-sized disposable lithium cell.

Common lithium grease contains lithium stearate to improve performance.

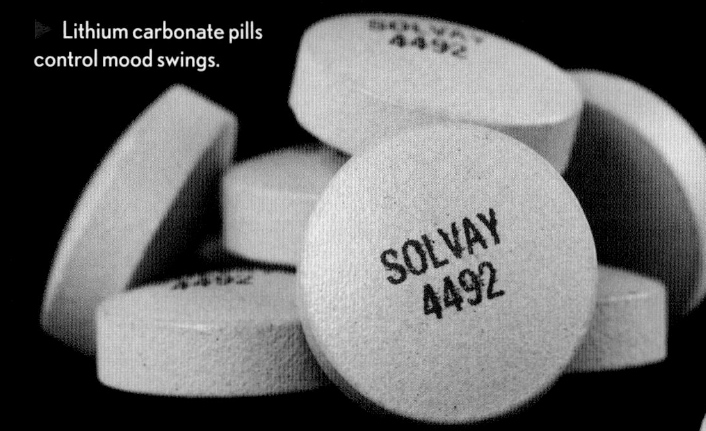

Lithium carbonate pills control mood swings.

The mineral elbaite, $Na(LiAl)_3Al_6(BO_3)_3Si_6O_{18}(OH)_4$, from Minas Gerais, Brazil.

Lithium is soft enough to cut with hand shears, which leave marks such as you see on this sample of the pure metal.

Beryllium

Be

4

Beryllium

A large aquamarine beryl ($Be_3Al_2Si_6O_{18}$) from the author's father's extensive collection.

Elemental

Atomic Weight
9.012182
Density
1.848
Atomic Radius
112pm
Crystal Structure

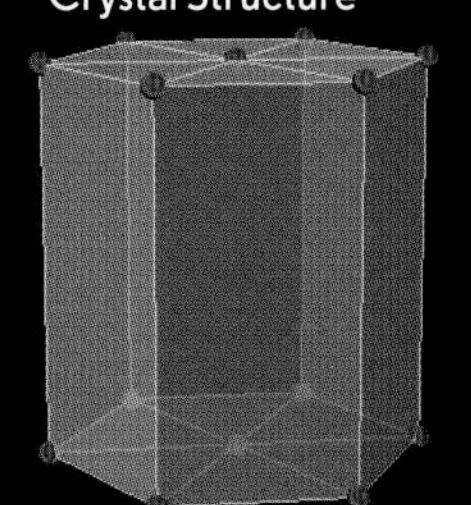

BERYLLIUM IS A LIGHT METAL (though three and a half times the density of lithium, it's still significantly less dense than aluminum, element 13). Where lithium is soft, low-melting, and reactive, beryllium is strong, melts at a high temperature, and is notably resistant to corrosion.

These properties, combined with its high cost and poisonous nature, account for the unique niche beryllium has carved out for itself: missile and rocket parts, where cost is no object, where strength without weight is king, and where working with toxic materials is the least of your worries.

Beryllium has other fancy applications. It is transparent to x-rays, so it's used in the windows of x-ray tubes, which need to be strong enough to hold a perfect vacuum, yet thin enough to let the delicate x-rays out. A few percent of it alloyed with copper (29) forms a high-strength, nonsparking alloy used for tools deployed around oil wells and flammable gases, where a spark from an iron tool could spell disaster, in great big flaming red letters.

In keeping with the sport of golf's tendency to use high-tech materials out of desperate hope that they may help get the ball where it's supposed to go, beryllium copper is also used in golf-club heads. Needless to say, it doesn't help any more than the manganese bronze or titanium (22) used for the same purpose.

Combining beauty with brawn, the mineral beryl is a crystalline form of beryllium aluminum cyclosilicate. You may be more familiar with the green and blue varieties of beryl, which are known as emerald and aquamarine.

Beryllium: A debonair, James Bond–style metal able to launch rockets one minute and charm the ladies the next. Then there's boron.

This pure broken crystal of refined beryllium ordinarily would be melted down and turned into strong, lightweight parts for missiles and spacecraft.

Beryllium oxide high-voltage insulator.

Beryllium copper non-sparking gas-valve wrench.

Complex beryllium missile gyroscope.

Beryllium foil windows mounted in an x-ray tube.

Beryllium copper golf club.

Electron Filling Order
1s 2s 2p 3s 3p 4s 3d 4p 5s 4d 5p 6s 4f 5d 6p 7s 5f 6d 7p

Atomic Emission Spectrum

State of Matter
0 500 1000 1500 2000 2500 3000 3500 4000 4500 5000 5500

Boron
B
5

Boron

POOR BORON—with a name like that, how can it get any respect? It doesn't help [...]on's most commonly found in bora[...]he laundry aid. But boron is more glam[...]ous than you might think.

Combine boron (5) with nitrogen (7), and you get crystals similar to those of their average, carbon (6), the element that forms diamond. Cubic boron nitride crystals are very nearly as hard as diamond, but much less expensive to creat[...] and more heat resistant, making them popular abrasives for industrial steelworking.

Recent theoretical calculations indicate that the alternate wurtzite-crystal form of boron nitride, as yet never created [...]ngle-crystal form, might actually be harder than diamond under certain conditions, and for certain technical definitions of "hard." Unseating diamond from its long reign as the hardest known material would be quite a coup, but for the time being "wurtzite" boron nitride's only accomplishment is causing an annoying footnote you now have to put next to any claim that diamond is the hardest known substance.

Boron carbide, also one of the hardest known substances, even has a genuine secret-agent application: Granules of it poured into the oil-fill hole of an internal combustion engine will destroy the engine by irreparably scoring the cylinder walls. Of slightly less interest to the CIA is the fact that boron is critical in cross-linking the polymers that gives Silly Putty its amazing ability to be both soft and moldable in your hand, yet hard and bouncy when you throw it against the wall.

B[...]hile boron is not quite the frump you [...] expect from its name, it's really not in the same league as carbon.

Boric acid was recommended for everything from eye washing to ant poison.

Cubic boron nitride is used in machine tool inserts for cutting hardened steel.

Boron carbide engine sabotage solution.

Silly Putty®.

Boron is rarely seen in pure form, as in these polycrystalline lumps. While extremely hard, boron is too brittle in pure form to have any practical applications.

Elemental

Atomic Weight
10.8111
Density
2.460
Atomic Radius
87pm
Crystal Structure

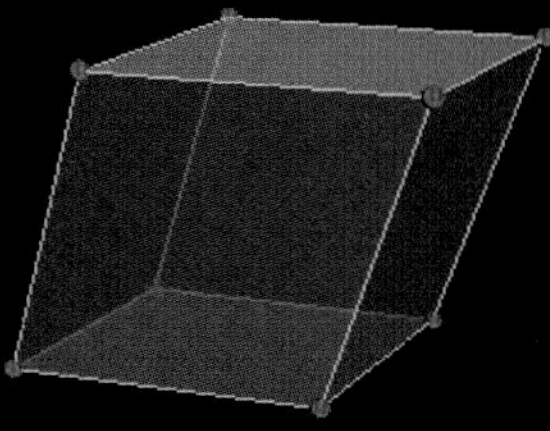

Electron Filling Order
Atomic Emission Spectrum
State of Matter

Carbon
C
6

Carbon

▷ Computer model of C_{60} "bucky ball"

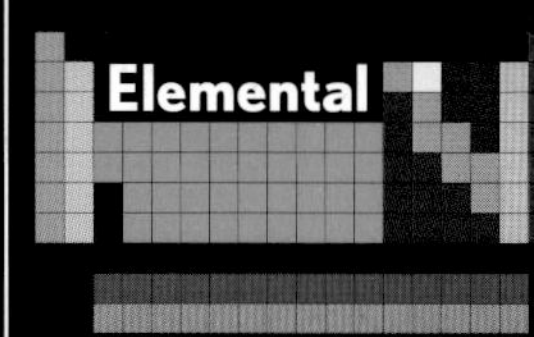

Atomic Weight **12.0107**
Density **2.260**
Atomic Radius **67pm**
Crystal Structure

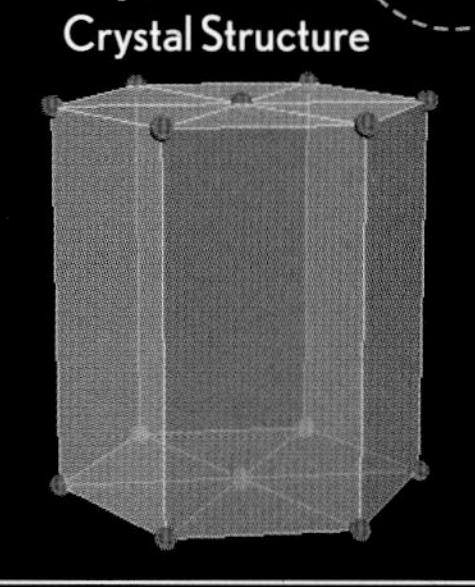

CARBON IS *THE* MOST IMPORTANT element of life, period. Sure, there are many others without which life would not exist, but from the spiral backbone of DNA to the intricate rings and streamers of the steroids and proteins, carbon is the element whose unique properties tie it all together. The very term "organic compound" refers exclusively to chemicals containing carbon.

Not content to be the foundation of all life on earth, carbon also forms diamond, the hardest known substance (at least for now; challengers are discussed under boron, element 5). But contrary to popular belief, diamonds are not particularly rare, nor are they unusually beautiful, nor are they forever: all three are myths created by the DeBeers diamond company. Diamonds would cost a tenth as much but for DeBeers's monopoly control. Cubic zirconia or crystalline silicon carbide are just as pretty. And at high enough temperatures, diamonds burn up into nothing but carbon dioxide.

If I were writing these words twenty-five years or so ago, I would probably have been doing it with carbon. The "lead" in pencils is actually graphite, a form of carbon, and has been since the 16th-century discovery in the English Lake District of the great mine at Borrowdale, the first source of pure graphite.

Carbon atoms like to form sheets, like a honeycomb with a carbon atom at each corner. Stack the sheets and you have graphite. Fold them into a sphere and you have a C_{60} "buckyball," named for Buckminster Fuller who invented the geodesic dome. Roll the sheets into tubes and you have the strongest material known to science: carbon nanotubes.

Carbon has now become a focus of political controversy centered on the fact that our civilization is pumping carbon dioxide back into the atmosphere at about 100,000 times the rate it was put away by the dinosaurs and their swamps. Interestingly, the situation with nitrogen is exactly reversed.

△ Tiny industrial diamonds embedded in this steel disk turn it into a powerful grinding wheel.

▽ A "Congo cube," natural cheap polycrystalline diamond clusters.

▷ Coal (roughly speaking C_nH_{2n}) carvings are found everywhere that coal is.

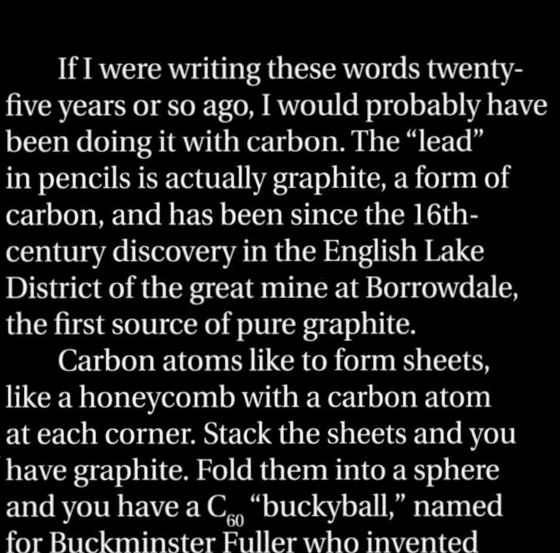

▽ A block of graphite (pure carbon) from the first atomic pile, described under fermium, element 100.

GRAPHITE FROM CP-1
FIRST NUCLEAR REACTOR
DECEMBER 2, 1942
STAGG FIELD – THE UNIVERSITY OF CHICAGO

◁ A diamond is forever, unless you heat it too much, in which case it burns up into carbon dioxide gas.

▷ Copper-clad graphite welding eletrodes are available in any welding shop

△ Coal as you buy it for heating and blacksmithing.

Electron Filling Order
1s 2s 2p 3s 3p 3d 4s 4p 4d 4f 5s 5p 5d 5f 6s 6p 6d 7s 7p

Atomic Emission Spectrum

State of Matter
0 500 1000 1500 2000 2500 3000 3500 4000 4500 5000 5500

Nitrogen

N

7

Nitrogen

AT THE SAME TIME that modern civilization has been pumping carbon dioxide into the atmosphere, we've been pulling out nitrogen and eating it.

Nitrogen as N_2 in the air is inert and largely useless, but when it's converted to a more reactive form, such as ammonia (NH_3), it becomes a vital fertilizer. Only some plants, beans for example, aided by microorganisms residing in their roots, are able to draw the nitrogen they need directly from the air. This is one reason that, before the advent of cheap nitrogen fertilizer, corn, which cannot "fix" nitrogen, was alternated in the fields with beans or alfalfa, which leave the soil with more nitrogen than it started with.

Just before World War I, Fritz Haber invented a practical process for converting nitrogen from the air into ammonia, one of the most important discoveries in human history. Ammonia fertilizer now feeds a third of the world (the rest being fed mainly by phosphate fertilizers). His work with chlorine (17) was less benevolent, as you can read about under that element.

And since plant growth absorbs carbon dioxide from the air, nitrogen fertilization even helps, at least a bit, with alleviating the effects of global warming.

Liquid nitrogen is a cheap and readily available cryogenic cooling liquid. With a boiling point of -196°C it is cold enough to freeze almost anything. It is used to preserve biological samples, to amuse children by freezing and shattering flowers, and occasionally to make ice cream in record time.

There's a lot of nitrogen around: Over 78 percent of the atmosphere is nitrogen. What's the other 22 percent? Most of it is the oxygen we need to breathe.

Silicon nitride (Si_3N_4) ceramic ball bearing for very expensive skateboards.

Elemental

Atomic Weight **14.0067**
Density **0.001251**
Atomic Radius **56pm**
Crystal Structure

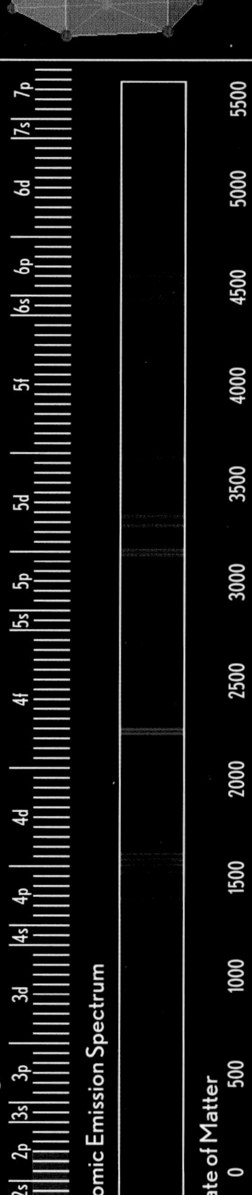

Nitrogen-gas canister for a wine-preservation gadget. The claim of 100% purity is suspect: Nothing is ever 100%.

Silicon nitride (Si_3N_4) is so hard it is used to make cutting tools, such as this milling bit insert.

The mineral nitratine ($NaNO_3$)

Nitroglycerine ($C_3H_5N_3O_9$) pills for angina.

A Dewar flask filled with boiling liquid nitrogen at -196°C (-320°F).

Oxygen

O

8

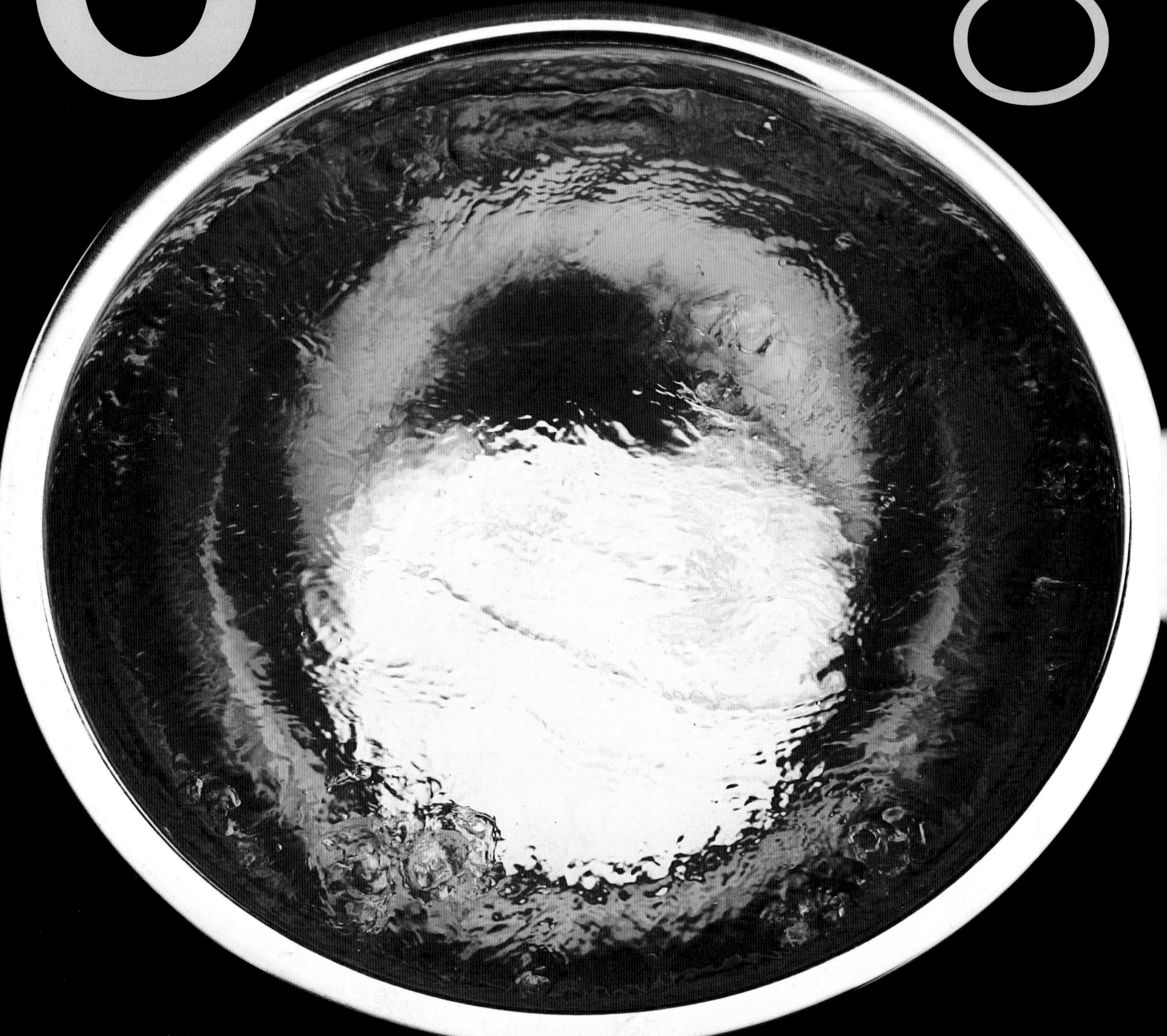

Oxygen

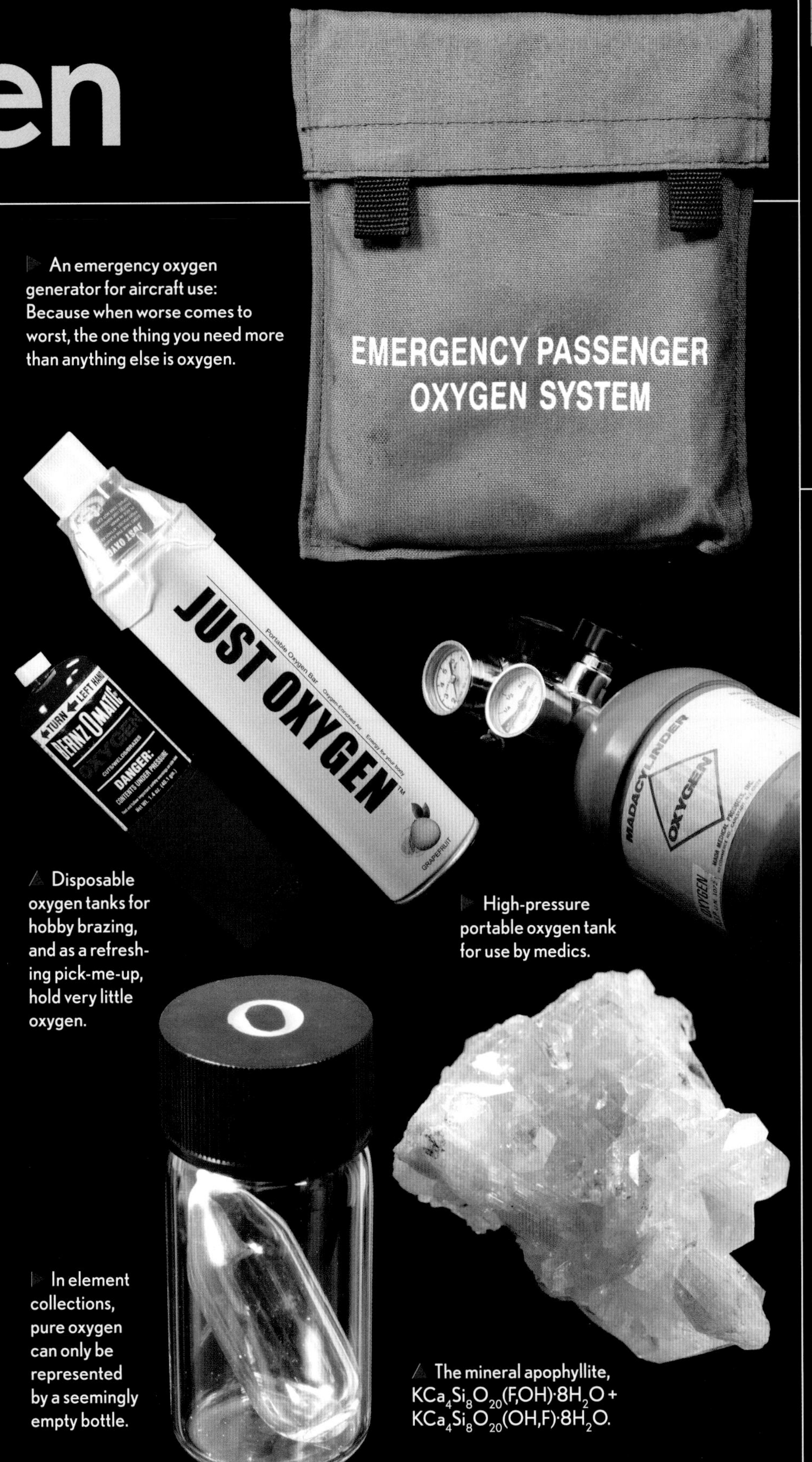

An emergency oxygen generator for aircraft use: Because when worse comes to worst, the one thing you need more than anything else is oxygen.

Disposable oxygen tanks for hobby brazing, and as a refreshing pick-me-up, hold very little oxygen.

High-pressure portable oxygen tank for use by medics.

In element collections, pure oxygen can only be represented by a seemingly empty bottle.

The mineral apophyllite, $KCa_4Si_8O_{20}(F,OH) \cdot 8H_2O$ + $KCa_4Si_8O_{20}(OH,F) \cdot 8H_2O$.

Elemental

Atomic Weight
15.9994
Density
0.001429
Atomic Radius
48pm
Crystal Structure

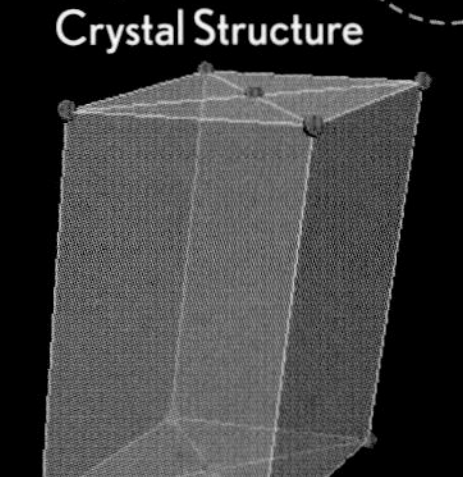

Electron Filling Order

Atomic Emission Spectrum

State of Matter

IF CARBON (6) is the foundation of life, then oxygen is the fuel. Oxygen's ability to react with just about any organic compound is what drives the processes of life. Combustion with oxygen also drives your car, your furnace, and if you work for NASA, your rockets. (Actually, the term "fuel" usually refers to the thing that is burned by an "oxidizer," so I'm speaking metaphorically when I say oxygen is the fuel of life. Technically speaking, oxygen is the oxidizer of life.)

The fact that you can light and burn wood, paper, or gasoline has less to do with what those things are made of, and more to do with the fact that our atmosphere is over 21 percent oxygen, providing a ready source of highly reactive oxidizer. Jet airplanes can travel great distances with far less fuel than a comparable rocket would require, because unlike jets that travel in air, rockets must function in the vacuum of space and must therefore carry their oxygen supply with them.

Concentrated into liquid form, oxygen goes from being gently life-giving to life-threateningly fierce. It's fair to say that the real power for most rockets comes not from the fuel they burn, but from their oxygen supply. The Saturn V moon rocket, for example, ran on kerosene. (Yes, we made it to the moon on diesel fuel.) But it wasn't the kerosene that was special, it was the ten cubic yards *per second* of liquid oxygen the Saturn V consumed at full thrust.

Given how intense oxygen is, it might surprise you to learn that it is the most abundant element on earth, accounting for nearly half the weight of the earth's crust and 86 percent of the weight of the oceans. But the crust and the oceans are made not of pure oxygen but of its compounds, and as we will learn from fluorine, the fiercer the element, the more stable its compounds.

At -183°C, oxygen is a beautiful pale blue liquid.

Fluorine

F

9

Fluorine

Beautiful purple fluorite with hydrocarbon impurities that tint the center yellow.

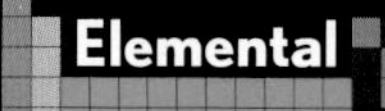

Atomic Weight
18.9984032
Density
0.001696
Atomic Radius
42pm
Crystal Structure

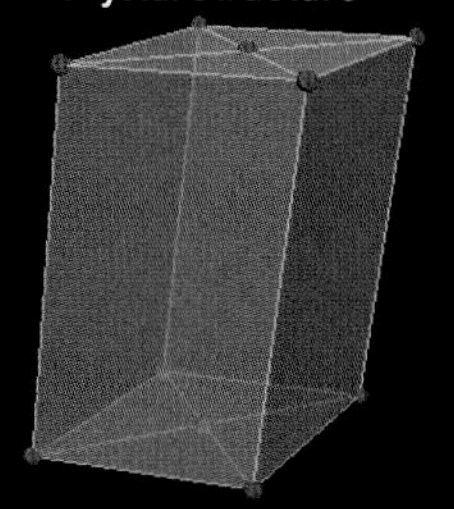

Electron Filling Order
Atomic Emission Spectrum
State of Matter

FLUORINE IS AMONG the most reactive of all the elements. Blow a stream of fluorine gas at almost anything, and it will burst into flame. That includes things not normally thought of as flammable, such as glass. Interestingly, the more reactive an element is, the more stable are its compounds.

When we say fluorine is highly reactive, we mean that a large amount of energy is released when it combines with other elements. The resulting compounds are very stable because of the large quantity of energy must be put back in if you want to tear them apart. This energy must be supplied by some yet-more-reactive substance, of which, in the case of fluorine, there are precious few.

The most famous highly stable fluorine compound is Teflon, which was discovered quite by accident. So many important chemicals have been discovered by accident that one has to wonder what a bunch of bumblers chemists are. Or maybe they are just exceptionally good at spotting serendipity when it ruins their day. Teflon was discovered when its unexpected formation completely ruined an attempt to create the first chlorofluorocarbon refrigerants, which have now been banned as ozone-depleting menaces. Not a bad trade, I'd say.

Teflon is almost completely resistant to chemical attack, and coincidentally also very slippery, which makes it useful in everything from nonstick pans to acid-storage bottles. Fluorine is important primarily because of the stable compounds it forms, while neon forms no stable compounds whatsoever.

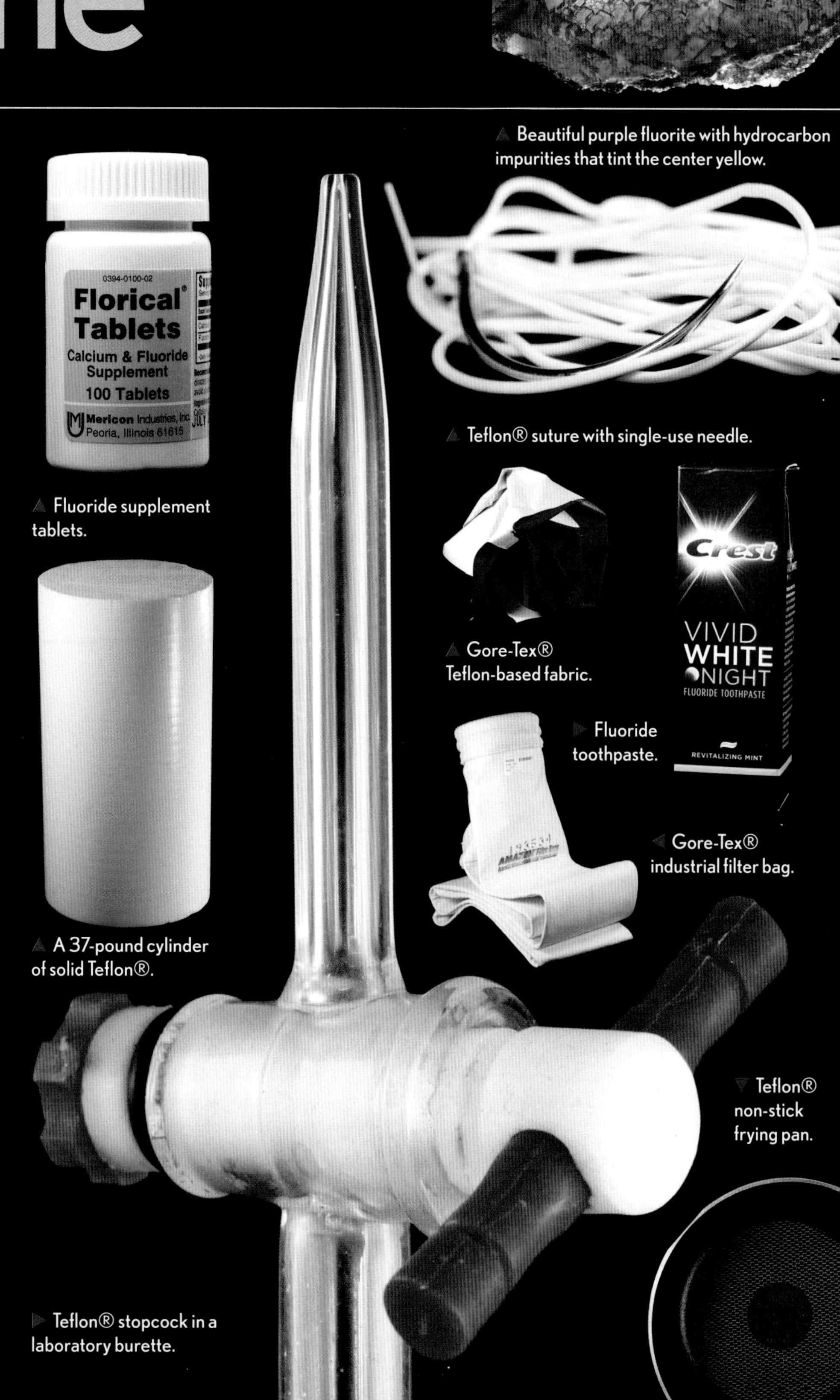

Fluoride supplement tablets.

A 37-pound cylinder of solid Teflon®.

Teflon® suture with single-use needle.

Gore-Tex® Teflon-based fabric.

Fluoride toothpaste.

Gore-Tex® industrial filter bag.

Teflon® non-stick frying pan.

Teflon® stopcock in a laboratory burette.

Fluorine is a pale yellow gas that reacts violently with virtually everything, including glass. This pure quartz ampoule probably held it for a while anyway.

10

Neon

NEON IS LITERALLY up in lights. As in, up there, in those lights, there is neon. So close is the association between the element and its most common application that Times Square and Las Vegas are described as being "awash with neon."

Unlike "platinum" credit cards that contain no platinum, some "neon" lights—the orange-red ones—really do contain neon. When a high-voltage electric discharge is run though a tube filled with low-pressure neon, the gas glows bright orange-red in a fuzzy line down the center of the tube. (Any other color, and it's not neon. And if you see a tube where the light comes from an opaque coating on the inside surface of the glass, rather than from inside the tube itself, you've got yourself a mercury vapor or krypton tube with a phosphor coating.)

Oliver Sacks, in his delightful book *Uncle Tungsten*, describes walking through Times Square with a pocket spectroscope, enchanted by the great variety of spectral lines he could see. That's another way to tell a genuine neon light—by its unique spectrum, unlike that of any other element or phosphor.

Helium-neon lasers were the first continuous-beam lasers in commercial use, and while they have been replaced in many applications by incredibly cheap laser diodes, HeNe lasers remain an important application for this element. There are very few things you can do with neon that don't rely in one way or another on the light it emits when stimulated with electricity. That neon has so few applications is masked by the fact that neon lights are so vivid and so widespread they make it seem like an important element, even though it would be one of the least missed.

The least reactive of all the elements, neon completely refuses to react with any others. That's something you definitely can't say about sodium, as we jump back to the left side of the periodic table.

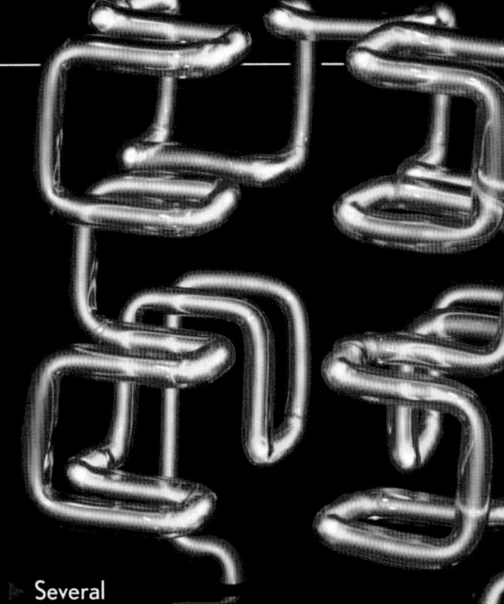

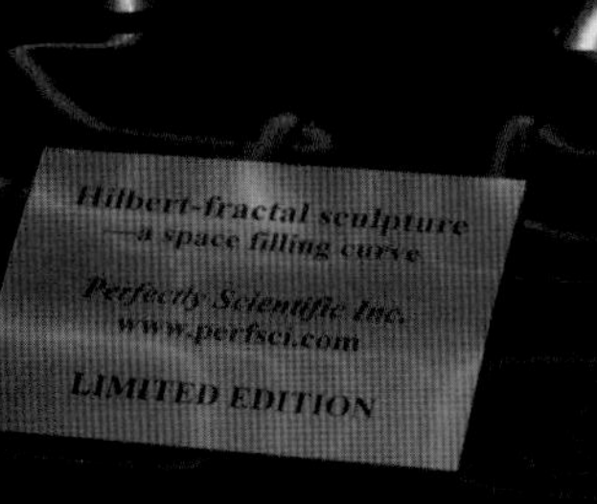

Several thousand volts illuminate this neon sculpture in the shape of a Hilbert fractal.

A tiny indicator light, no more than 1/8 inch across, glows from applied 120V AC.

Pure neon is an invisible gas, seen here in an antique sample ampoule.

Neon signs really are made with neon, like this Ne tube. An electric current runs through it, creating the light.

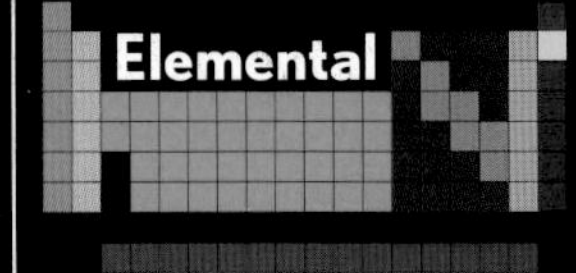

Atomic Weight
20.1797
Density
0.000900
Atomic Radius
38pm
Crystal Structure

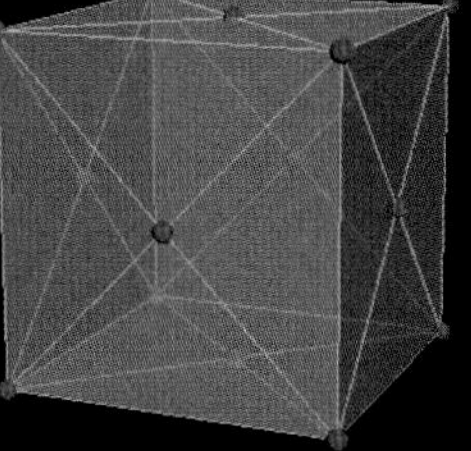

Electron Filling Order
Atomic Emission Spectrum
State of Matter

Sodium
Na
11

Sodium

SODIUM IS THE MOST EXPLOSIVE, and the best tasting, of all the alkali metals (the elements from the first column of the periodic table).

Explosive because if you throw it into water, it rapidly generates hydrogen gas, which seconds later ignites with a tremendous bang, throwing flaming sodium in all directions. (The other alkali metals react similarly with water, but sodium, overall, creates the most attractive explosions and is thus favored by mischief makers the world over for throwing into lakes and rivers.)

Best tasting because, together with chlorine (17), it forms sodium chloride, or table salt, widely considered the tastiest of the alkali metal chloride salts. Potassium chloride is sold as a salt substitute for people on a low-sodium diet, but it adds a bitter metallic note to its saltiness. Rubidium chloride and cesium chloride are less salty and more metallic in taste, while lithium chloride produces a burning sensation followed by an oily metallic aftertaste.

Pure sodium metal is used in large quantities in the chemical industry as a reducing agent, and while it might seem like a really bad idea, liquid sodium is used to move heat from the reactor core to the steam turbines in some nuclear reactors (yes, there have been spectacular sodium leaks). Closer to home, yellowish sodium vapor lamps create more light per unit of electricity than nearly any other type, while making people under them look dead.

Sodium is used only for its chemical properties. The next element, magnesium, is very useful both for its chemical and its structural properties.

These soft, silvery sodium chunks were cut with a knife and stored under oil. In air they turn white in seconds; exposed to water, they generate hydrogen gas and explode in flaming balls of molten sodium.

Low-pressure sodium vapor light, which produces horrible light very efficiently.

Sodium hydroxide, whose traditional name is lye, is commonly sold as a drain opener.

Sodium-filled valve stem from a high-performance racing engine, cut away to show the sodium.

High-pressure sodium vapor light, commonly used for efficient, not completely unpleasant light.

The mineral sodalite ($Na_4Al_3Si_3O_{12}Cl$).

A block of salt (sodium chloride) for horses to lick.

Elemental

Atomic Weight
22.989770
Density
0.968
Atomic Radius
190pm
Crystal Structure

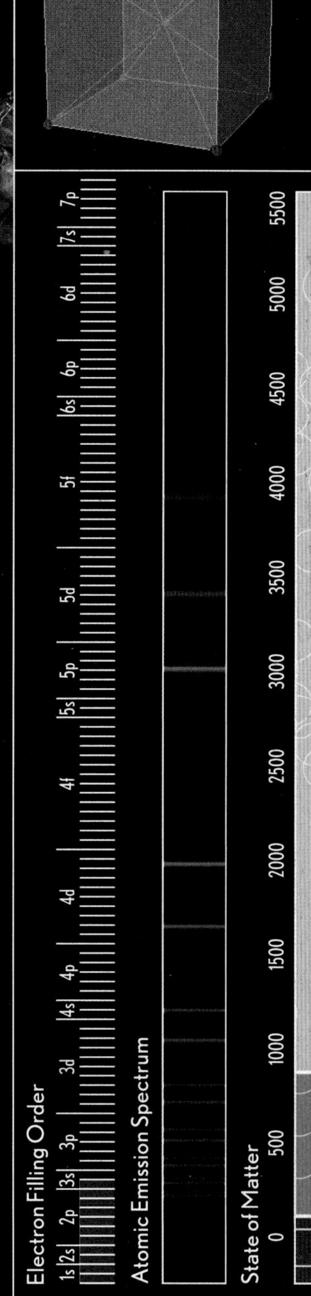

Magnesium

Mg

12

Magnesium

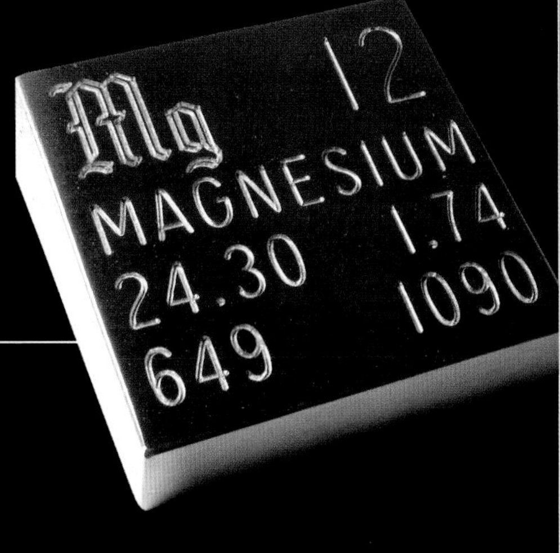

A block of magnesium engraved with that element's physical properties.

MAGNESIUM IS THE FIRST of the truly marvelous structural metals. (Beryllium, element 4, is a fine metal, but its high cost and toxicity keep it from being marvelous.) Magnesium is moderately priced, strong, light, and easy to machine. About the only downside is that it's highly flammable.

Magnesium is so flammable that you can light a ribbon of it with a match, and fine powders of it are positively explosive. Early photographic flashes were nothing more than a rubber bulb used to blow a puff of magnesium powder into a candle flame, and many modern pyrotechnic mixtures contain magnesium powder to create a bright and loud report.

The fact that it's flammable might seem like a deal breaker for using magnesium to make car parts, but in large solid pieces, it's surprisingly difficult to ignite. The bulk metal conducts heat away from the surface fast enough to keep it from lighting. Magnesium is used in race cars, airplanes, and bicycles, despite the occasional mass fatality when a magnesium race-car frame catches fire. (Eighty-one people died at Le Mans in 1955 when a flaming magnesium-bodied car crashed into the stands, an event not considered serious enough to stop the race.)

Much more common are alloys of aluminum (13) that contain a few percent magnesium. Confusingly, wheels made of this imposter are often referred to as "mag wheels," even though they are 60 percent heavier than true magnesium wheels (which are also available, at several times the price).

But as marvelous as magnesium is, for overall supreme goodness as a metal there is really no competition: aluminum wins hands down.

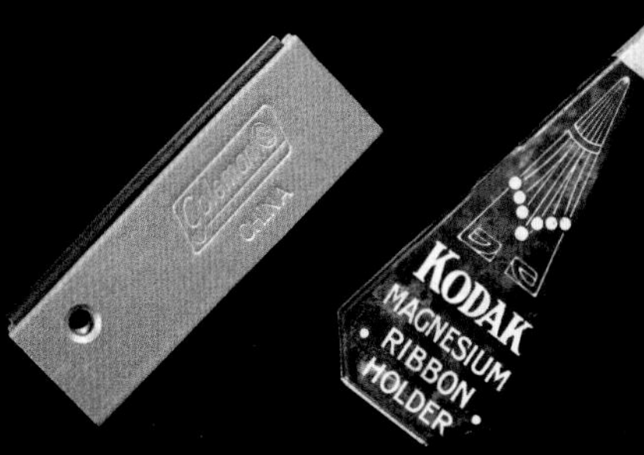

A magnesium block campfire starter.

Early ribbon holder used to expose contact prints.

Atomic Weight
24.3050
Density
1.738
Atomic Radius
145pm
Crystal Structure

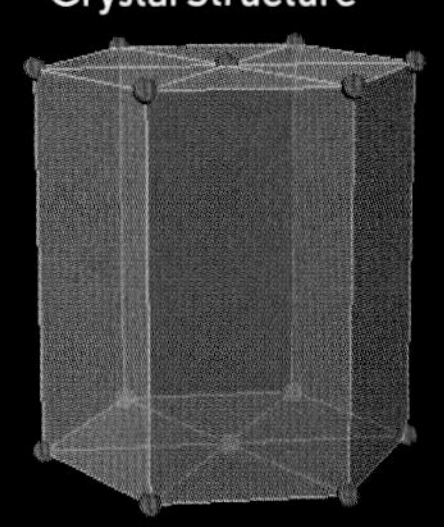

A magnesium printing block.

Magnesium powder photoflash kit from the 1920's.

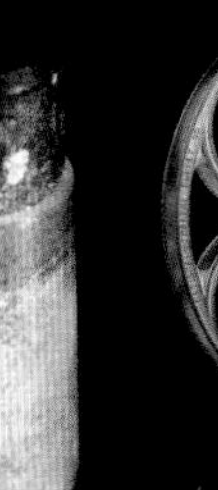

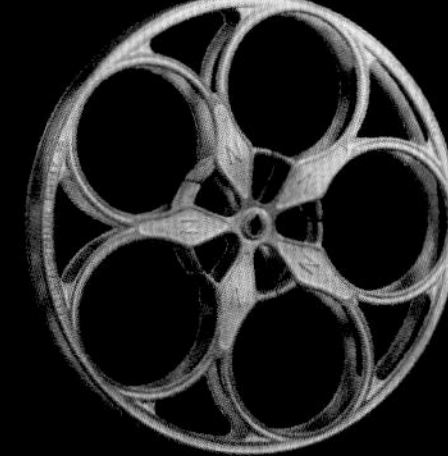

Magnesium film reel.

These magnesium nodules grow during the refining process and are usually melted down into useful products.

A solid magnesium car brake mounting hub.

Aluminum
Al
13

Aluminum

ALUMINUM IS PRETTY close to being the ideal metal, though it could be improved in a few ways: It could be as cheap and easy to weld as iron (26), or it could take to casting as well as zinc (30) or tin (50). But overall it's very fine stuff: light and strong enough to form the structure of most airplanes except the most exotic high-performance military aircraft, yet cheap enough to be in every kitchen. (It wasn't always cheap: When the pure metal was first produced, it was considered a noble metal alongside gold and silver. Napoleon III served his most important guests on aluminum plates; ordinary princes and dukes had to settle for mere gold.)

Aluminum's signature advantage over steel is that it doesn't rust, which makes it all the more surprising to learn that aluminum reacts with air even more rapidly than does iron. The difference is that aluminum "rust" is a tough, transparent oxide, also known as corundum, one of the hardest substances known. Exposed to air, aluminum instantly protects itself with a thin layer of this material, harder than the metal itself. Iron foolishly coats itself with a red flaky powder that soon falls off, exposing fresh metal to further oxidation.

But deep down, aluminum really is very reactive. Powdered aluminum is a basic ingredient in modern flash powder and rocket-fuel mixtures, and its sale below a certain particle size is restricted for this reason.

Aluminum minerals are extremely common, including such basics as corundum (the generic form of ruby and sapphire) and beryl (the generic form of emerald and aquamarine). Aluminum in minerals and rocks makes up a large part of the crust of the earth, as does its neighbor in the periodic table, silicon.

Solid block of aluminum for testing purposes.

Metalized (aluminum coated) Mylar emergency blanket.

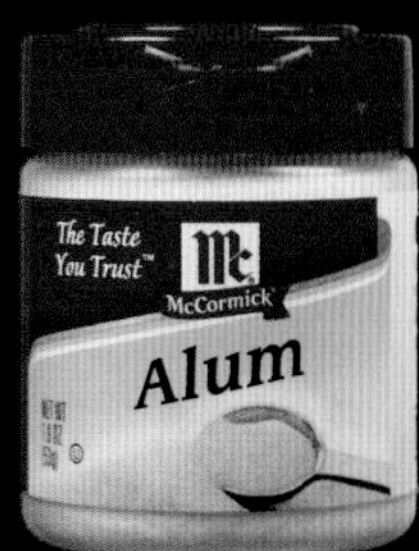

Antique and modern medical Alum (potassium aluminum sulfate).

Heat sinks use aluminum's high thermal conductivity.

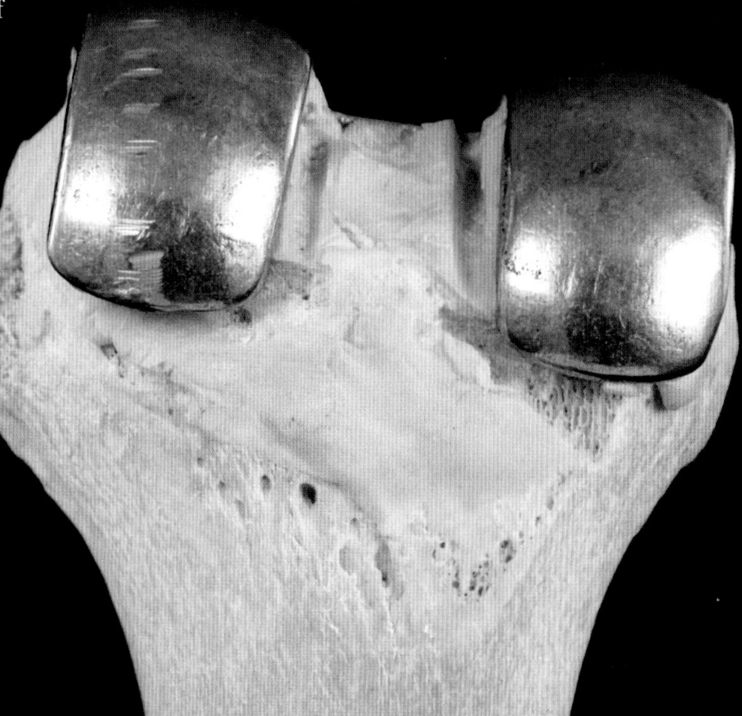

Aluminum is never used for medical implants, but this one was made for doctors to practice on. The bone is real but the aluminum implant is just pretend.

Nodules created by pouring molten aluminum into a bucket of water.

Etched high-purity aluminum bar showing internal crystal structure.

Elemental

Atomic Weight
26.981538
Density
2.7
Atomic Radius
118pm
Crystal Structure

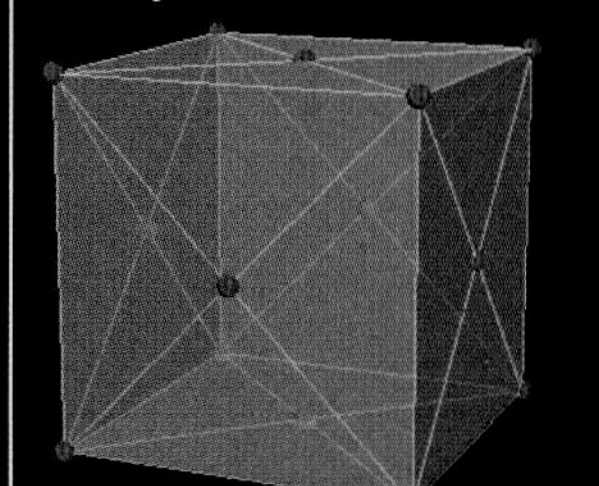

Electron Filling Order: 1s 2s 2p 3s 3p 3d 4s 4p 4d 4f 5s 5p 5d 5f 6s 6p 6d 7s 7p

Atomic Emission Spectrum

State of Matter: 0 500 1000 1500 2000 2500 3000 3500 4000 4500 5000 5500

Aluminum 13

Charming hourglass dodecahedron rendered in rings of 5356 alloy aluminum.

"Firefly" aluminum has a mixture of fine powder and coarse flakes to give a random sparkle effect to fireworks.

This bumpy surface is formed mechanically when a cylinder of very pure aluminum is crushed under tremendous pressure to a fraction of its original height.

A bad day at the factory sent this massive block of machined aluminum to the Boeing surplus store in Seattle.

An aluminum cannon makes no sense, unless it's just a model. This one was made by the author in shop class back when high schools still had shop classes.

Aluminum oxide grinding disks are very common.

A colleague's gift teases the author with chocolate encased in aluminum hard disk platters.

High purity aluminum sputtering targets the size of dinner plates.

Common aluminum cookware is typically made of fairly pure aluminum for good thermal conductivity.

Silicon

Si

14

Silicon

This silicon boule was pulled out of the melting pot prematurely: We're seeing the underside, where molten silicon dripped off.

Elemental

Atomic Weight
28.0855
Density
2.330
Atomic Radius
111pm
Crystal Structure

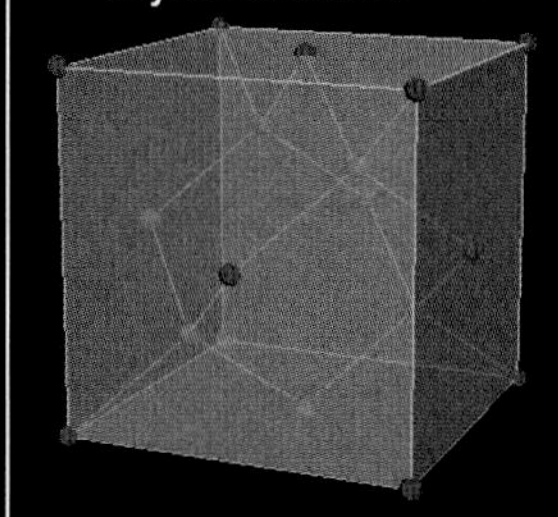

SILICON-BASED LIFE-FORMS have been the subject of speculation in science fiction ever since chemists pointed out that silicon, of all the elements, is most like its neighbor, carbon (6), in its ability to form complex molecular chains, in some ways not unlike the long-chain carbon molecules that are reading this text. (That means you.)

But it now seems quite clear that when silicon-based life emerges, at least on this planet, it will not be due to silicon's ability to form molecular chains, but rather to its ability to form semiconducting crystals. Computer chips start out as common white silica beach sand (silicon dioxide) and end as nearly perfect single crystals of hyperpure silicon, etched with patterns beyond the resolution of visible light. That this can be done at all is remarkable; that your average child's toy today contains more computing power than the Apollo moon rockets is the kind of thing that turns civilization on its head. Or replaces it.

The bones of the earth—the rock, sand, clay, and soil—consist in very large part of silicate minerals, combinations of silicon and oxygen (8) with smaller amounts of aluminum (13), iron (26), calcium (20) and others. (Only oxygen occurs in a larger quantity than silicon in the earth's crust. So if the computers take over, they'll have plenty of raw material to breed with, or whatever it is they do.)

About the only thing that doesn't have a lot of silicon in it is you: While some sea sponges grow bones of silica glass, your bones, assuming you are not a sea sponge, are calcium phosphate, in the form of rigid hydroxyapatite foam with almost no silicon. It is unclear why most earthly life evolved using our ubiquitous silicon only in incidental ways (unlike those clever sea sponges and computers), opting instead for phosphorus—which, as you can read on the next page, is in tragically short supply.

A bowlful of diced silicon chips.

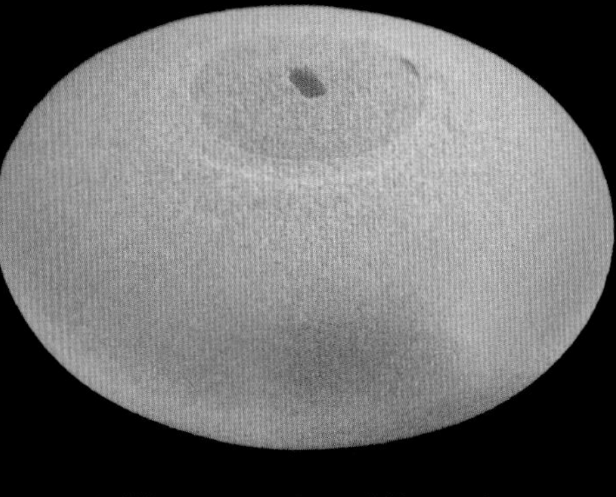

Silicon is not silicone! This implant is made of soft silicone rubber, not hard crystalline silicon.

Large silicon crystal boules rejected for chip making.

A Venus's Flower Basket's glass (silica) skeleton.

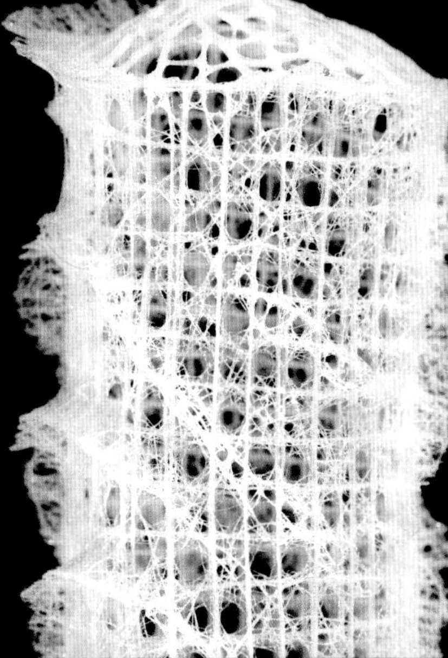

High-purity refined silicon.

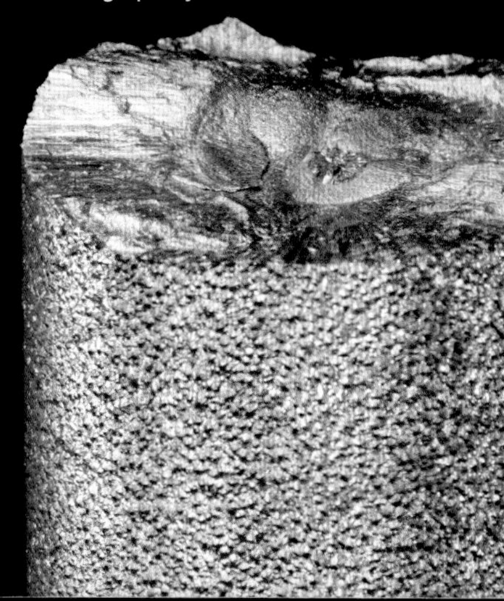

Low-purity—but pretty—molten blob of silicon from the first step of refining from sand.

Electron Filling Order: 1s 2s 2p 3s 3p 4s 3d 4p 5s 4d 5p 6s 4f 5d 6p 7s 5f 6d 7p

Atomic Emission Spectrum

State of Matter: 0 500 1000 1500 2000 2500 3000 3500 4000 4500 5000 5500

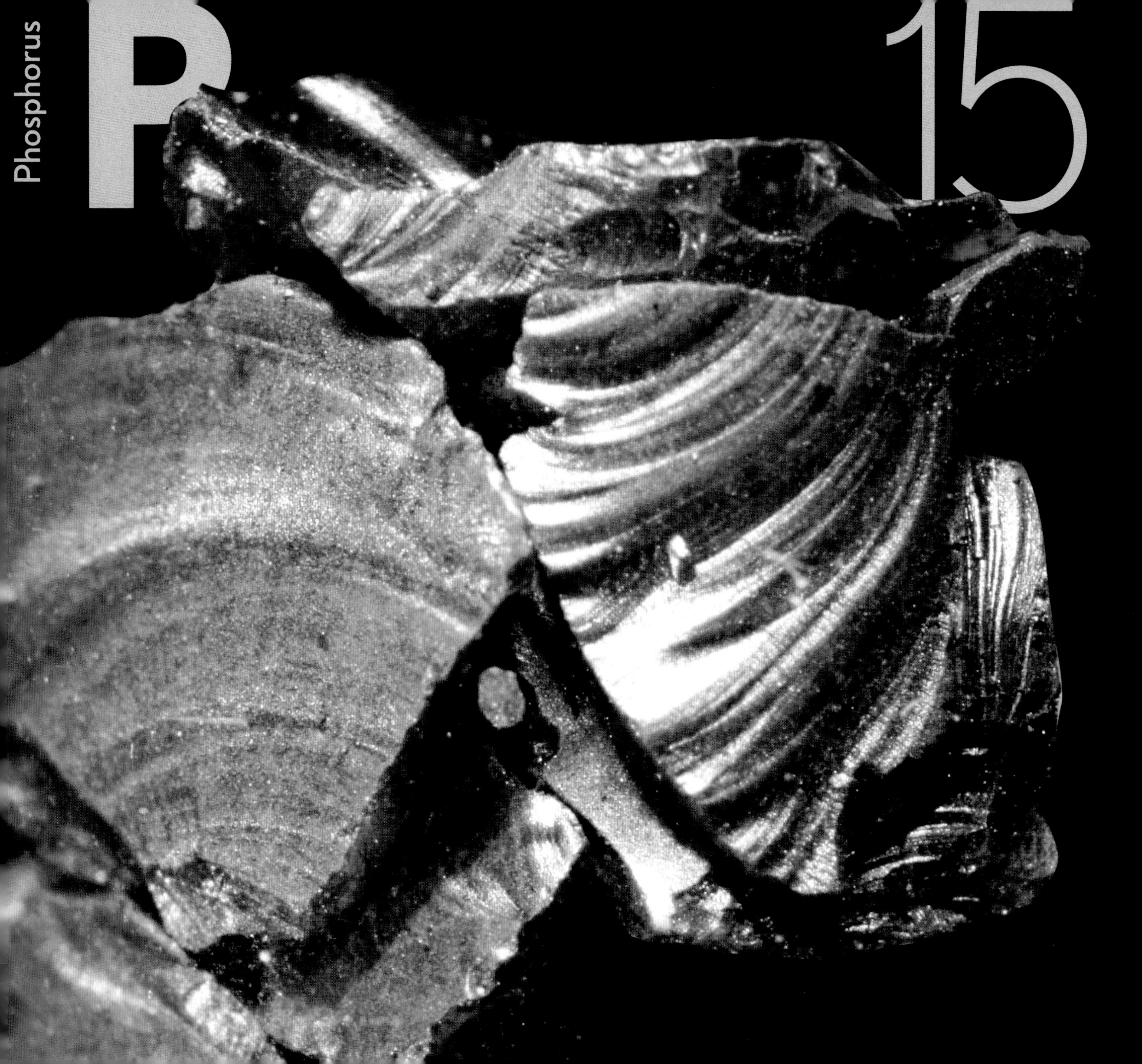

Phosphorus

P

15

Phosphorus

IN ELEMENTAL FORM, phosphorus is nasty stuff, particularly the white phosphorus allotrope, discovered in 1669 in Hamburg and responsible in 1943 for helping burn that city to the ground in one of the great firestorms of World War II (magnesium incendiary bombs leveled the buildings; white phosphorus burned the people driven outside). Even today white phosphorus artillery and mortar shells are used in warfare with horrific results.

Even in the form of phosphates (compounds containing the PO_4^{3-} group), phosphorus is vital, and was for most of human history *the* limiting factor in the growth of food crops. Depletion of phosphorus in the soil has caused mass starvations throughout history, and the search for its replenishment though guano, bonemeal, or other fertilizers determined the fate of civilizations.

Not until we learned in the mid-1800s how to create fertilizer from phosphate rocks was a technical solution to this shortage found. Phosphate fertilizers are arguably responsible for the explosion of human populations to the point where water, not phosphorus, is now the limiting factor in many places.

Phosphorus in pure form exists in several allotropes, or molecular forms. Red phosphorus is relatively stable and widely used in matches as an igniter. Black phosphorus is hard to make and rarely seen, as it has no important applications. White phosphorus—toxic, pyrophoric, and used mainly in war—is fairly close to pure evil, though if it were judged purely on the basis of smell that contest would be won by sulfur.

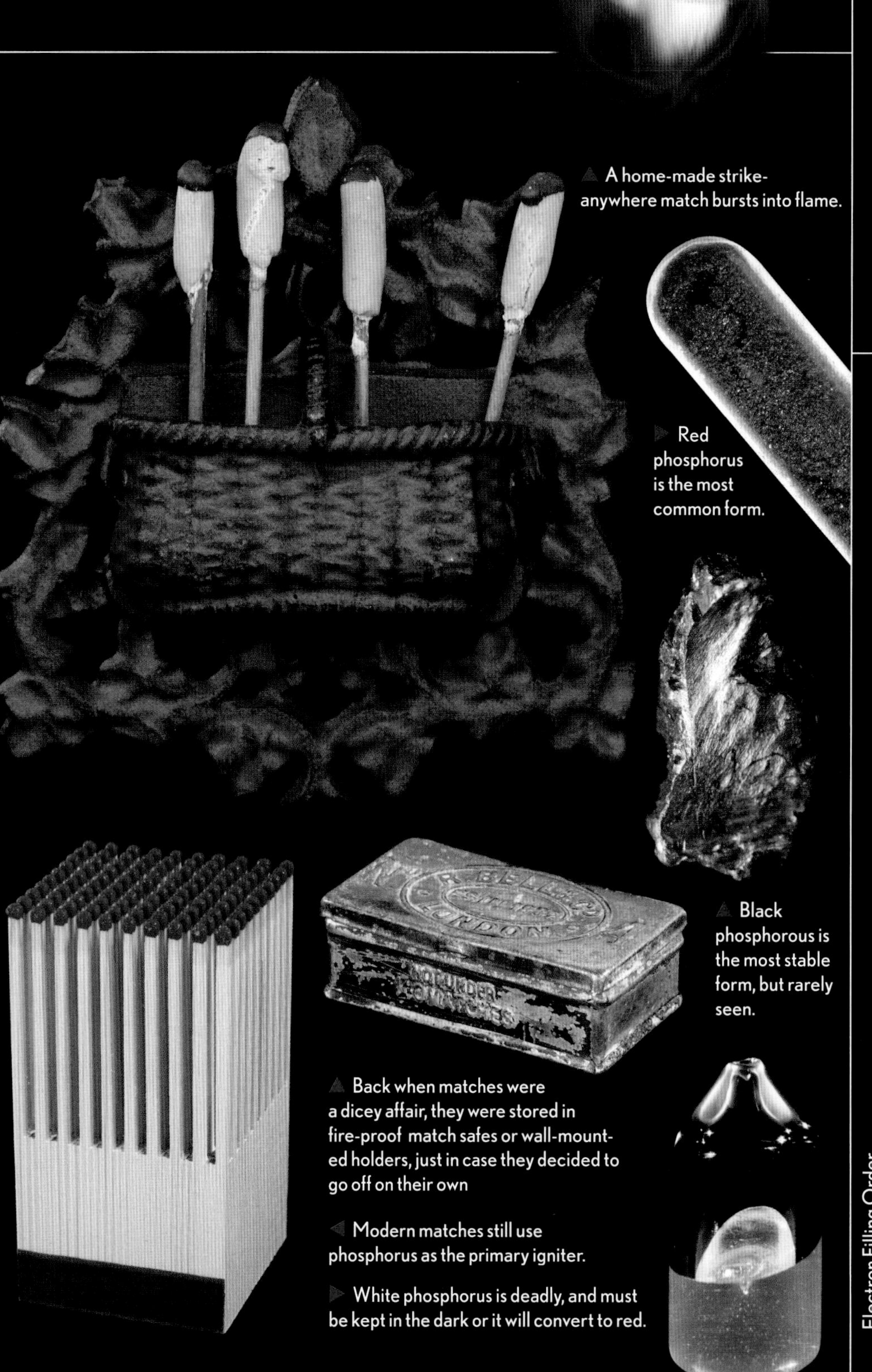

◀ This rare violet phosphorus is considered to be a mixture of red and black phosphorus, not a true allotrope in its own right.

▲ A home-made strike-anywhere match bursts into flame.

▶ Red phosphorus is the most common form.

▲ Black phosphorous is the most stable form, but rarely seen.

▲ Back when matches were a dicey affair, they were stored in fire-proof match safes or wall-mounted holders, just in case they decided to go off on their own

◀ Modern matches still use phosphorus as the primary igniter.

▶ White phosphorus is deadly, and must be kept in the dark or it will convert to red.

Elemental

Atomic Weight
30.973761
Density
1.823
Atomic Radius
98pm
Crystal Structure

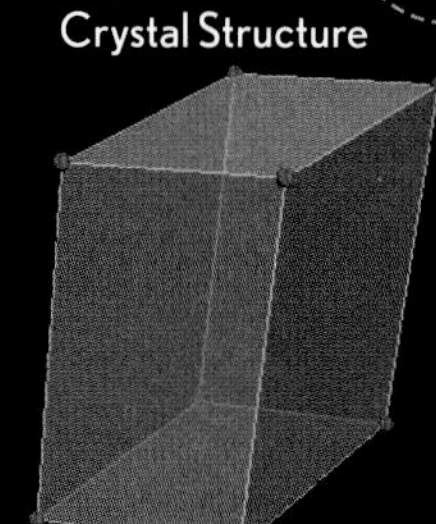

Electron Filling Order: 1s 2s 2p 3s 3p 4s 3d 4p 5s 4d 5p 6s 4f 5d 6p 7s 5f 6d 7p

Atomic Emission Spectrum

State of Matter: 0 500 1000 1500 2000 2500 3000 3500 4000 4500 5000 5500

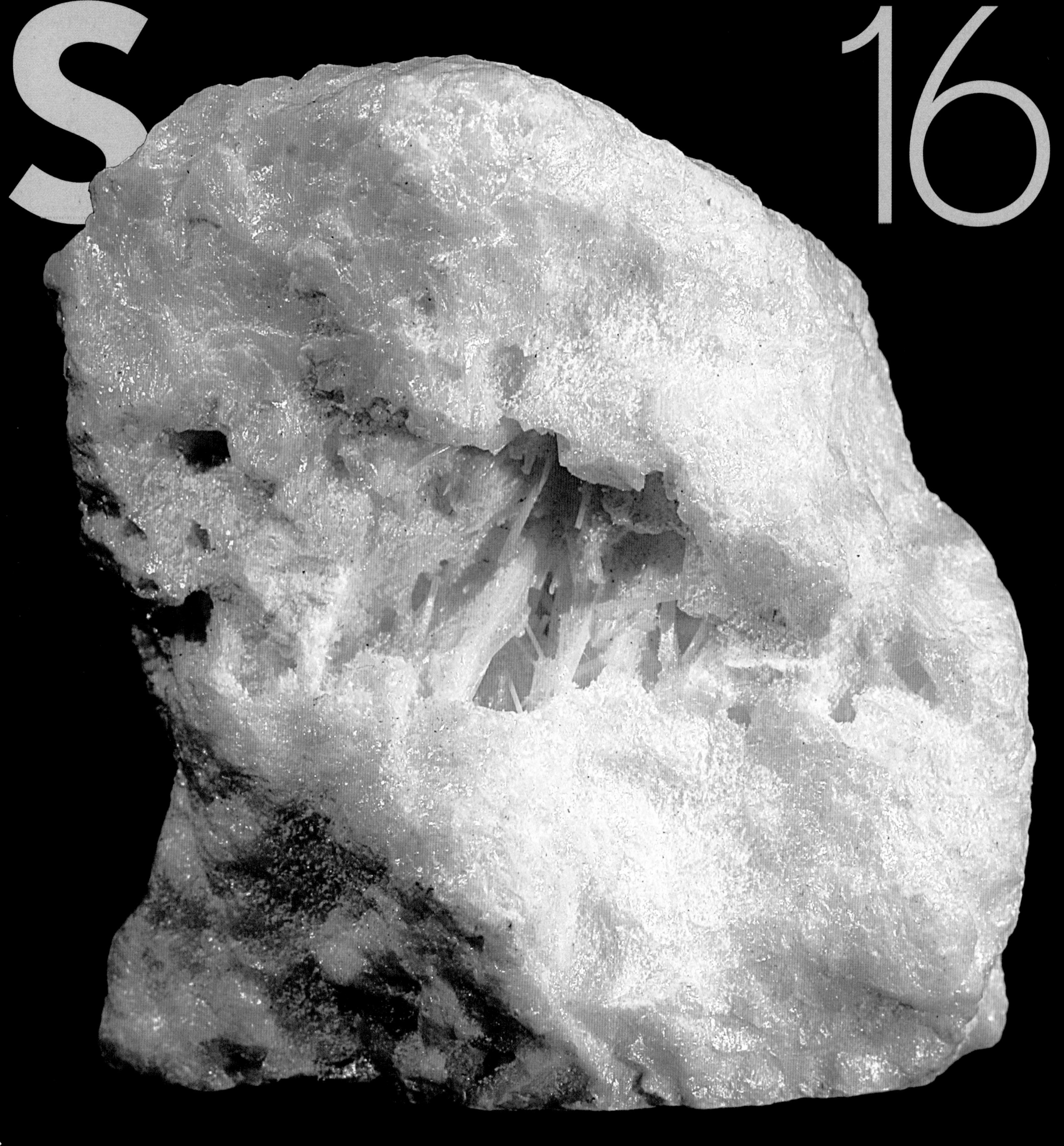
Sulfur
S
16

Sulfur

Elemental

Atomic Weight
32.065
Density
1.960
Atomic Radius
88pm
Crystal Structure

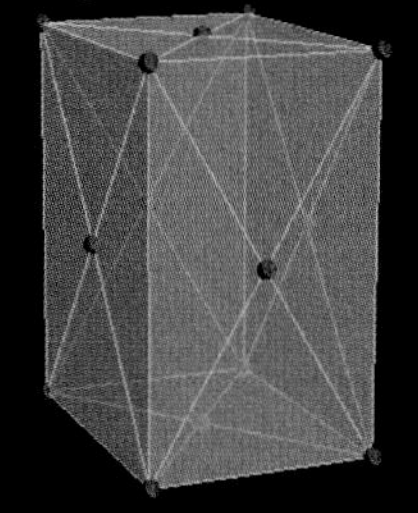

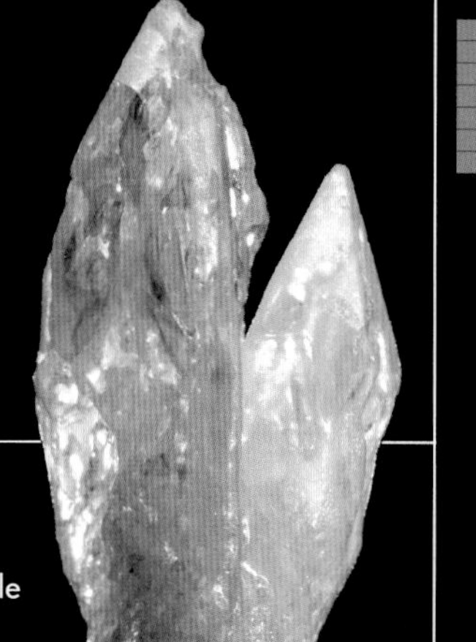

Large crystal of pure native, or naturally occurring, sulfur.

SULFUR IS SMELLY STUFF, there's just no two ways about it. It's smelly as a powder, it's smelly as a solid crystal, and when it's burning you understand why many traditions fill their hell with it (sulfur's historical name is "brimstone").

Many sulfur compounds are similarly unpleasant, chief among them hydrogen sulfide, the smell of rotten eggs. Sulfur compounds released from burning coal, oil, and diesel fuel are major components of urban smog, and cleaning it from exhaust streams and fuels is now mandatory.

Sulfur is also one of the three basic ingredients of gunpowder, and thus has the blood of millions on its hands.

Is there anything positive to say about sulfur? Well, it cannot be denied that sulfur is very *useful*. Vast quantities of it are produced and consumed in the chemical industry, primarily in the form of sulfuric acid, the workhorse acid for countless manufacturing processes.

Smelly as it is, you can buy bags of powdered sulfur in any garden center for use in adjusting soil pH. (For some reason, sulfur is generally considered an "organic" material, not like those nasty "chemical" alternatives, though frankly I find that assessment a bit hard to fathom.)

Sulfur smells bad, but you can handle large amounts of it safely. Chlorine, on the other hand, has an almost pleasant smell in low concentrations, reminding people of the pleasure of swimming pools. But watch out if there's more than a trace around.

Ninety-percent pure sulfur is available cheap in any garden center.

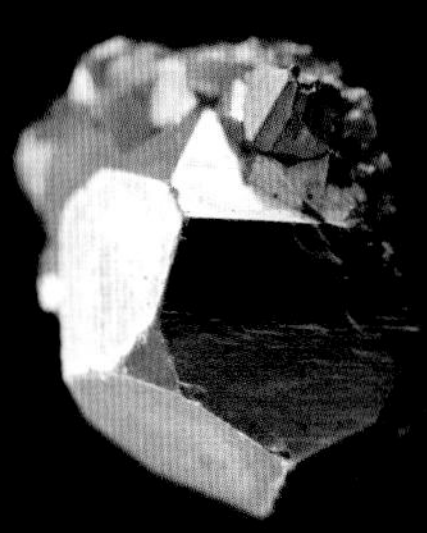

The mineral pyrite (FeS).

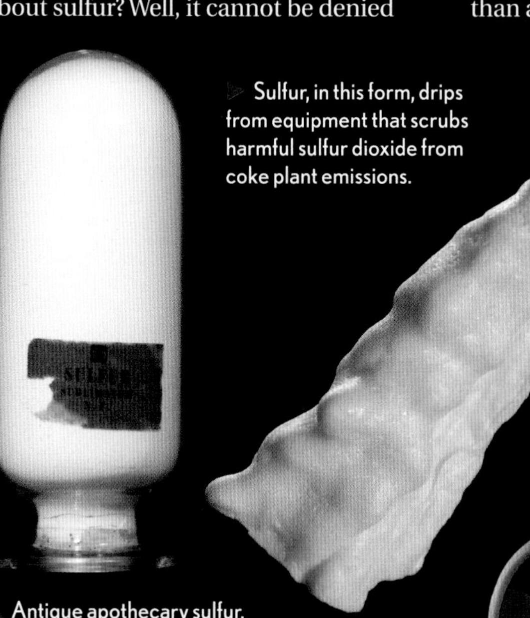

Sulfur, in this form, drips from equipment that scrubs harmful sulfur dioxide from coke plant emissions.

Antique apothecary sulfur.

The characteristic smells of garlic and onion both come from sulfur compounds.

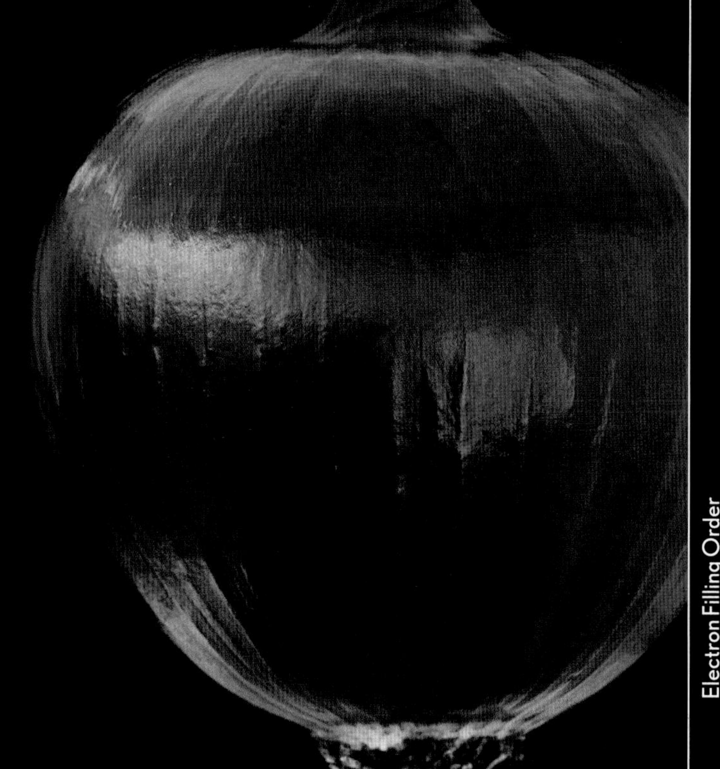

Penicillin ($C_{16}H_{18}N_2O_4S$) was once so rare, they collected the urine from patients to reuse it. This 100ml bottle, for treating horses, cost $7.

Sulfur occurs naturally in fairly pure form around volcanoes and geothermal vents.

Electron Filling Order: 1s 2s 2p 3s 3p 3d 4s 4p 4d 4f 5s 5p 5d 5f 6s 6p 6d 7s 7p

Atomic Emission Spectrum

State of Matter: 0 500 1000 1500 2000 2500 3000 3500 4000 4500 5000 5500

Chlorine
Cl
17

Chlorine

CHLORINE WAS USED in World War I as a poison gas during the grueling trench-warfare phase. Soldiers would position a line of gas cylinders at the front lines, wait for the wind to shift toward the enemy, then open the valves and run like hell. This practice—sometimes overseen personally by Fritz Haber, a man whose positive contributions to humanity are discussed under nitrogen (7)—was slowly phased out as experience showed that roughly equal numbers of soldiers on both sides died regardless of who set off the gas.

I have inhaled whiffs of pure chlorine, not enough to cause injury but probably close to the edge. The sensation is one of pure, instant agony, as if someone is pointing a blowtorch at your sinuses. Death by chlorine gas must be unimaginably awful.

On the other hand, chlorine in small amounts is one of the cheapest, most effective, and least harmful of disinfectants, saving millions upon millions of lives through treatment of drinking and wastewater, with no long-lasting environmental effects. On balance, chlorine has saved vastly more lives than it has taken.

Chlorine is found in many common household chemicals. Chlorine bleach is a solution of sodium hypochlorite (NaClO), and can release chlorine gas, with its characteristic odor, when combined with any acidic material. Common table salt is sodium chloride (NaCl), and the main component of stomach acid is hydrochloric acid (HCl).

Chlorine is a diverse element, widely distributed in nature, and chlorine ions participate in manifold functions of living organisms, from nerve conduction to digestion. While chlorine is an element of the world, argon earns its title as a noble gas by staying above the fray.

◀ Chlorine gas has a pale yellow color, just visible against a white background.

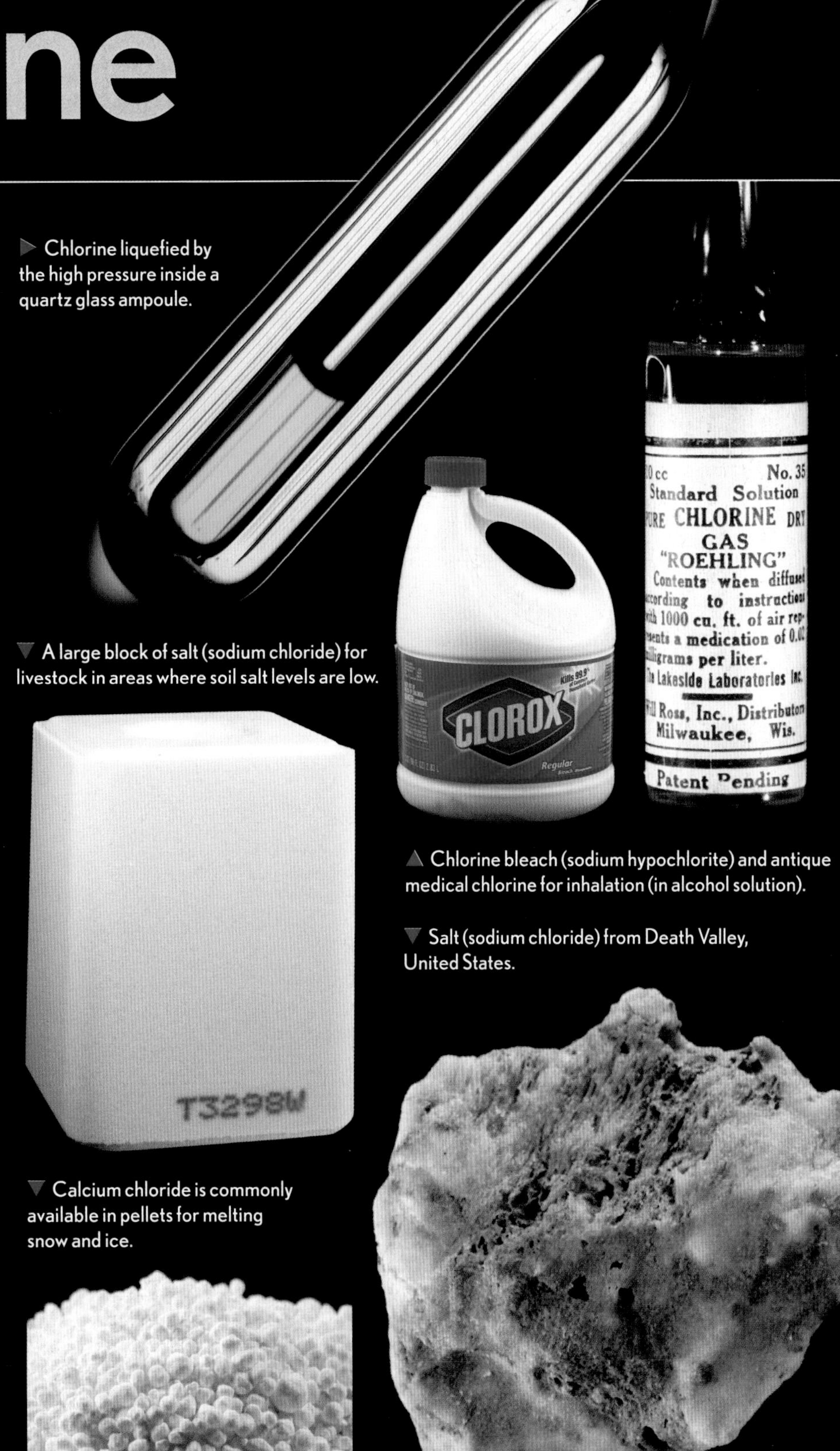

▶ Chlorine liquefied by the high pressure inside a quartz glass ampoule.

▼ A large block of salt (sodium chloride) for livestock in areas where soil salt levels are low.

▲ Chlorine bleach (sodium hypochlorite) and antique medical chlorine for inhalation (in alcohol solution).

▼ Salt (sodium chloride) from Death Valley, United States.

▼ Calcium chloride is commonly available in pellets for melting snow and ice.

Elemental

Atomic Weight
35.453
Density
0.003214
Atomic Radius
79pm
Crystal Structure

Electron Filling Order

Atomic Emission Spectrum

State of Matter

0 500 1000 1500 2000 2500 3000 3500 4000 4500 5000 5500

18

Argon

ARGON, FROM THE GREEK for "inactive," is exactly that. Nearly all its applications relate to the fact that it is the cheapest totally inert gas. Nitrogen, N_2, is even cheaper, and inert enough for many applications, but at high temperatures it can break down, while argon remains inherently and unshakably uninterested in chemical combination (except for a few highly unstable compounds of purely academic interest).

Edison's first lightbulbs used a vacuum to protect the filament from oxidation, but modern incandescent bulbs are instead filled with a mixture of nitrogen and argon at near atmospheric pressure, allowing them to have paper-thin glass walls. (Fancy, smaller bulbs are filled with krypton (36), xenon (54), and/or halogen gases to allow their filaments to burn hotter and thus brighter.)

The fact that you can buy small metal cylinders of argon gas in retail stores is thanks to the existence of gadgets used to protect opened bottles of wine from oxidation by topping off the bottle with argon. (As far as I'm concerned, it would be much easier just to drink the grape juice before it goes bad; by that simple expedient, we could avoid an awful lot of wine snobbery.)

Argon is surprisingly abundant in our atmosphere, nearly one percent by weight, which accounts for its relatively low price. Commercial argon is a by-product of the production of liquid oxygen (8) and liquid nitrogen (7), both of which are produced in huge quantities.

Moving on to potassium, we are back to elements that are intimately tied to worldly things, in this case radioactive bananas.

Pale glow from electric discharge in an indicator light.

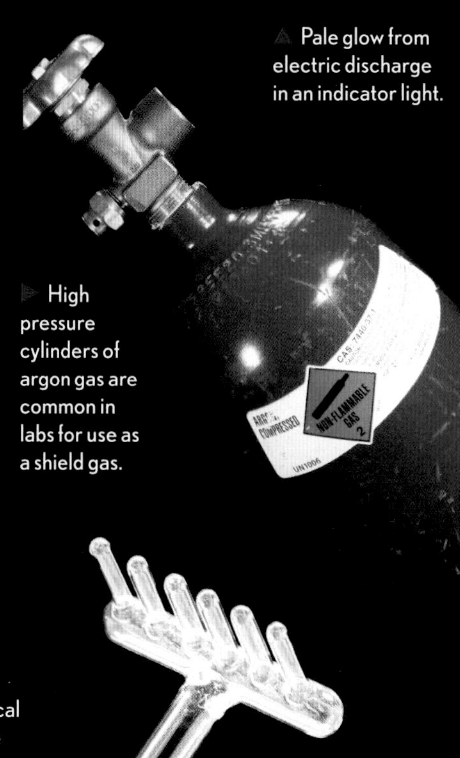

High pressure cylinders of argon gas are common in labs for use as a shield gas.

Elemental

Atomic Weight
39.948
Density
0.001784
Atomic Radius
71pm
Crystal Structure

Electron Filling Order

Atomic Emission Spectrum

State of Matter

Small disposable argon cylinder for a wine-protection gadget.

Pure argon is an invisible gas.

This quack medical "violet ray" machine created an impressive argon violet-colored electric discharge of no medicinal value.

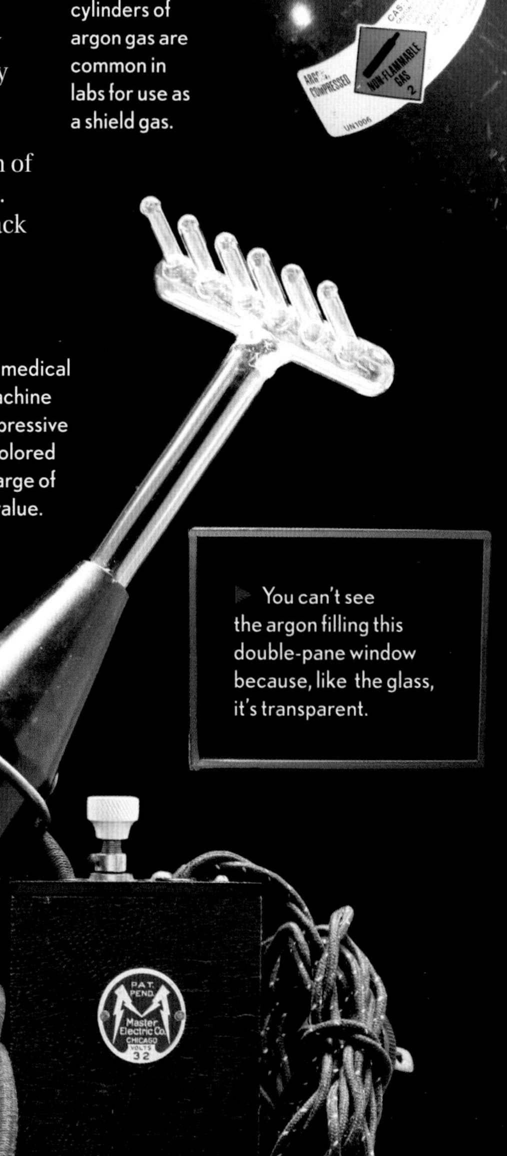

You can't see the argon filling this double-pane window because, like the glass, it's transparent.

A noble gas, argon is inert and colorless until an electric current excites it to a rich sky-blue glow.

Potassium K 19

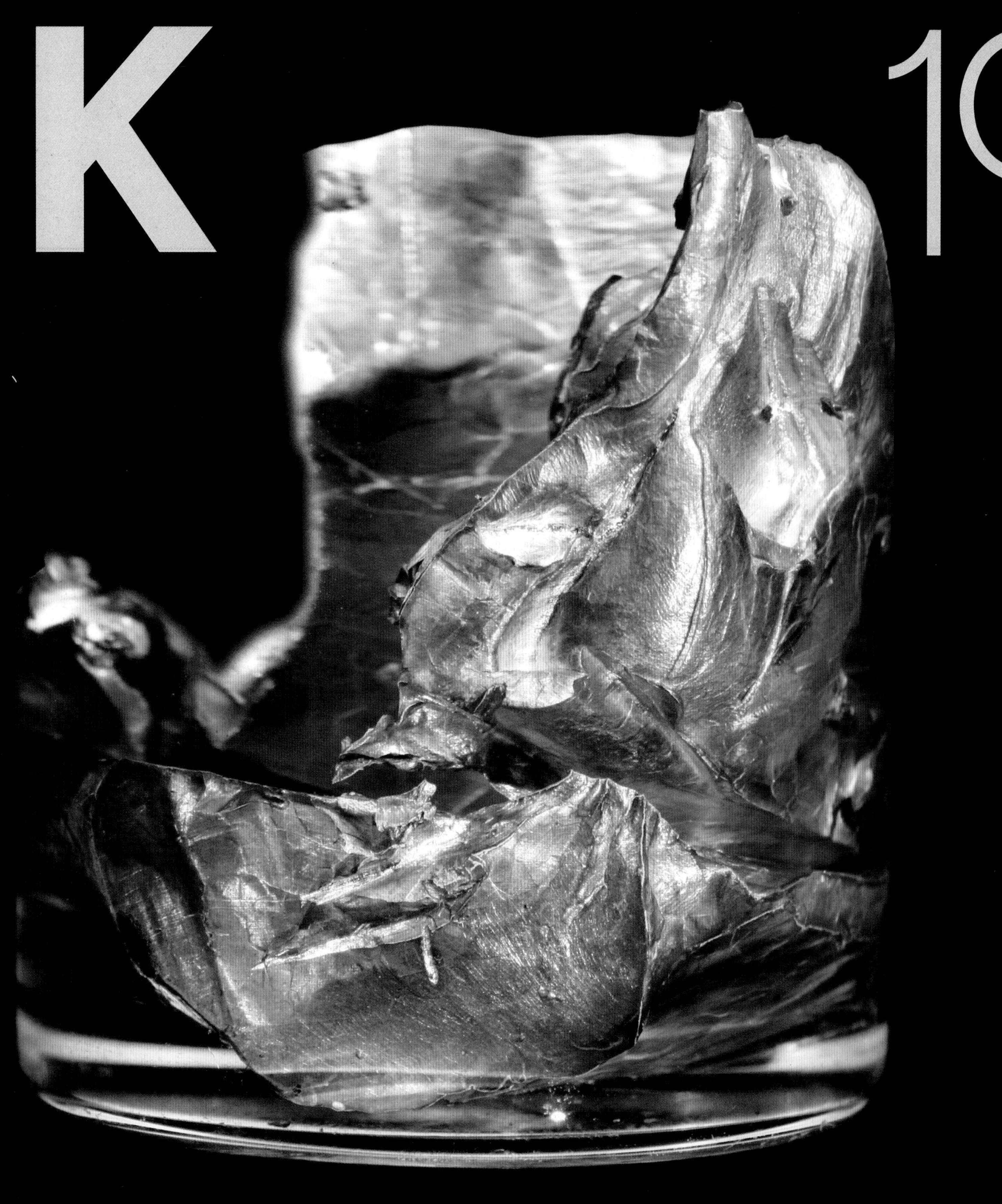

Potassium

Very pure, un-oxidized potassium is a bright shiny metal.

RADIOACTIVE BANANAS! At least that's how the headline might read if a reporter got hold of half the facts. The reassuring truth is that virtually everything you eat is radioactive, bananas just a bit more so. Bananas are rich in the important nutrient potassium, and about one hundredth of one percent of the potassium atoms in the world are the radioactive isotope ^{40}K.

This trace accounts for a significant fraction of the natural background radiation we are all exposed to every day. Intriguingly, the writer Isaac Asimov speculated that the level of ^{40}K radiation, which has been decreasing on a billion-year timescale since the formation of the earth, determined the window of opportunity for intelligent life to evolve. Too much ^{40}K in the early earth prevented the formation of fragile long genomes; too little ^{40}K later on would make the rate of mutation, and thus the rate of evolution, too slow to accomplish much.

Pure speculation of course, but it is interesting to reflect on the idea that without radiation-induced mutation, we might not be here to speculate.

Potassium, radioactive or otherwise, is one of the alkali metals, and therefore fun to throw into water. More reactive than sodium (11), potassium bursts into beautiful violet flames the instant it hits water, typically with such explosive force that the fire is spread some distance in all directions.

In the body, potassium (in the form of the K^+ ion) is critical to nerve transmission: If levels get too low, fingers start to freeze in place, and death follows if the deficiency reaches the heart. The cure, if medical care is not immediately available, is to eat bananas.

Potassium keeps things moving in the body, but calcium is what keeps the body in shape.

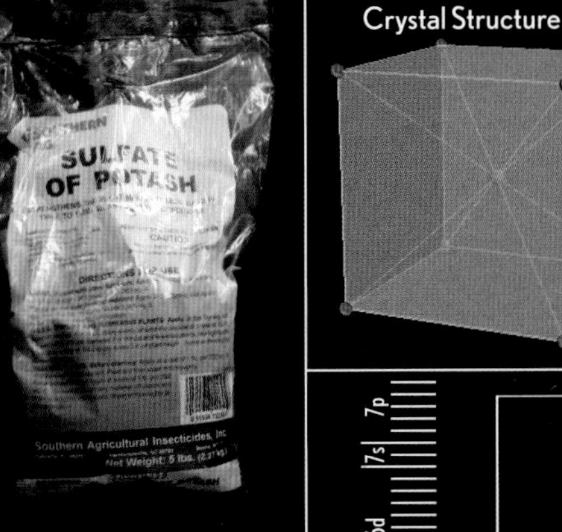

Potash (potassium carbonate) and sulfate of potash (potassium sulfate) are common fertilizers.

Sodium-free salt (potassium chloride) is very slightly radioactive.

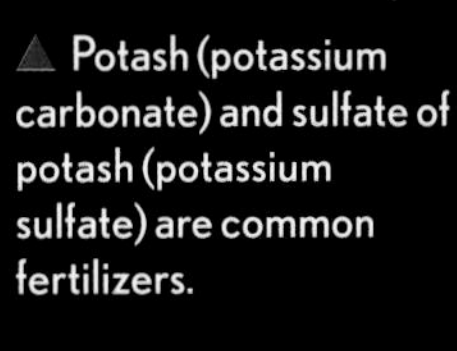

Admirably shiny potassium prepared by a collector in Germany. Keeping the metal free of oxidation is extremely difficult.

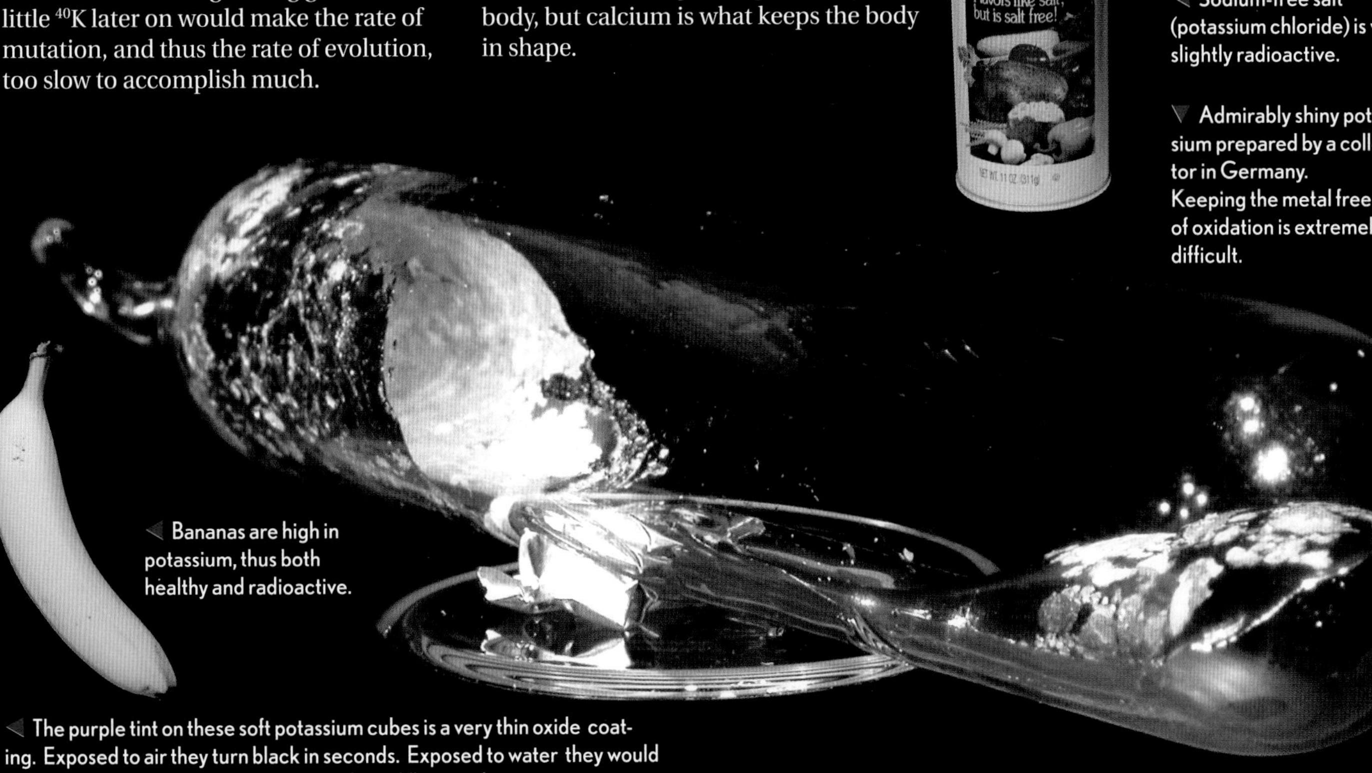

Bananas are high in potassium, thus both healthy and radioactive.

The purple tint on these soft potassium cubes is a very thin oxide coating. Exposed to air they turn black in seconds. Exposed to water they would explode, sending off characteristic purple-red flaming drops.

Elemental

Atomic Weight
39.0983
Density
0.856
Atomic Radius
243pm
Crystal Structure

Electron Filling Order: 1s 2s 2p 3s 3p 4s 3d 4p 5s 4d 5p 6s 4f 5d 6p 7s 5f 6d 7p

Atomic Emission Spectrum

State of Matter: 0 500 1000 1500 2000 2500 3000 3500 4000 4500 5000 5500

Calcium

Ca

20

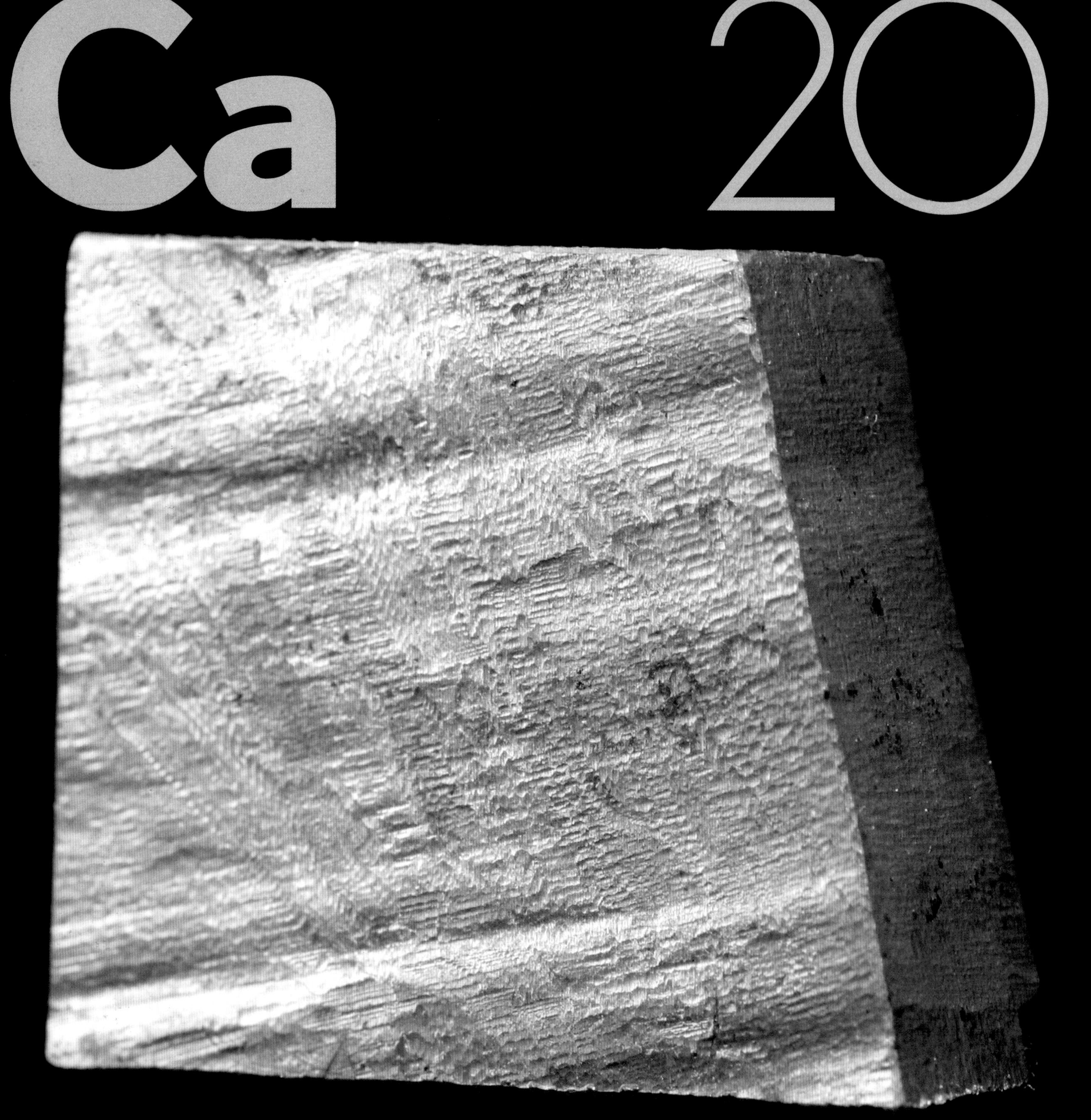

Calcium

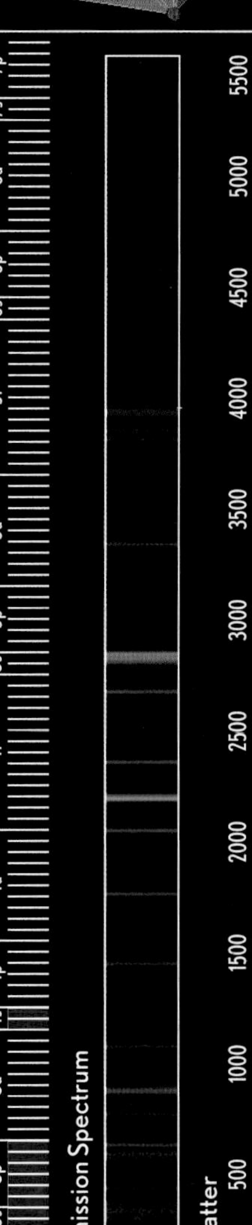

SAY "CALCIUM" and most people think of white, chalky things, or perhaps milk. The stone called "chalk," as in the white cliffs of Dover, is calcium carbonate, while chalk of the blackboard variety is today made of calcium sulfate, properly known as gypsum. (The "lead" in pencils is not made of lead and common "chalk" is not made of chalk; what is it with writing instruments and misleading names?)

Both kinds of chalk, and the calcium found in milk, are compounds of calcium. The pure element itself is a shiny metal, similar in appearance to aluminum. You rarely see it in metallic form because it is unstable in air, decomposing fairly rapidly into calcium hydroxide and calcium carbonate, which, as you might expect, are chalky white substances. On contact with water or acid, calcium metal generates hydrogen gas much as the alkali metals do, but at a slower, controlled pace that makes it a useful source of small amounts of hydrogen.

We're always told that calcium is important for strong bones, and it is indeed a major component of bone mineralization (mammal bones are a rigid hydroxyapatite foam, a form of hydrated calcium phosphate). But while it's possible to imagine bones made of something else (such as glass—see silicon, element 14), the calcium ion's function in the biochemistry of the cell is more fundamental. Calcium is constantly moving in and out of cells, mediating the action of nerves and muscles in such important ways that the body will begin to dissolve bones rather than allow blood calcium levels to fall. (In fact, one theory posits that bones evolved initially as a way to store calcium for just such an eventuality, taking on structural functions only as an afterthought.)

Calcium is among the elements that life requires in substantial quantities. Others, like selenium (34), are required in tiny amounts for just a few specialized enzymes. Still others, like scandium, have absolutely no function whatsoever in the body.

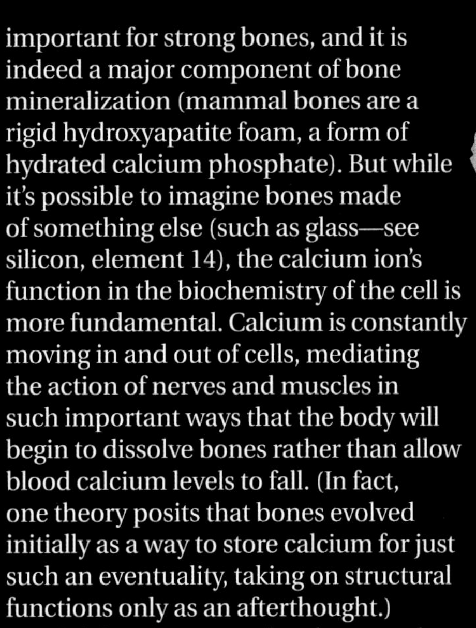

▲ Chalk is made of gypsum (calcium sulfate).

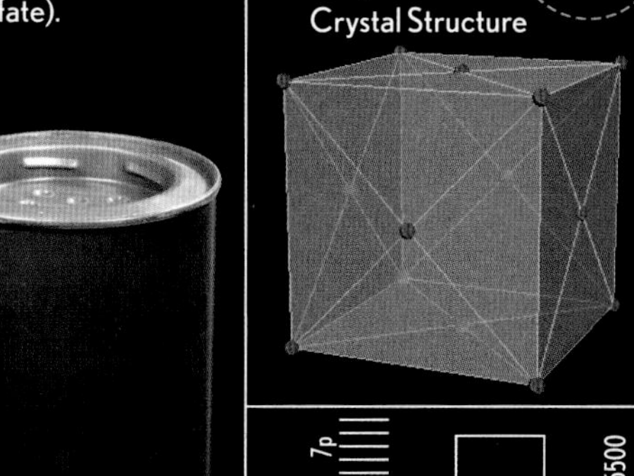

▲ The mineral calcite (calcium carbonate).

▼ Seashells are made of calcium carbonate.

▲ A canister of calcium hydride, used to generate hydrogen to inflate weather balloons.

▲ Calcium metal in a canister used to generate hydrogen for an obscure military purpose.

◄ Surprisingly, pure calcium is a firm silvery metal. Only in compounds is it characteristically chalky.

◄ Rare Hawaiian coral made of calcium carbonate.

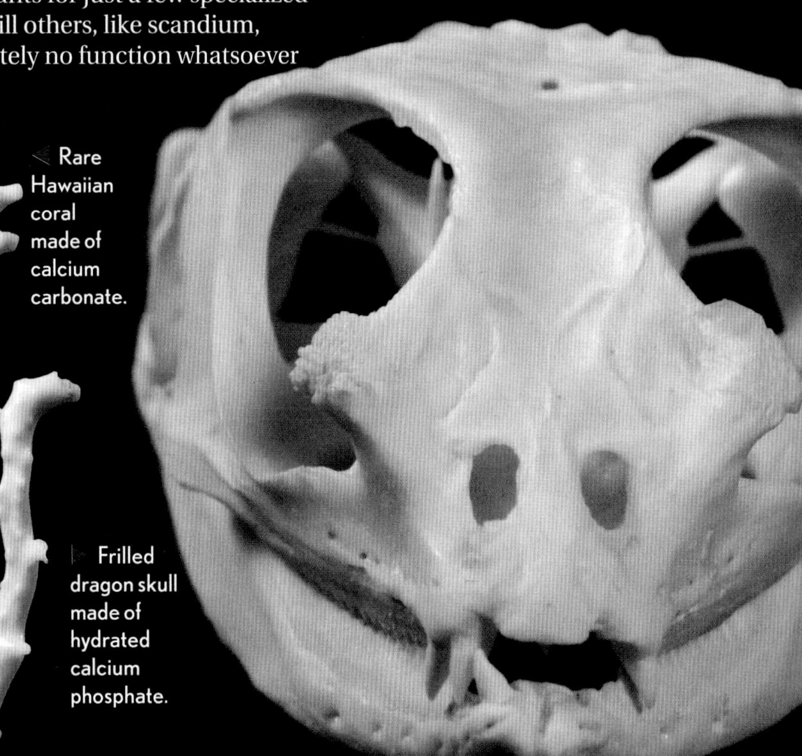

► Frilled dragon skull made of hydrated calcium phosphate.

Scandium

Sc 21

Scandium

SCANDIUM IS THE FIRSTof the elements you've never heard of. The total world trade in scandium in pure metallic form is less than a hundred pounds a year, so it's safe to say that very few people have ever seen this element in pure form. (The trade in scandium oxide is around ten tons per year, still a tiny amount by world standards.)

Scandium is an example of an element that is expensive not because it's particularly rare in the earth's crust, but because there's no place where it's concentrated. For most other elements, even those much rarer overall, there is some ore somewhere in which it can be found at much higher concentration; but scandium is spread thinly all over, which makes it expensive to gather and purify.

Scandium is used to make strong metals and bright lights. A tiny amount mixed with aluminum creates some of the strongest aluminum alloys known, which are used in fighter jets, baseball bats, and bicycle frames (all of the expensive variety). Scandium iodide in high-intensity metal-halide discharge lighting transforms what would otherwise be harsh light into a more pleasant sun-like spectrum.

Metal-halide lighting is used where very large amounts of light are needed, on streets and in warehouses and megastores. It is more efficient than any other common light source except sodium vapor, whose yellow color has a tendency to make people look like zombies and is thus too unpleasant for anything but highway lighting. While LED lighting may one day dominate our homes, the sheer volume of light you can get from metal-halide bulbs will continue to ensure its place in the public eye. Literally.

Scandium light is something millions see without ever hearing the name. Titanium, on the other hand, is a name millions hear, even when there is no actual titanium in sight.

Scandium-aluminum alloy is used in high-end bicycle frames for its great strength.

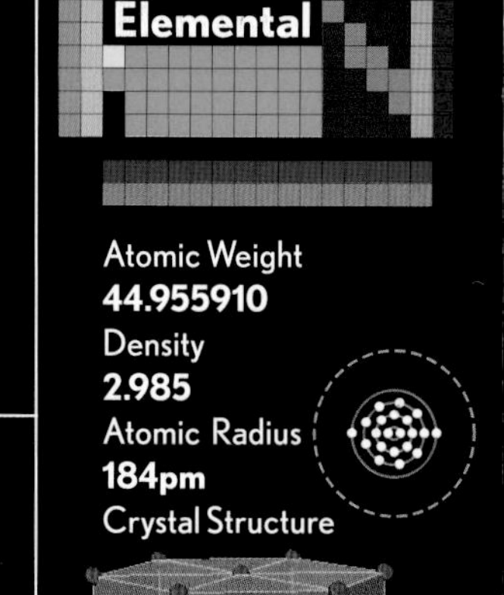

Scandium-aluminum master alloy ingot, the form in which much of the world's scandium is traded.

The mineral kolbeckite ($ScPO_4 \cdot 2H_2O$).

These vacuum distilled scandium crystals are destined for use in daylight spectrum metal halide arc lights.

Scandium in metal halide lights helps produce a particularly pleasing spectrum of light.

Titanium

Ti

22

Titanium

Atomic Weight
47.867
Density
4.507
Atomic Radius
176pm
Crystal Structure

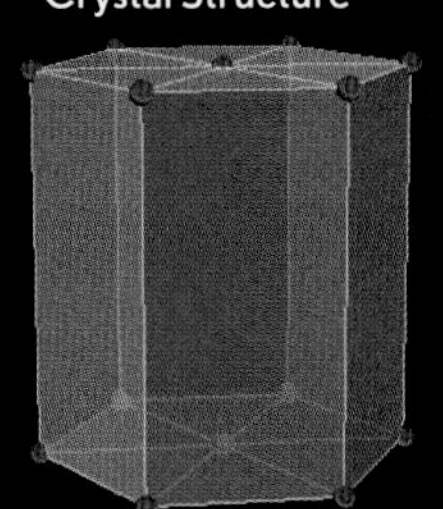

TITANIUM IS ONE OF the most popular element names, so popular that marketers apply it to thousands of products whether they contain actual titanium or not.

If you have a golf club with the word "TITANIUM" prominently molded into the metal of the club, think twice before concluding that it's actually *made* of titanium. Some are, some are not. An easy test is to hold the club up against a grinding wheel: If you don't see the characteristic bright white sparks of genuine titanium, then you haven't damaged anything of value.

Titanium stands for strength, both in name (after the Titans, gods of Greek legend) and in fact (it's used in jet engines, tools, and rockets for its tremendous strength). It's also completely nonrusting and nonallergenic, so much so that it's popular for use inside the body in the form of artificial hip joints, dental implants, and body jewelry (i.e., tongue studs, eyebrow rings, and other such items commonly found stuck in teenagers).

Though titanium metal is expensive, its ore is actually quite abundant. The high cost comes from the difficulty of refining the metal, not its scarcity. Titanium dioxide is everywhere. It is the white in white paint, and in all other colors of paint TiO_2 is the opacity, the substance that prevents what's underneath from showing through. Even this book's paper contains titanium dioxide to keep print on one side from showing through to the other.

From missiles to razors, titanium is a popular superstar, and must be a source of great envy to its neighbor, vanadium, which labors in obscurity even while helping create an alloy of much greater strength than titanium.

A ring machined by the author from a solid bar of 99.999% pure crystal titanium.

Gold-colored titanium nitride coating on electric razor blades.

Top half of an artificial hip joint made of pure titanium.

Clockwise from top left, a gear wire-cut from solid titanium, titanium speaker cone, titanium ring, and titanium piercing barbell.

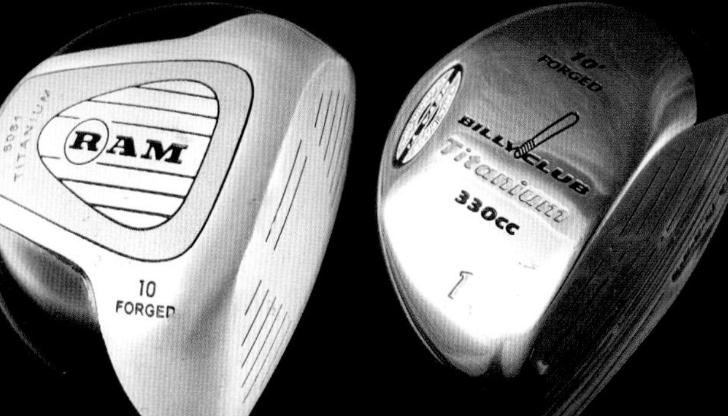

Two golf clubs, one real titanium, one fake. Hint: 6061 is a standard aluminum alloy.

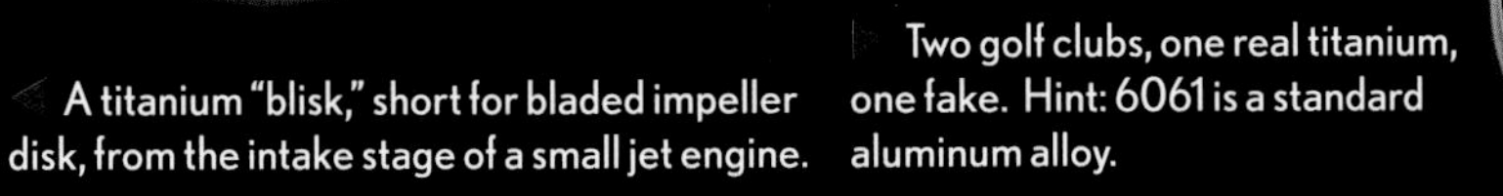

A titanium "blisk," short for bladed impeller disk, from the intake stage of a small jet engine.

Electron Filling Order
1s 2s 2p 3s 3p 3d 4s 4p 4d 4f 5s 5p 5d 5f 6s 6p 6d 7s 7p

Atomic Emission Spectrum

State of Matter
0 500 1000 1500 2000 2500 3000 3500 4000 4500 5000 5500

Titanium 22

The titanium beaded surface on this artificial hip joint encourages bone intergrowth.

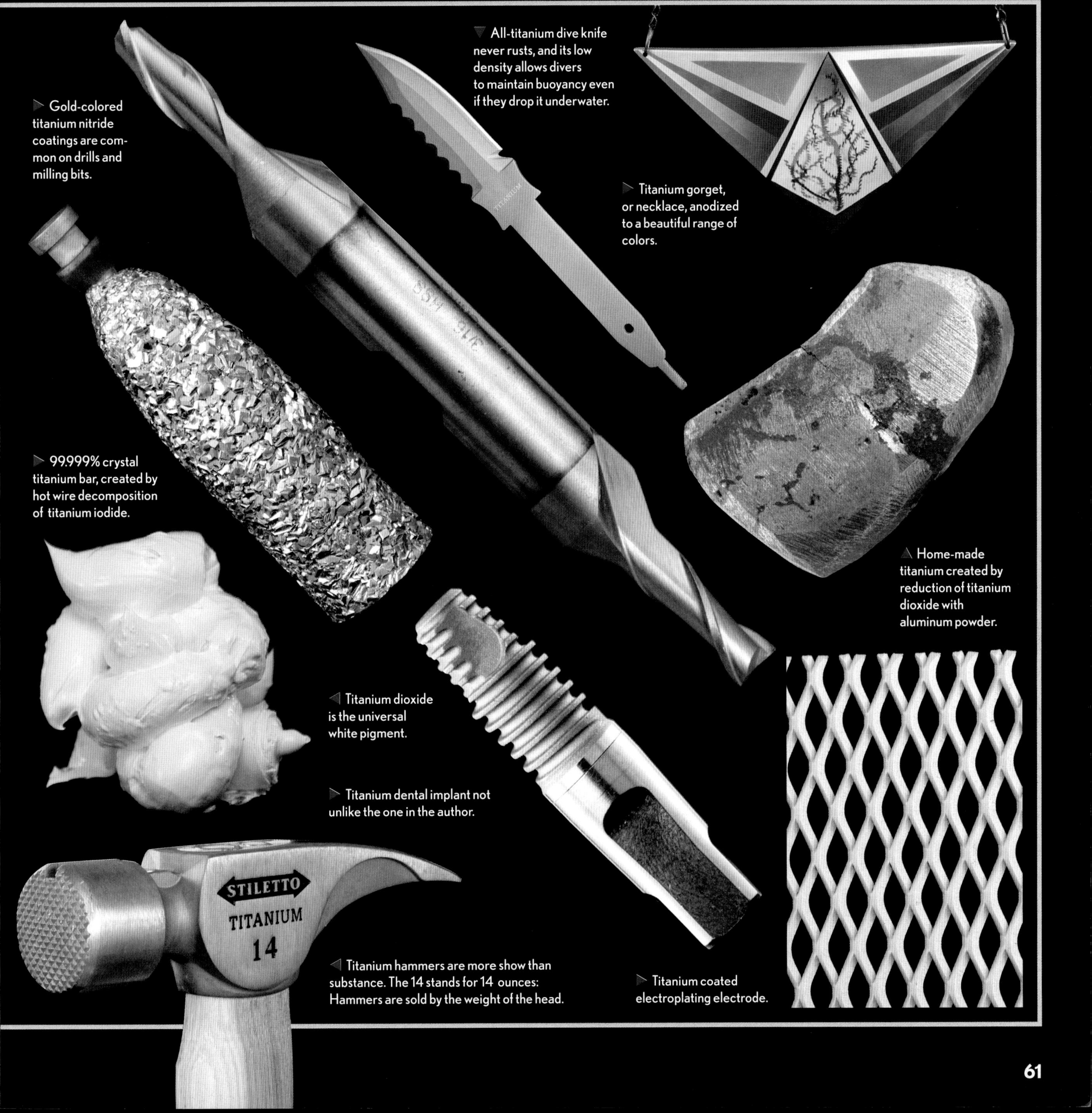

▶ Gold-colored titanium nitride coatings are common on drills and milling bits.

▼ All-titanium dive knife never rusts, and its low density allows divers to maintain buoyancy even if they drop it underwater.

▶ Titanium gorget, or necklace, anodized to a beautiful range of colors.

▶ 99.999% crystal titanium bar, created by hot wire decomposition of titanium iodide.

▲ Home-made titanium created by reduction of titanium dioxide with aluminum powder.

◀ Titanium dioxide is the universal white pigment.

▶ Titanium dental implant not unlike the one in the author.

◀ Titanium hammers are more show than substance. The 14 stands for 14 ounces: Hammers are sold by the weight of the head.

▶ Titanium coated electroplating electrode.

Vanadium

V

23

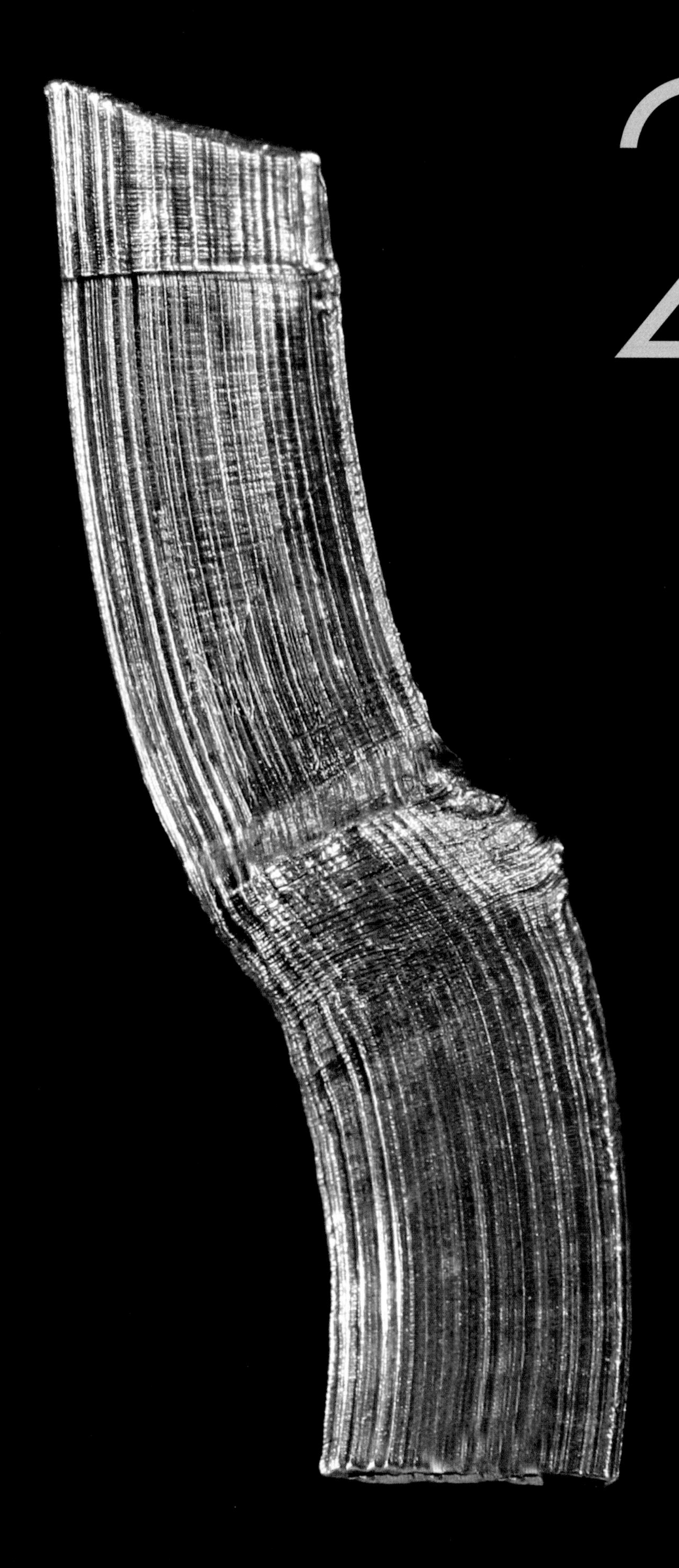

Vanadium

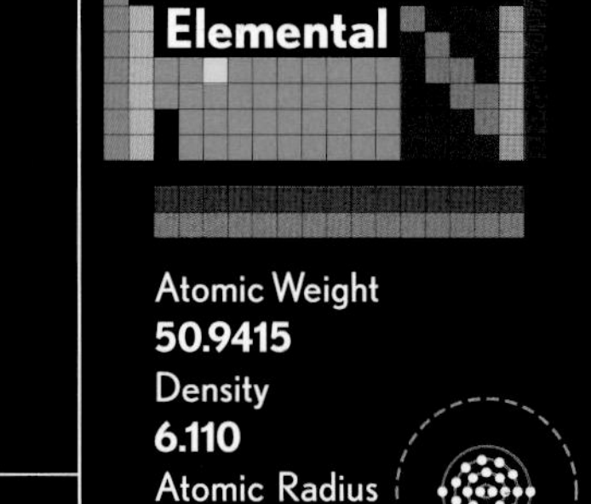

Atomic Weight
50.9415
Density
6.110
Atomic Radius
171pm
Crystal Structure

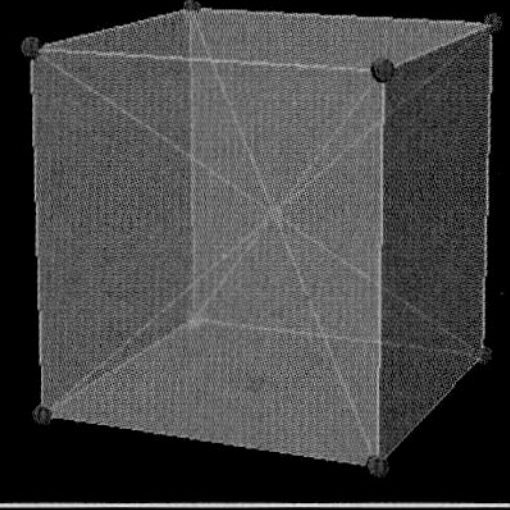

TOOL STEEL AND HIGH speed steel are families of iron (26) alloys distinguished by their supreme hardness, toughness, and wear resistance, properties contributed by a few percent of vanadium in the form of vanadium carbide. While heavier than titanium (22), vanadium steel is also *much* harder.

Because steel alloying is its main application, vanadium is most often sold in the form of ferrovanadium master alloy meant to be added to steel in the pot before casting. The master alloy contains a much higher percentage of vanadium than the final product, but melts easily when added to liquid iron—unlike pure vanadium, which has a much higher melting point.

Though not nearly as glamorous and popular a marketing term as titanium, you will find "vanadium" stamped prominently onto many tools. Unlike many instances of titanium branding, you can be fairly sure that those tools really are made of vanadium alloy steel. Though tungsten carbide cutting bits now provide a harder alternative, vanadium high speed steel remains a workhorse of industrial machining and a staple in every home workshop in the form of drill and router bits, socket wrenches, pliers, and so on.

Power and grit define vanadium's working life, but it has a dainty side as well: The green color of some emeralds comes from an impurity of vanadium. (Quite a few otherwise hardworking elements come together to form the beauty of emeralds, which are crystals of beryllium aluminum silicate, generically known as beryl.)

If vanadium makes *some* emeralds green, what about the other green emeralds? Their color comes from a close neighbor, chromium.

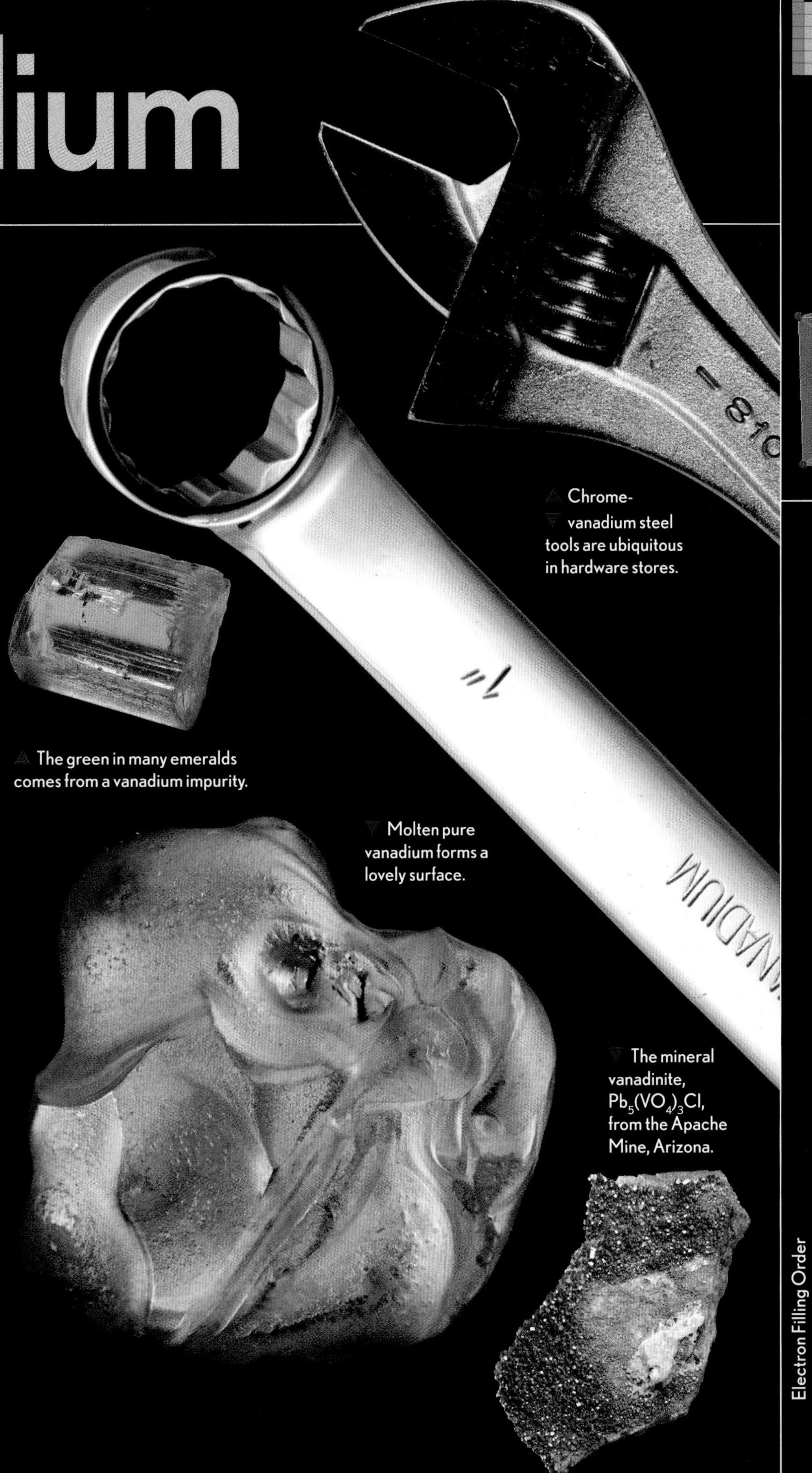

Chrome-vanadium steel tools are ubiquitous in hardware stores.

The green in many emeralds comes from a vanadium impurity.

Molten pure vanadium forms a lovely surface.

The mineral vanadinite, $Pb_5(VO_4)_3Cl$, from the Apache Mine, Arizona.

This elegant vanadium sculpture is actually a tiny chip cut from a vanadium cylinder on a lathe.

Electron Filling Order: 1s 2s 2p 3s 3p 4s 3d 4p 5s 4d 5p 6s 4f 5d 6p 7s 5f 6d 7p

Atomic Emission Spectrum

State of Matter: 0 500 1000 1500 2000 2500 3000 3500 4000 4500 5000 5500

Chromium

Cr 24

Chromium

THE CAR INDUSTRY went through its chrome phase in the 1950s and '60s, when cars were decked out with acres of blinding chrome from bumper to bumper (literally, since the bumpers were where much of the chrome resided). This type of chrome, like virtually all the pure chrome you will see in everyday life, is a very thin layer of chromium metal electroplated over a thicker layer of nickel (28), which in turn is electroplated over a base of iron (26), zinc (30), brass, or even plastic.

A microscopic layer is all you'll normally see of this element in pure form, but alloyed with iron and nickel, chromium is a key ingredient in stainless steel, making up as much as a quarter of the weight of some stainless alloys. Chromium is also very commonly used with its neighbor, vanadium (23) in chrome-vanadium steel. Go into any hardware store and it will not take you long to find a crescent wrench, socket set, or other tool with "Cr-V" stamped on it.

Extremely shiny, highly corrosion resistant, and beautiful in so many ways, the only reason chromium plating isn't used instead of silver (47) in jewelry is that it's just too cheap to be taken seriously. One place where chromium has displaced silver is in "silverware"; our eating utensils these days are chrome-based stainless steel in all but the most pretentious settings.

Chromium is valued by artists for the rich green pigment it creates, known as chromium oxide green for fairly self-evident reasons. (Not to be confused with Paris green, which is made with arsenic, element 33.)

One of the first pigments, found in cave paintings tens of thousands of years old, is made not with chromium but with manganese.

◀ These chips show the result of plating chromium until a thick slab is built up. This process, called electrowinning, is how high-purity chromium is obtained from solution.

▲ Crystal structure is visible in this high-purity sputtering target.

◀ Chrome-vanadium socket wrench proudly advertises its chemical composition.

▶ There isn't anything you can't get chrome-plated.

▶ Chromium oxide green is a common pigment in paints and glazes.

▲ Super-high-purity vapor-deposited chromium crystals.

▶ Common stainless steel is around 20 percent chromium.

Elemental

Atomic Weight
51.9961
Density
7.140
Atomic Radius
166pm
Crystal Structure

Electron Filling Order: 1s 2s 2p 3s 3p 4s 3d 4p 5s 4d 5p 6s 4f 5d 6p 7s 5f 6d 7p

Atomic Emission Spectrum

State of Matter: 0 500 1000 1500 2000 2500 3000 3500 4000 4500 5000 5500

Manganese

Mn

25

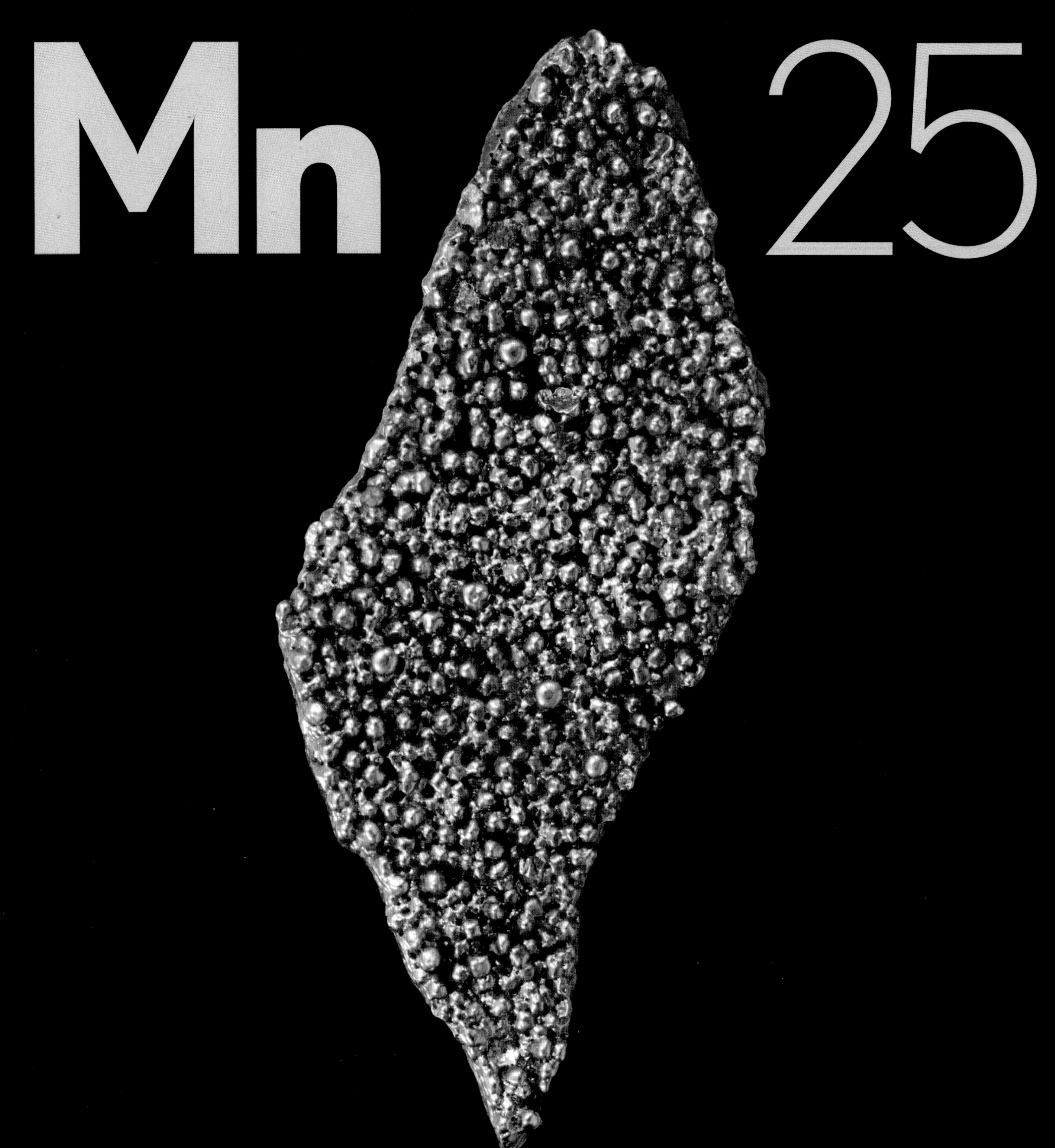

Manganese

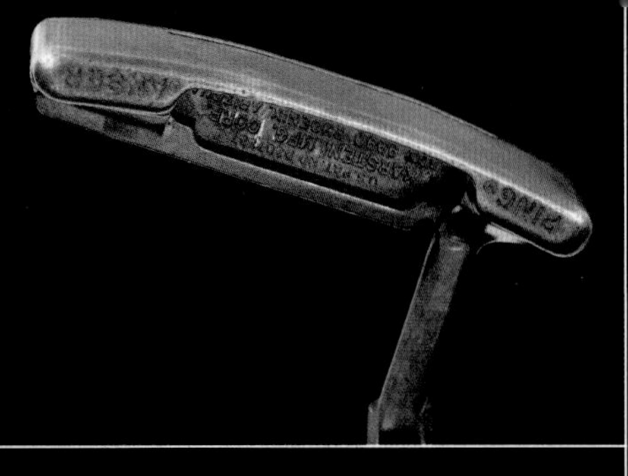

▲ Manganese in the bronze of this putter won't help your score any more than the dozen other exotic elements tried for golf clubs.

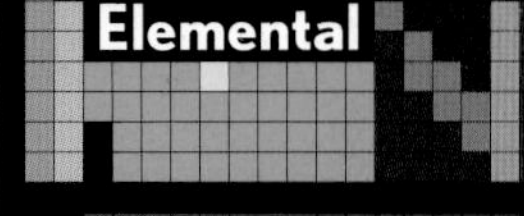

Atomic Weight
54.938049
Density
7.470
Atomic Radius
161pm

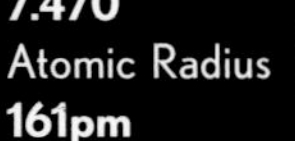

Crystal Structure

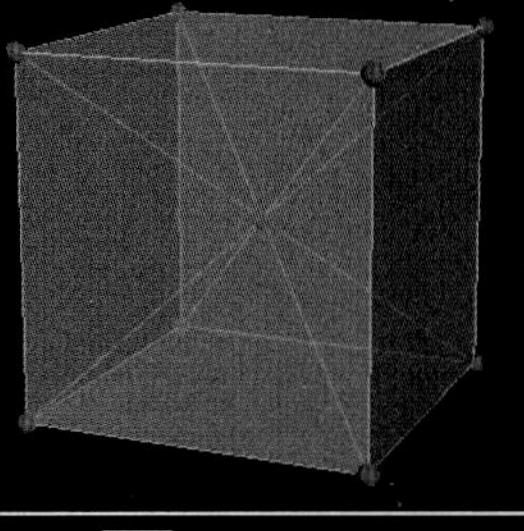

Electron Filling Order
Atomic Emission Spectrum
State of Matter

ALONG WITH RED IRON OXIDE, black manganese oxide pigments are among the earliest known, having been found in cave paintings dating from at least 17,000 years ago. But the most interesting episode in manganese history is much more recent.

In the mid 1970s, there was much excitement about the riches to be made collecting manganese nodules from the deep ocean. The eccentric billionaire Howard Hughes set off a manganese rush by commissioning a special ship, the *Hughes Glomar Explorer*, to probe the ocean floor northwest of Hawaii and harvest its bounty of manganese nodules.

The whole thing was a complete lie. Hughes had been hired by the CIA to participate in an elaborate cold war ruse. The real purpose of the *Glomar Explorer* was to raise a sunken Russian ballistic missile submarine, the K-129.

The CIA knew that any sort of deep-sea exploration in that part of the ocean would immediately raise suspicions, unless it had an airtight cover story of such sweeping scope and detail that no one but a conspiracy nut would imagine it was a cover-up. But that's what it was.

Yes, there are manganese nodules on the ocean floor, but no one ever made any money from them, and most likely no one ever will. The CIA didn't get their prize either; the section of the submarine containing the codebooks broke off as it was being raised, and in the end they got nothing more than a few torpedoes and six dead Russian crewmembers, who were buried at sea with military honors.

I guess we should mention that manganese is actually quite useful, its main application being in alloys made mainly of our next element, iron.

▲ These antique glazed tiles show the use of manganese oxide as a black pigment.

▼ The author traded a mineral dealer this magnificent rhodochrosite (manganese carbonate) crystal for several hundred lesser minerals.

◀ These rough slabs are created by electroplating manganese out of a solution until enough metal builds up to break off. The bumpy surface occurs naturally as the current finds the path of least resistance.

▶ Manganese steel is prized for the sharp edge it can take, as in this straight razor.

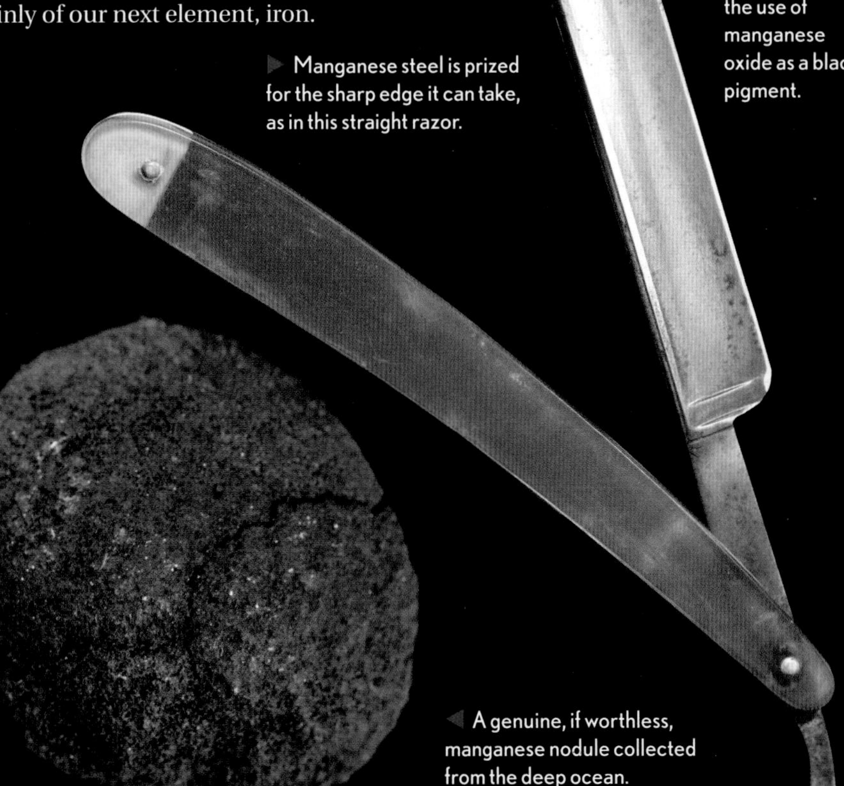

◀ A genuine, if worthless, manganese nodule collected from the deep ocean.

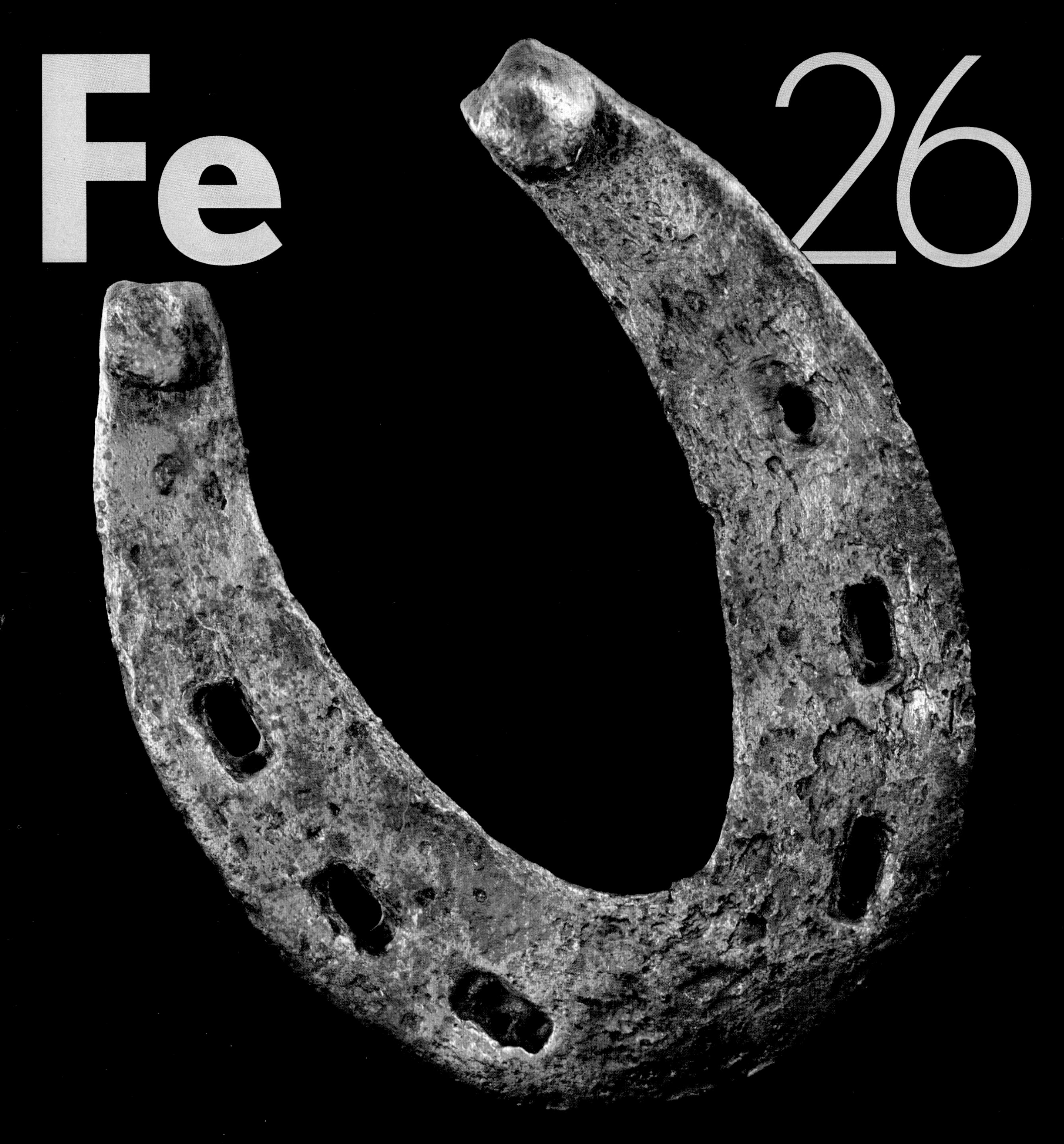
Iron
Fe
26

Iron

IRON IS THE ONLY ELEMENT to have an Age named after it (the other Ages being Stone and Bronze, which are, respectively, varied mixtures of compounds and an alloy). Iron richly deserves the honor; if you're going to name Ages after their primary toolmaking material, iron is absolutely without equal. You could argue we're still very much in the Iron Age.

When people describe metals such as aluminum (13) or titanium (22) as being lighter, stronger, or more corrosion resistant, they are always comparing them to one thing, iron, because to this day iron in the form of steel remains *the* metal of industry. When push comes to shove, if you want to build something really big, or really strong, iron is what you use (the one exception being when it's supposed to fly; then weight settles the argument in favor of a more expensive, lighter metal).

The fact that iron rusts so readily is one of the great lousy breaks of chemistry, responsible for untold billions in costs every year. But in iron's favor are its very low overall cost and its ability to form an astonishing range of alloys whose properties can be finely tuned, from super-hardness to extreme tensile strength to high vibration damping. The ease with which iron can be welded, cast, machined, forged, cold-worked, tempered, hardened, annealed, drawn, and generally persuaded to take on unlikely shapes and tempers is unequaled by any other metal.

So important is iron as a metal that it's easy to forget that many forms of life depend crucially on iron atoms, such as those trapped in the core of the protein hemoglobin and responsible for transporting oxygen in our blood. Iron is thus one of the most critical of the minor constituents of the human body.

Metal ions are often found at the core of important enzymes. For hemoglobin it's iron, while for the very similar chlorophyll molecules in plants it's magnesium (12), and in the blue blood of spiders and horseshoe crabs it's copper (29). For vitamin B_{12} the core is cobalt.

Iron is synonymous with tools, but not all are as fabulous as this Chinese marvel.

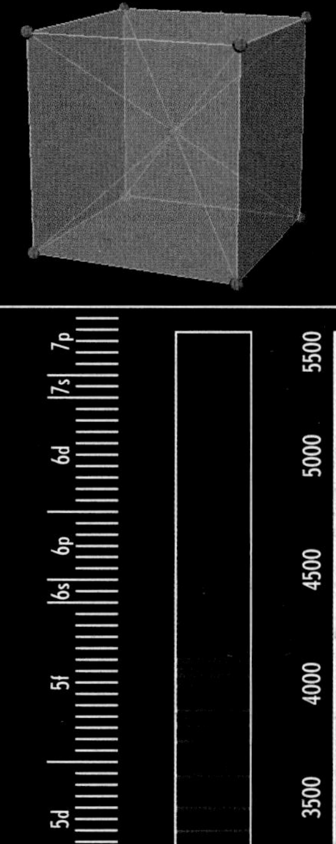

Common iron horseshoe magnets are weak in comparison to modern alternatives.

Cute salesman's sample of a cast-iron stove, made of real cast iron.

Stainless steel chain-mail glove used by butchers.

A section of the steel cable that pulls visitors up the St. Louis Gateway Arch.

Medieval horseshoe shows pitting from centuries of slow rusting.

High speed steel milling bit.

Iron 26

This two and a half inch diameter iron ball was one of many in a load of canister shot, fired from a canon during the Civil War and found over a century later in a forest in Pennsylvania.

Iron meteorites are often sliced and polished to show off their internal structure.

Iron coins have an obvious problem: They rust.

A huge iron wrench head for 4-inch nuts.

This ammonite fossil formed out of pyrite (iron sulfide, also known as fool's good). The gold color is not painted on, it's completely natural.

Hand-cut iron nails were so valuable they were carefully recovered from burned buildings, but with mass production came cheap, ubiquitous steel nails.

Fifty pounds of iron in the form of a fifty-pound weight.

Cast iron cookware is heavy, but indestructible.

Cobalt

Co 27

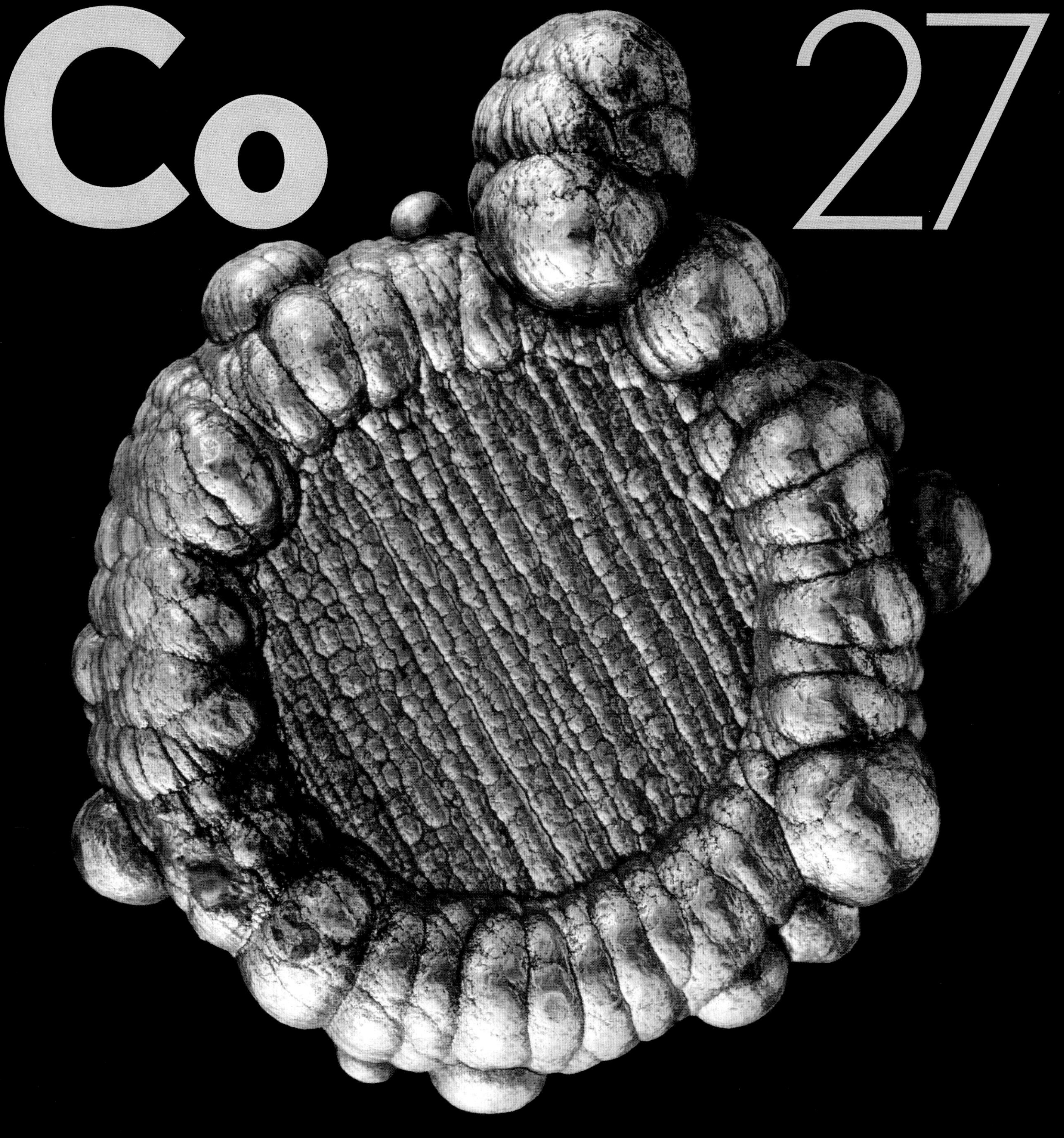

Cobalt

I'M NOT SURE HOW common this sentiment is, but for years cobalt was an element that made me feel nervous. Its main association in my mind, and doubtless in that of many others, was with nuclear fallout. But that's just a particular isotope, ^{60}Co. Although this isotope is indeed highly radioactive, and was a deadly component of the fallout from atmospheric nuclear bomb tests in the 1950s, ordinary cobalt is not radioactive at all.

In fact, cobalt is a quite ordinary metal, similar in appearance to nickel (28), and used, like many of the other elements in this neighborhood of the periodic table, as a component in steel alloys. Cobalt steel is one of the hardest and toughest alloys used for drills and milling machine bits.

Aficionados of glass trinkets will know the deep blue color of cobalt glass, used in everything from bottles to electrical insulators. (For some reason, glass insulators from old telephone wires, power lines, and railroad signals are avidly collected, commanding shockingly high prices on eBay.)

The blue comes from a trace of cobalt compounds added to the glass, and actually has a more serious application beyond cheap bottles and overpriced antique insulators. When the very strong yellow emission lines from sodium (11) interfere with spectroscopic measurements, cobalt-blue filters are used to block it out selectively, allowing other colors of light through.

Cobalt and its neighbor, nickel, are chemically quite similar, but nickel has a far higher public profile, primarily due to its frequent presence in American pockets.

A cobalt electrowinning button, built up by extended electroplating.

Cobalt-aluminum oxide has been an important pigment for centuries.

Cobalt-steel is widely used for milling bits.

Rare pale cobalt blue glass insulator.

Cobalt glass telephone-wire insulator.

Cobalt electrowinning nodules.

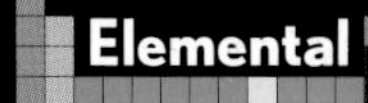

Atomic Weight
58.9332
Density
8.9
Atomic Radius
152pm

Crystal Structure

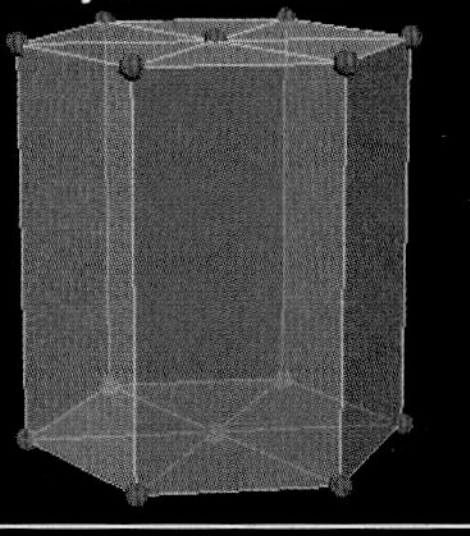

Electron Filling Order: 1s 2s 2p 3s 3p 3d 4s 4p 4d 4f 5s 5p 5d 5f 6s 6p 6d 7s 7p

Atomic Emission Spectrum

State of Matter: 0 500 1000 1500 2000 2500 3000 3500 4000 4500 5000 5500

Ni
28

Nickel

NICKEL IS WIDELY USED in coins, as reflected by the fact that there is a U.S. coin called "a nickel," without further qualification, making the word "nickel" both an element and a denomination. And that's information I wouldn't give you a nickel for.

Pure nickel is everywhere in daily life. It's plated over iron (26) to prevent rust, and over brass to make it colorless rather than yellow. Huge quantities are used to plate automobile bumpers. Because of its high value, the metal is stored in special warehouses under armed guard before being applied to the bumpers. (About a pound is plated onto each bumper, worth between $5 and $25 depending on the highly fluctuating market price of nickel.)

Sometimes, but not always, the nickel layer is covered with a thinner layer of chromium (24). Plain nickel plating is found in utilitarian applications because the layer of chrome affects only the appearance, giving a brighter, more perfectly mirrorlike shine than nickel. Rustproofing is provided entirely by the nickel layer.

Nickel is also a component of stainless steels, and more exotically it is a key ingredient in the nickel-iron superalloys used in jet engines. These superalloys maintain high strength even at very high temperatures, for example in the exhaust gas path of a jet engine. Titanium (22) can be (and is) used in cooler parts of the engine, because of its lighter weight; but for the truly hellish jobs, nickel-iron alloys are the metals of choice.

U.S. nickel coins are actually only about 25 percent nickel; the rest is copper, historically the most popular coinage metal of them all.

Antique nickel-plated dollhouse scale.

Nickel-metal hydride batteries are losing out to lithium varieties.

Chemical mixing propeller made of Hastelloy C nickel alloy.

Nickel/chrome nodules like this grow up where insulation is cracked on electroplating racks. They are a beautiful nuisance to the plating industry.

Square-cut chunks of nickel electrowinning plate are used in the anodes of electroplating lines.

Nickel-plated handcuffs show off the beauty of this metal when properly plated.

Elemental

Atomic Weight
58.6934
Density
8.908
Atomic Radius
149pm
Crystal Structure

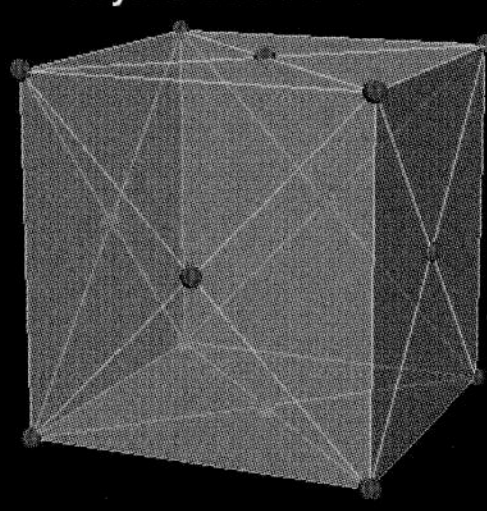

Electron Filling Order: 1s 2s 2p 3s 3p 3d 4s 4p 4d 4f 5s 5p 5d 5f 6s 6p 6d 7s 7p

Atomic Emission Spectrum

State of Matter: 0 500 1000 1500 2000 2500 3000 3500 4000 4500 5000 5500

Copper

Cu 29

Copper

COPPER IS WONDERFUL STUFF. Just wonderful. Many other elements have some kind of a gotcha about them: maybe they are great in every way except they're poisonous, or they would be perfect except they explode when they touch water. Copper has no gotcha—it's just nice stuff all around.

Copper can be toxic, but it takes special effort—eating large amounts of copper sulfate, or routinely eating acidic foods that have been stored in copper containers for a long time. Extended contact with copper objects rarely causes harm. In fact, copper has antimicrobial properties that make it useful in hospitals for doorknobs and other surfaces on which infections may be passed (though claims of the mystical healing powers of copper bracelets are, of course, nonsense).

Copper is soft enough to be worked using hand tools or modest power tools, yet hard enough to be made into very useful things, especially when alloyed with tin (50) or zinc (30) to create, respectively, bronze or brass. You can even find copper in native metallic form in several places around the world, making it one of the first useful metals (hence "the Bronze Age," which I guess sounds better than "the Copper Alloy Age").

Copper is the only reasonably priced metal that isn't more or less gray, quite a remarkable fact if you think about it. Every single one of the hundred-odd metallic elements is some shade of silvery gray, except cesium (55), gold (79) and copper. Not surprisingly, copper has been used in jewelry since antiquity, where its only real disadvantage is that it tarnishes slowly, while gold remains bright forever, at six thousand times the price. (The main disadvantage of cesium as a metal for jewelry is that it explodes on contact with skin.)

Unbeknownst to the ancients, copper has another nice attribute: the second-highest electrical conductivity of any metal. Vast quantities of copper are used for electrical wiring, making it as vital in the modern age as it was in the Bronze Age.

It may not be as pretty as copper, but I will forever hold a special place in my heart for the next element, zinc.

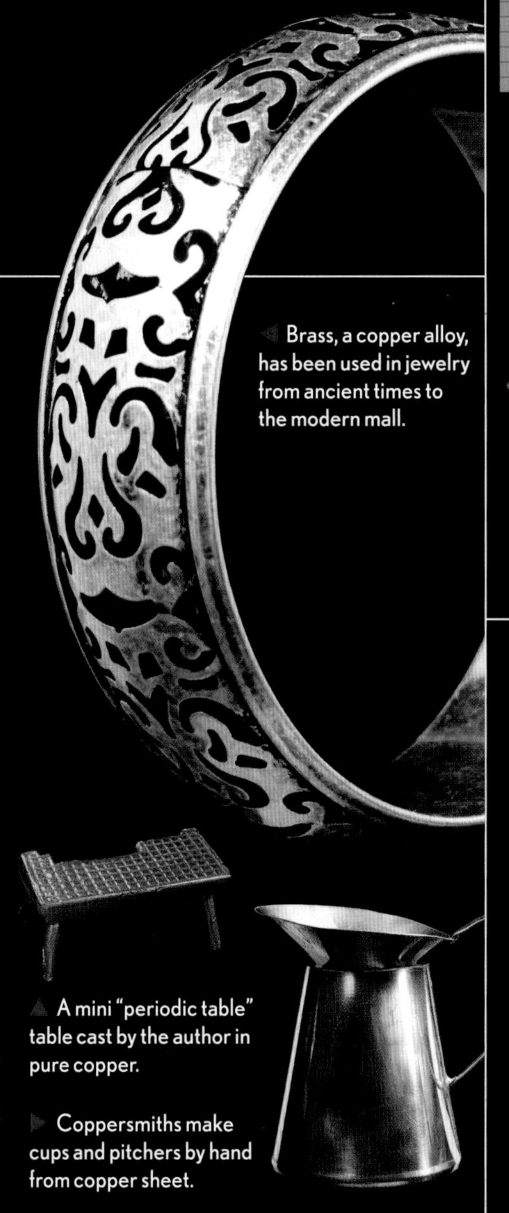

Brass, a copper alloy, has been used in jewelry from ancient times to the modern mall.

A mini "periodic table" table cast by the author in pure copper.

Coppersmiths make cups and pitchers by hand from copper sheet.

Half-Persian 4-in-1 weave chain made from copper electrical wire.

Solid copper heat sink for a CPU chip.

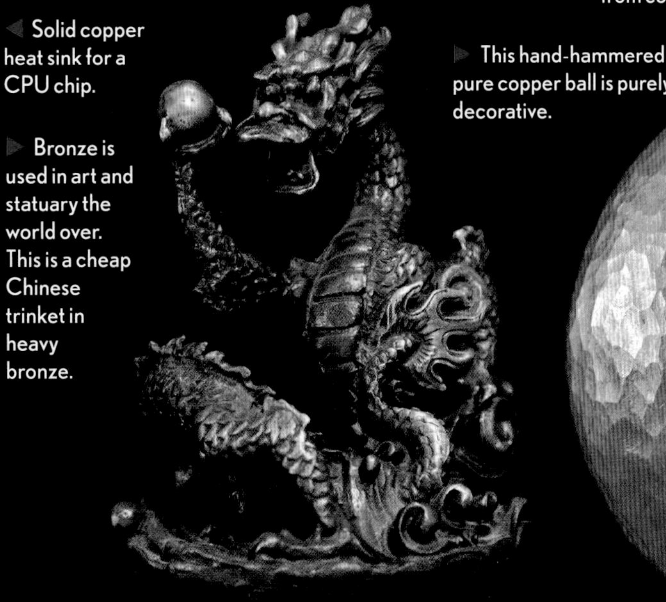

Bronze is used in art and statuary the world over. This is a cheap Chinese trinket in heavy bronze.

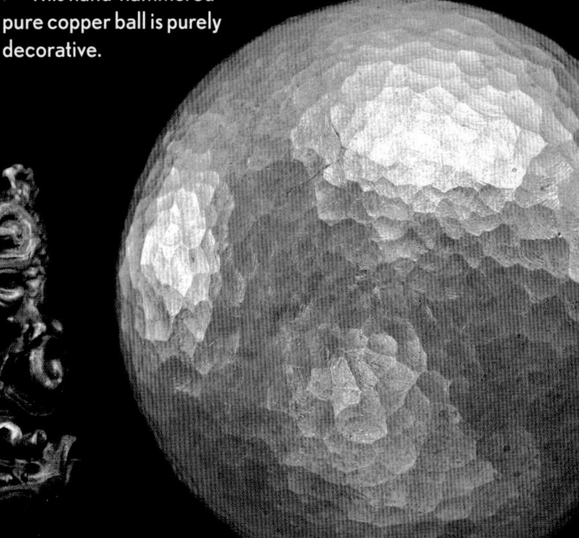

This hand-hammered pure copper ball is purely decorative.

Elemental

Atomic Weight
63.546
Density
8.920
Atomic Radius
145pm
Crystal Structure

Electron Filling Order
1s 2s 2p 3s 3p 4s 3d 4p 5s 4d 5p 6s 4f 5d 6p 7s 5f 6d 7p

Atomic Emission Spectrum

State of Matter
0 500 1000 1500 2000 2500 3000 3500 4000 4500 5000 5500

Copper 29

▲ Copper electrowinning nodule.

Large Japanese commemorative medal made of copper.

A pig? The pig on this penny from Bermuda celebrates the importance of this animal to the island.

Copper pipes, joined with solder, are common targets of thieves stripping abandoned buildings, due to the high value of scrap copper.

Copper electrical cable thick enough to carry 400 amps.

Copper earring: the copper hanging loop may be a problem as some people are allergic.

Copper has become expensive enough that some people are offering it in bullion form for investment purposes.

Zinc

Zn 30

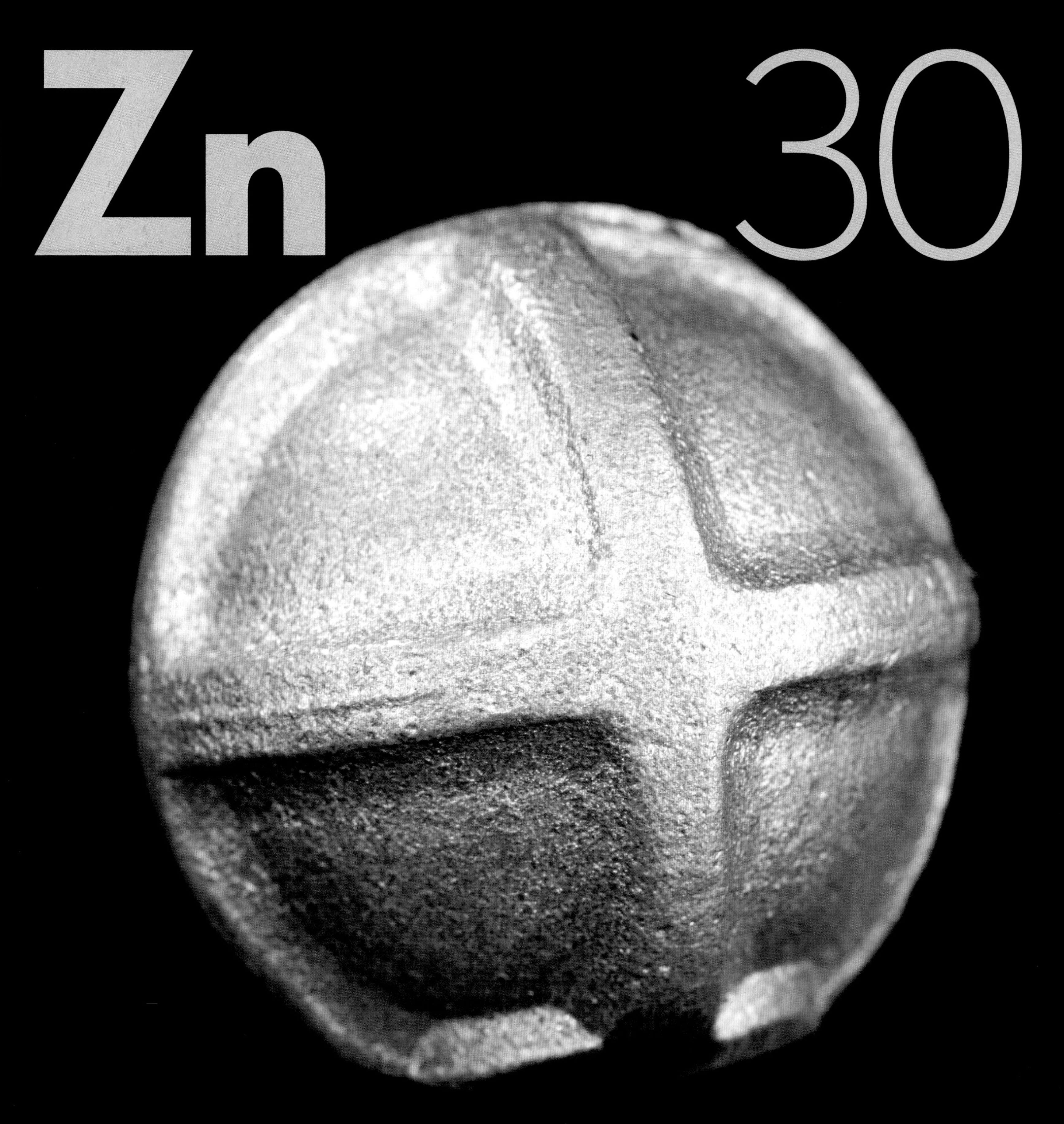

Zinc

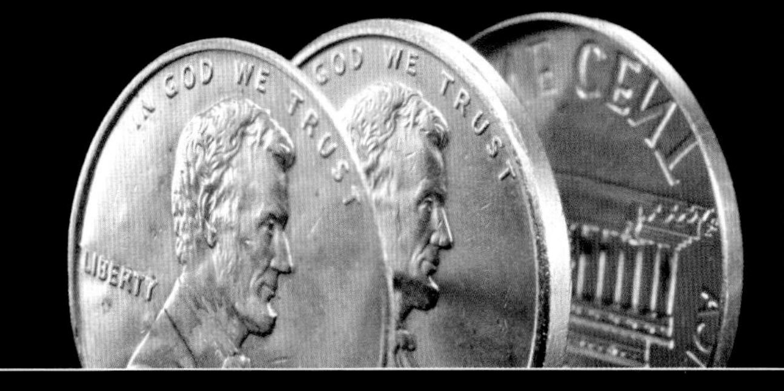

▲ A post-1982 US penny split open to show its zinc core.

Elemental

Atomic Weight
65.409
Density
7.140
Atomic Radius
142pm
Crystal Structure

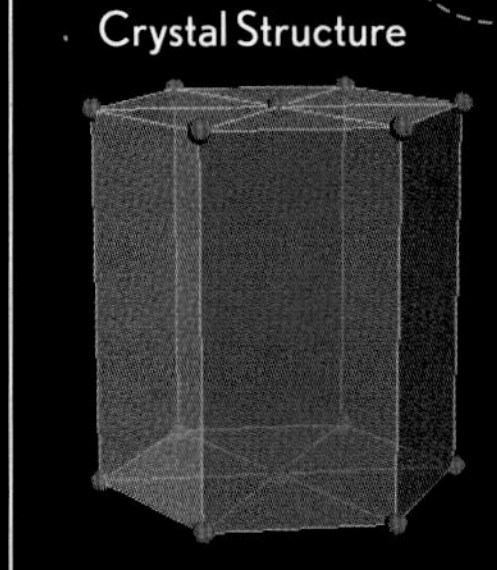

THE FIRST METAL that the ancients learned to cast was probably either lead (82) or perhaps the copper (29) alloy known as bronze. But the first metal *I* learned to cast was zinc. In recent history it would have been more common for a child to start by casting lead or tin (50): that is what "tin soldiers" were made of before they all turned plastic. Casting them at home was a common hobby in the youth of my father and grandfather.

But I arrived too late for that, and the only metal I could find with a low enough melting point for the kitchen stove was zinc (from scrap roof flashing, and later, starting in 1983, from pennies). As a casting metal, zinc is quite practical; indeed, it is the primary component of inexpensive "pot metal" alloys used for casting parts that don't need to be particularly strong.

The reason pennies are now made primarily of zinc instead of copper is simple: the value of a penny sank until the copper contained in it was worth more than a penny, a completely untenable situation for obvious reasons. More recently, around 2008, the price of zinc in turn threatened to exceed one penny per penny's worth, prompting serious discussion of converting pennies to aluminum, the last refuge of worthless coinage. (A far better solution would of course be to eliminate pennies entirely.)

Cheap zinc-based pot metals get no respect, but consider the indignity suffered by zinc when used in sacrificial anodes. These blocks or slabs of solid zinc are electrically connected to steel structures such as bridges, railroad tracks, and the hulls of large ships. Zinc's job is to slowly dissolve away to nothing, its sacrifice generating a small electrical potential that protects the more worthy iron (26) from rusting. When the anode has given all it has to give, a new one is unceremoniously bolted in its place.

Moving right along, gallium is a *much* more interesting element! (Sorry, not even in a book of the elements can zinc get much respect.)

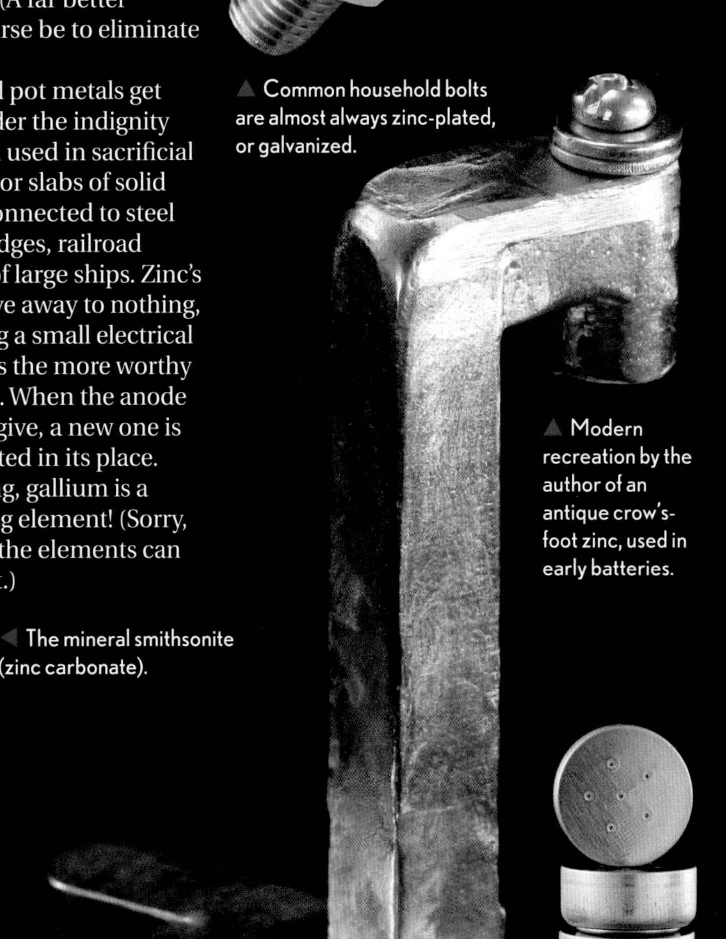

▲ Common household bolts are almost always zinc-plated, or galvanized.

▲ Modern recreation by the author of an antique crow's-foot zinc, used in early batteries.

◀ The mineral smithsonite (zinc carbonate).

◀ A crude zinc casting made by the author as a child.

◀ Sacrificial zinc anodes are used to protect steel tanks, rails, and ship hulls from rusting. Since zinc oxidizes more easily than iron, it corrodes first.

▲ Note the air holes in these zinc-air hearing-aid batteries.

Electron Filling Order: 1s 2s 2p 3s 3p 4s 3d 4p 5s 4d 5p 6s 4f 5d 6p 7s 5f 6d 7p

Atomic Emission Spectrum

State of Matter: 0 500 1000 1500 2000 2500 3000 3500 4000 4500 5000 5500

Gallium

Ga 31

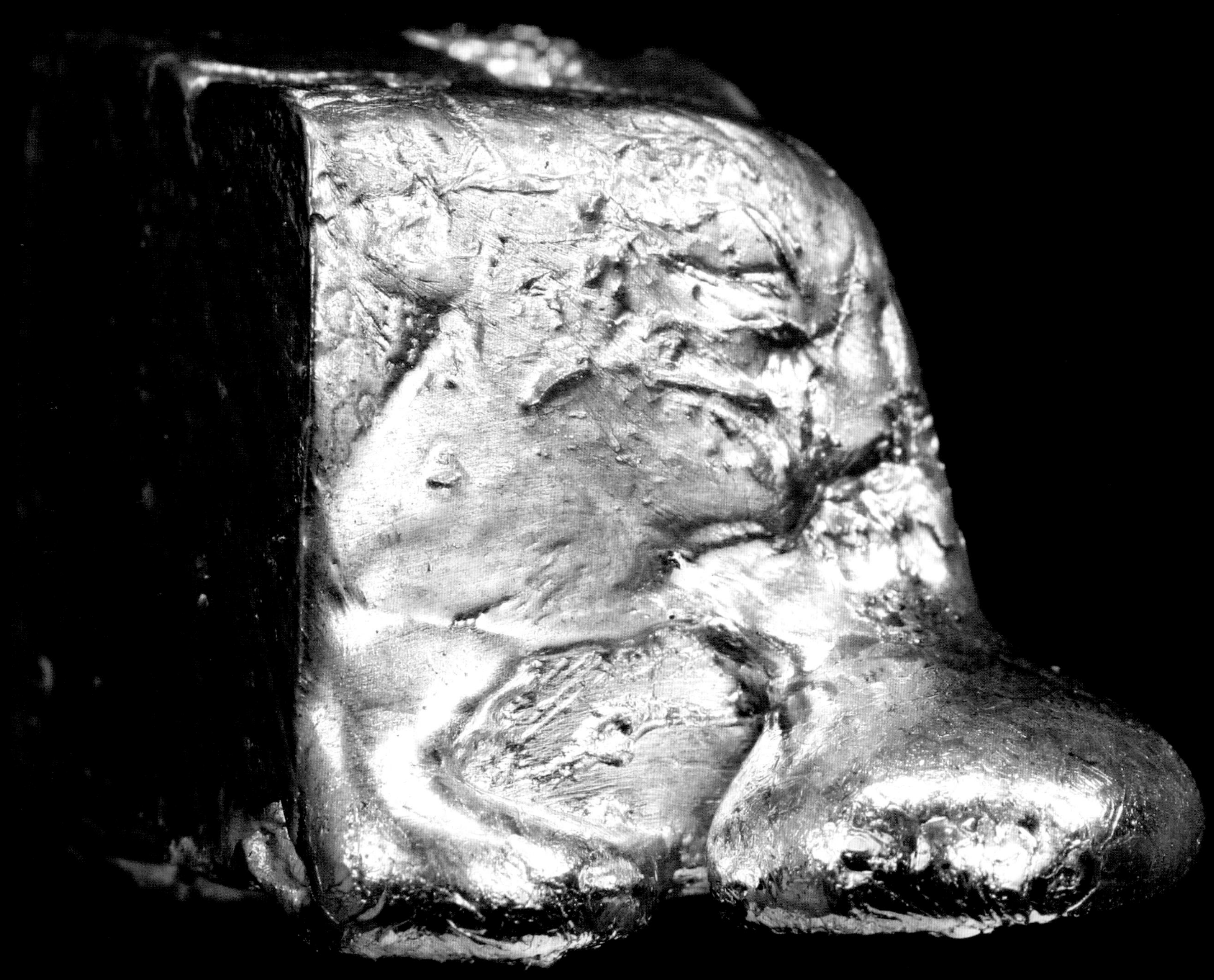

Gallium

MERCURY (80) IS OFTEN CITED as the only metal element liquid at room temperature, but this reveals a climatological bias. In more tropical, less air-conditioned parts of the world, gallium and cesium (55) would be on the list too; they melt at a comfortable 85.57°F (29.76°C) and 83.2°F (28.44°C) respectively.

Even in Alaska gallium will literally melt in your hand—quite an unusual experience, though one you're not likely to repeat: While gallium is not known to be toxic, it tends to stain skin dark brown and is best played with in a plastic bag.

Gallium's low melting point finds a practical application in a patented alloy known as Galinstan—its name from the starting syllables of gallium, indium (49) and *stannum*, the Latin name for tin (50)—which is liquid down to -2.2°F (-19°C). If you buy a fever thermometer today that looks like a mercury thermometer, it most likely contains Galinstan, because mercury has been banned in that application for years.

But gallium's most important modern use is in semiconductor crystals. (This is true of most of the elements on or near the diagonal line of elements known as metalloids.) Silicon semiconductors stop working above a few gigahertz, but gallium arsenide circuits operate as high as 250 gigahertz, which is at the upper end of the microwave frequency range.

Gallium is also present in nearly all light-emitting diodes (LEDs), in the form of gallium arsenide, gallium nitride, indium gallium nitride, aluminum gallium nitride, and numerous variations thereof.

But gallium's usefulness in semiconductors pales in comparison to the fundamental and ubiquitous role of silicon (14), and the historical role of their mutual neighbor

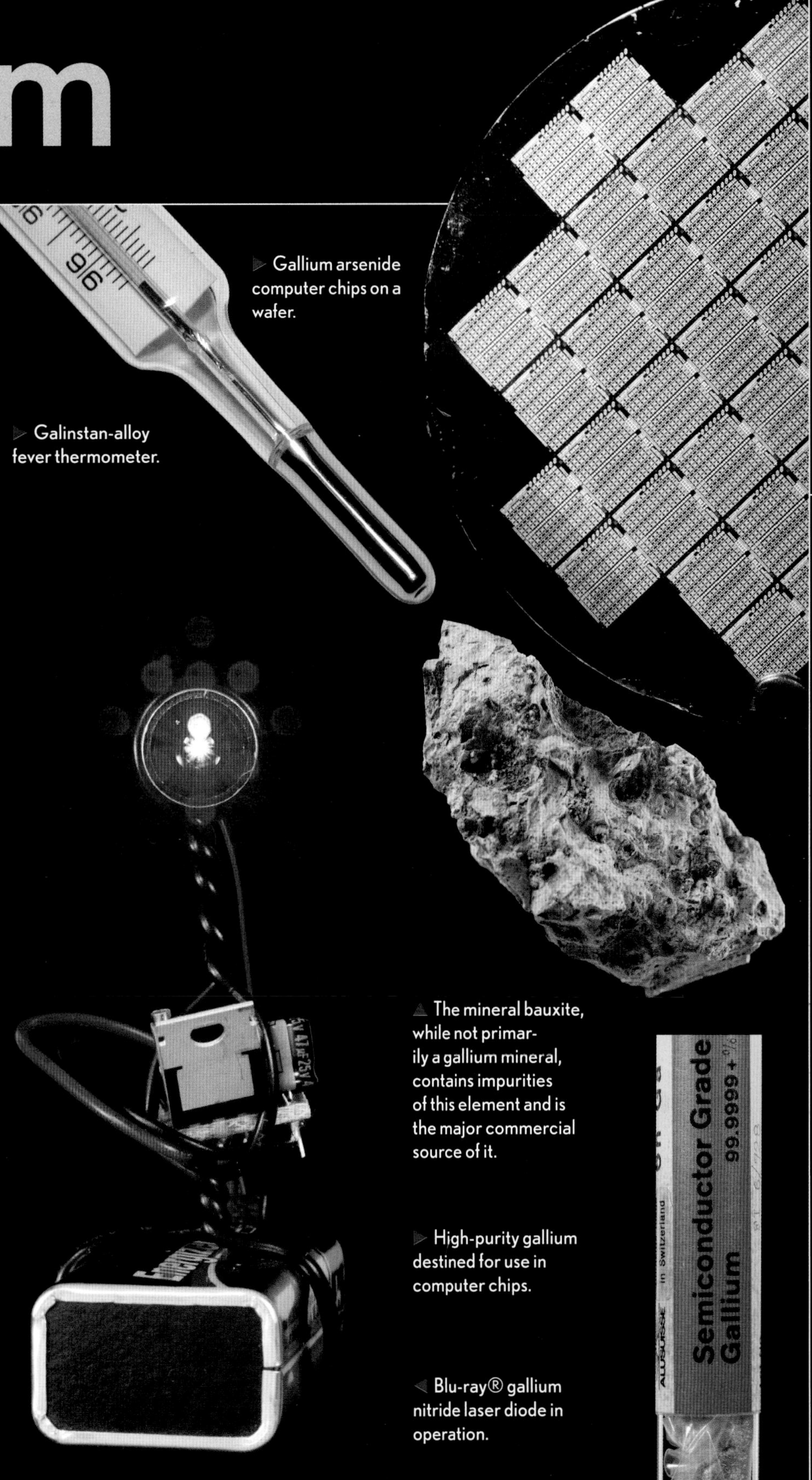

▷ Gallium arsenide computer chips on a wafer.

▷ Galinstan-alloy fever thermometer.

▲ The mineral bauxite, while not primarily a gallium mineral, contains impurities of this element and is the major commercial source of it.

▷ High-purity gallium destined for use in computer chips.

◁ Blu-ray® gallium nitride laser diode in operation.

◁ Gallium melts at just above room temperature. A blow-dryer turned this perfect cube surreal.

Elemental

Atomic Weight
69.723
Density
5.904
Atomic Radius
136pm
Crystal Structure

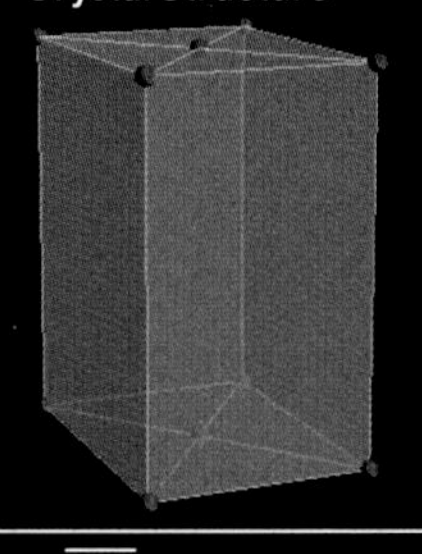

Electron Filling Order: 1s 2s 2p 3s 3p 4s 3d 4p 5s 4d 5p 6s 4f 5d 6p 7s 5f 6d 7p

Atomic Emission Spectrum

State of Matter: 0 500 1000 1500 2000 2500 3000 3500 4000 4500 5000 5500

Germanium

32

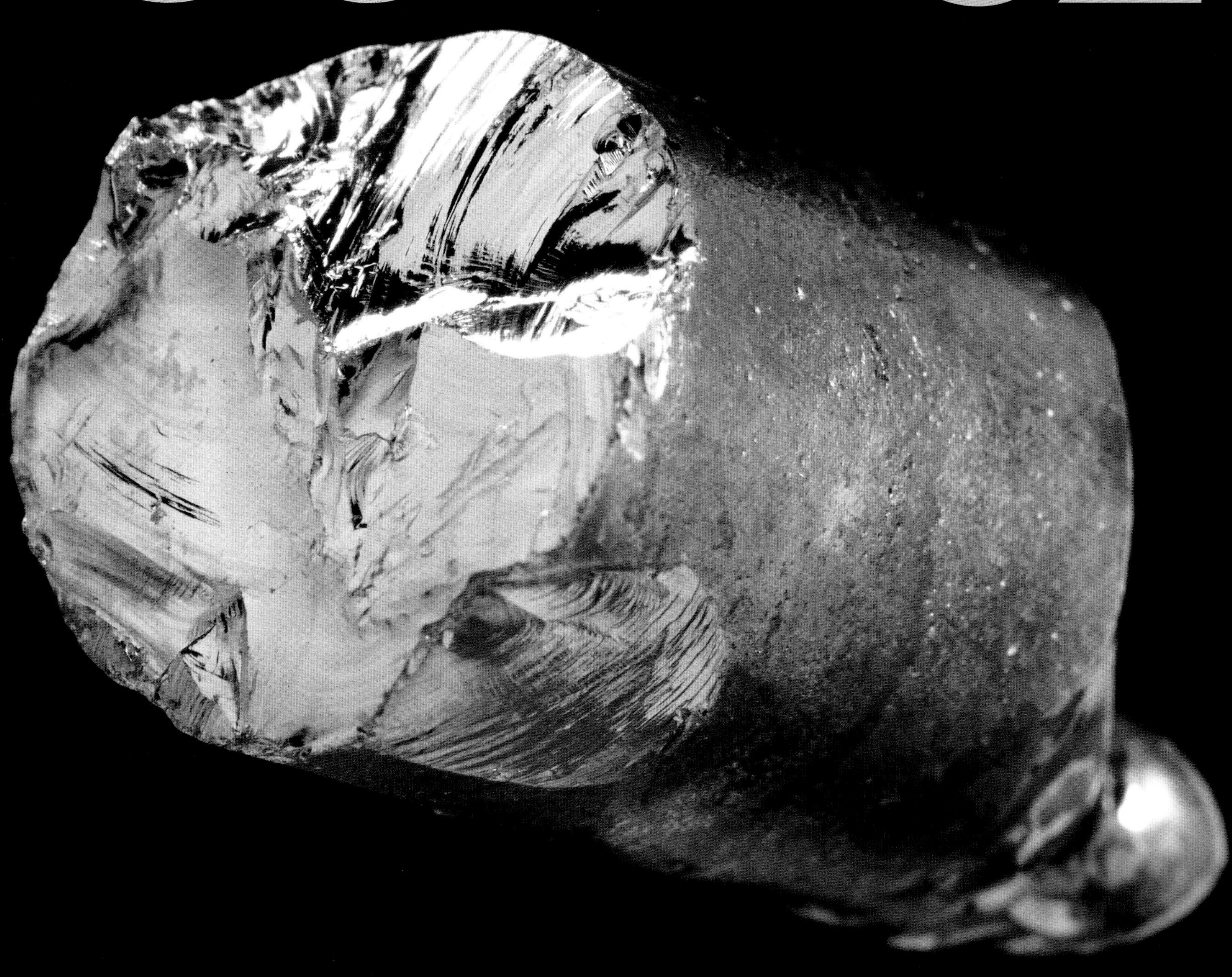

Germanium

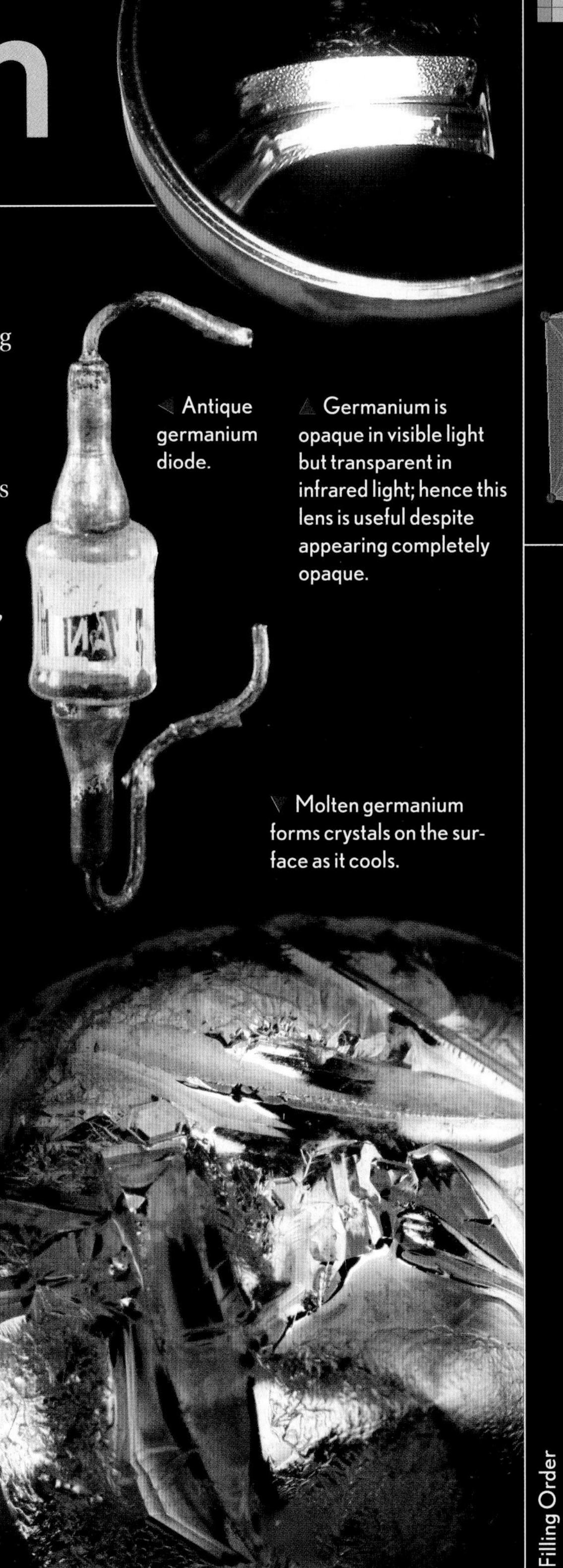

GERMANIUM BEARS the English name for a modern country, and is the only such element that is stable and common. All of the others—francium (87), polonium (84), and americium (95)—are radioactive, were discovered much later, and are not found in nature to any appreciable degree. In grabbing element names for your country, it pays to get there first.

When Dimitri Mendeleev in 1869 created his first systematic arrangements of the elements, which would eventually become the modern periodic table, he bravely left gaps where he thought as-yet-undiscovered elements must exist. Germanium, discovered nearly twenty years later with properties that matched Mendeleev's predictions almost exactly, filled one of those holes and helped establish his periodic table as one of the most important scientific discoveries of all time.

Germanium's importance extends to the history of technology as well, as the first diodes and transistors were made not with silicon but with semiconducting germanium. Silicon transistors, while superior in some ways to germanium ones, work only when the material is incredibly pure, while germanium transistors work at the lower purity levels available in the mid 20th century, at the dawn of the semiconductor age.

Germanium is still used in specialized semiconductor applications, but these days its main uses are in fiber optics and infrared optics, where lenses made of germanium appear completely opaque to the eye but are transparent to invisible infrared light. Its use in bath salts for medicinal purposes is popular in Japan, as are a fair number of other surprisingly nutty ideas.

Also surprising is that arsenic, besides being a notorious poison, has health benefits—though for chickens, not humans.

◄ Antique germanium diode.

▲ Germanium is opaque in visible light but transparent in infrared light; hence this lens is useful despite appearing completely opaque.

▼ Molten germanium forms crystals on the surface as it cools.

◄ High-purity germanium crystal bar.

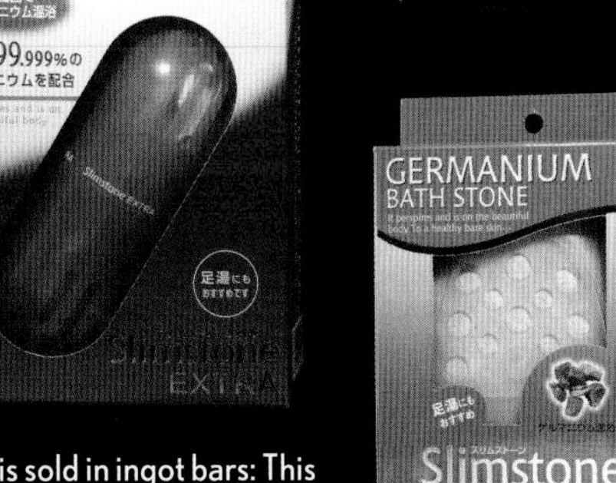

► Germanium dietary supplements and bath salts from Japan are largely just silly.

◄ Bulk commercial germanium is sold in ingot bars: This is the broken end of one, showing the internal crystals.

Elemental

Atomic Weight
72.64
Density
5.323
Atomic Radius
125pm
Crystal Structure

Electron Filling Order: 1s 2s 2p 3s 3p 4s 3d 4p 5s 4d 5p 6s 4f 5d 6p 7s 5f 6d 7p

Atomic Emission Spectrum

State of Matter: 0 500 1000 1500 2000 2500 3000 3500 4000 4500 5000 5500

Arsenic
As
33

Arsenic

PARIS GREEN, or copper acetoarsenite, is one of the few chemicals that can be used both as an artist's pigment and as a rat poison.

Given how strongly arsenic is associated with poison, it may come as a surprise that it is intentionally added to the feed of chickens raised for human consumption. It turns out that organic arsenic compounds are less toxic than pure arsenic, and actually promote the growth of chickens. There is some evidence that very low concentrations may be required for optimal health in chickens, possibly even in humans. (It should not really surprise anyone, however, that under some conditions the arsenic in chicken feed can end up getting converted back to its toxic inorganic form. Generally speaking, when an idea sounds as stupid as intentionally feeding arsenic to chickens, it probably is.)

Another idea that turns out to be as stupid as it sounds is using arsenic as a pigment. Paris green, also known as emerald green, was popular in the 19th century—William Morris himself, the great arbiter of taste in Victorian England, promoted its use in wallpaper over new-fangled synthetic pigments. The trouble was that in damp English winters mold growing on the wallpaper converted the arsenic to gas form, which sickened and even killed those living in the room. The more green your wallpaper, and the damper the winter, the sicker you got. Could it be that the common belief that damp weather is unhealthy originated with green wallpaper? Move to a nice, dry climate for a few months, and you feel much better! But was it the pleasant weather or the fact that you weren't breathing arsenic vapor anymore? Unaware of the latter possibility, people at the time naturally concluded it was the former. Besides, who's going to argue with a doctor who orders you to spend a month at the beach?

While it is an unsettled question whether arsenic in very low concentrations is an essential nutrient, the next element is well known for having a dual nature as nutrient and poison.

Glass ampoule filled with pure arsenic metal granules.

Paris green, copper (II) acetoarsenite, is equally useful in pigments and rat poison.

Chromated copper arsenate–treated wood is now banned, but still found everywhere.

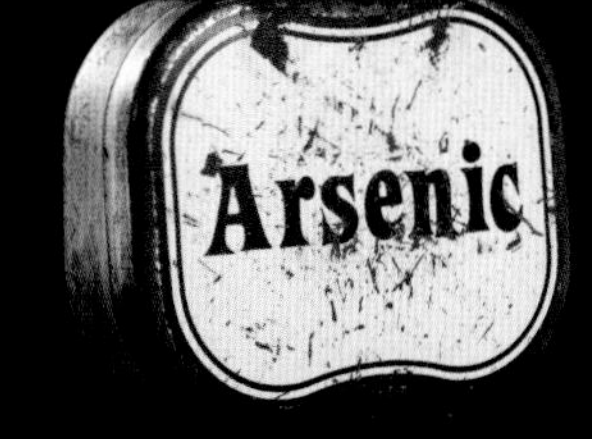

Why someone would carry a small tin of arsenic, I do not know.

A mixture of realgar (As_4S_4) and orpiment (As_2S_3).

This gallium arsenide microwave amplifier is like a city unto itself.

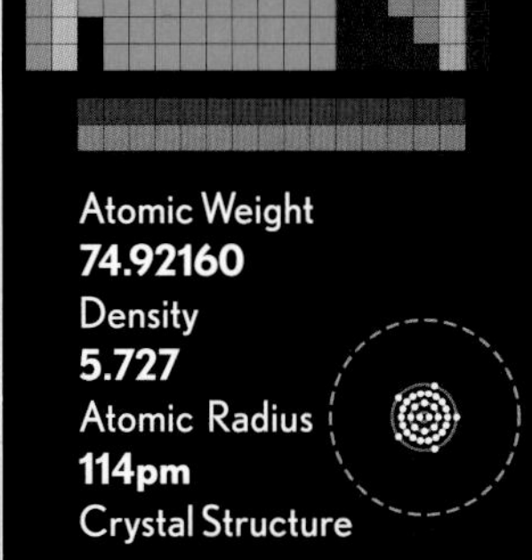

Atomic Weight
74.92160
Density
5.727
Atomic Radius
114pm
Crystal Structure

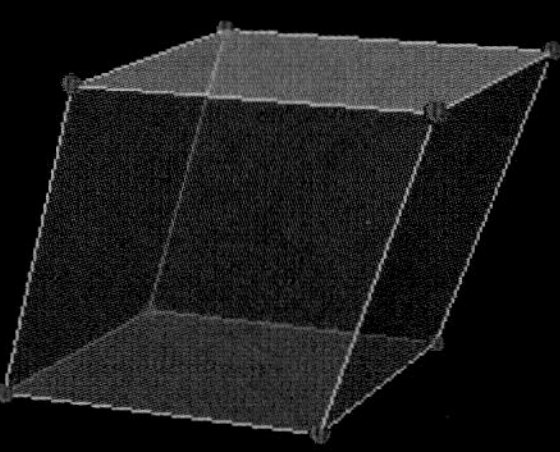

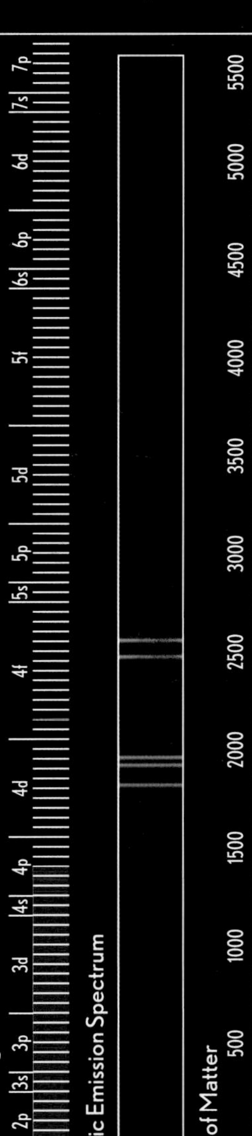

Selenium

Se

34

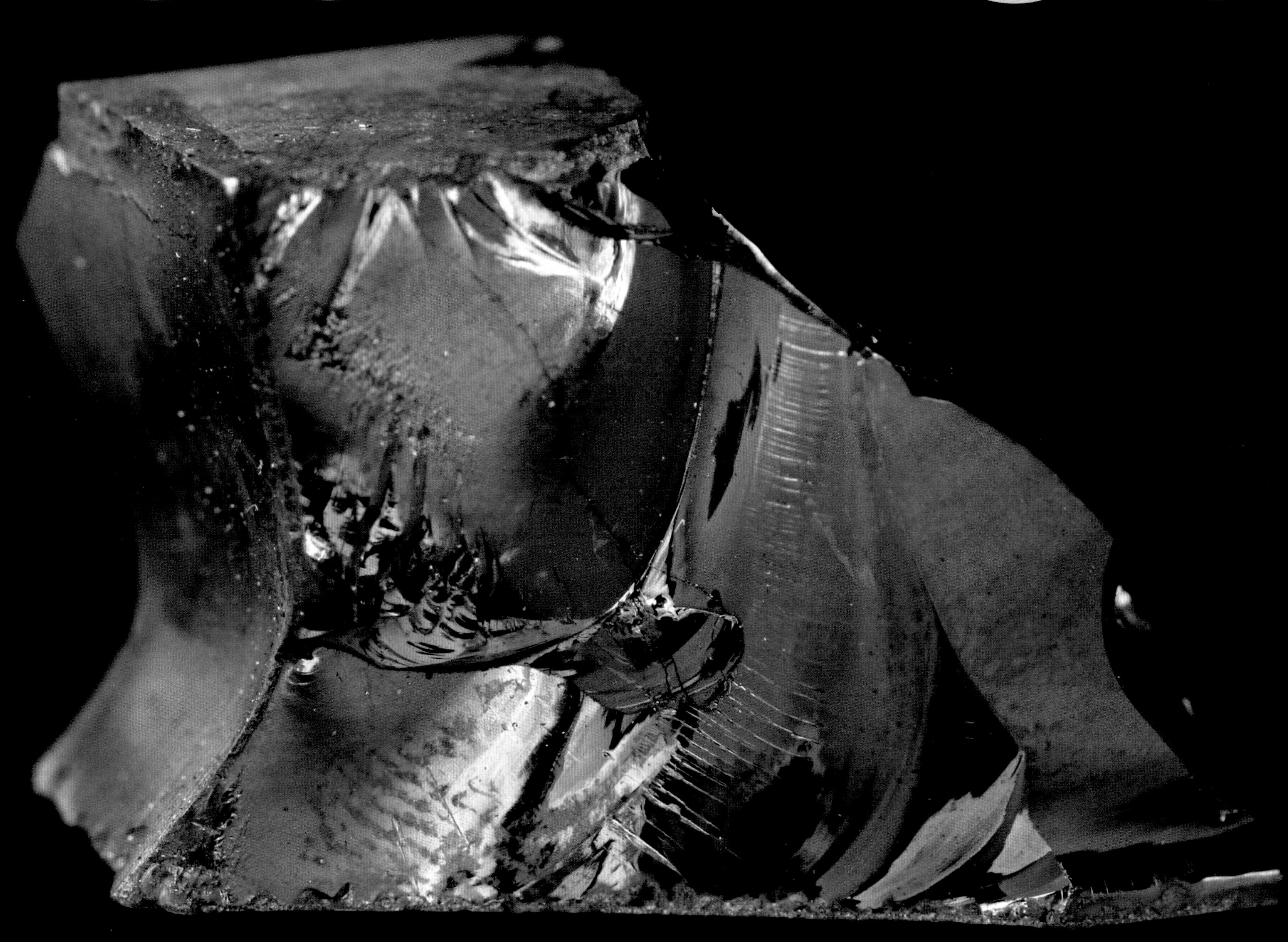

Selenium

SELENIUM IS AN ESSENTIAL nutrient in small amounts, but too much of it is toxic. This is true of quite a few substances, but it's particularly relevant for selenium because people, animals, and plants commonly suffer both from too much of it and from too little, depending on the concentration in the soil where they live.

Some plants, locoweed in particular, seem to require more of it than most, and the presence of large amounts of locoweed indicates high soil selenium levels, and a potential danger to livestock (both from the selenium and from the fact that locoweed itself produces a neurotoxin unrelated to selenium).

Loco livestock aside, the principal modern interest in selenium revolves around its response to light. Xerographic photocopiers and laser printers contain a cylinder coated with selenium in a form that acts as an insulator in the dark but a conductor when exposed to light. A static charge is spread evenly over the cylinder, which is then exposed to an image. Where the image is bright, the coating becomes conductive and the static charge drains off. Where the image is dark, the static charge remains. Then a very fine black powder is dusted over the drum and sticks only where there is a static charge, forming a copy of the original image in black powder. Paper is rolled by the cylinder and picks up the powder, which is then fused to the paper with heated rollers. Yes, it all sounds very finicky, and it's kind of amazing that xerography works at all. Before the selenium drum was invented, mostly it didn't.

Selenium light meters were once an essential tool for any serious photographer, but the rise of digital cameras has by and large rendered a separate light meter unnecessary. A digital camera is in effect several million individual light meters (pixels) that display their results in the form of an image. The image itself is a far more comprehensive measure of whether you got the lighting right than any individual meter reading would be.

Moving on from selenium, we reencounter the halogens: in the liquid shape, just barely, of bromine.

◄ A broken crystal of pure selenium.

Kodak Professional rapid selenium toner for films and papers Net Contents 946 mL 1 U.S. Quart CAT 146 4486

▲ Selenium glaze gives this vase its red color.

▼ Selenium rectifiers (diodes) predate silicon and germanium varieties, and were much larger.

▲ Interesting surface formed when selenium is cooled in a mold.

◄ Selenium sulfide medicated shampoo.

◄ Selenium is one of many chemicals once used to impart a tone or hue to photographs.

▲ Brazil nuts are notoriously high in selenium.

▼ Selenium photocells were widely used in photographers' light meters.

Elemental

Atomic Weight
78.96
Density
4.819
Atomic Radius
103pm
Crystal Structure

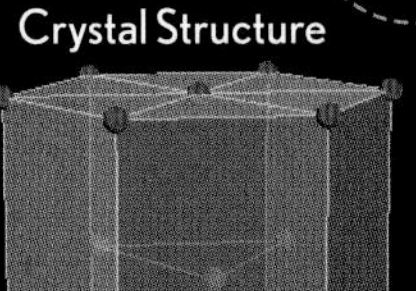

Electron Filling Order: 1s 2s 2p 3s 3p 4s 3d 4p 5s 4d 5p 6s 4f 5d 6p 7s 5f 6d 7p

Atomic Emission Spectrum

State of Matter: 0 500 1000 1500 2000 2500 3000 3500 4000 4500 5000 5500

Bromine

Br

Bromine

EXACTLY TWO STABLE elements are liquid at conventional room temperature, mercury (80) and bromine. But while mercury is firmly liquid—it melts at -37.8°F (-38.8°C) and does not boil until 675°F (357°C)—bromine's boiling point of just 138°F (59°C) means that it is just barely still liquid at comfortable temperature. So close is it to boiling that a puddle of it at room temperature will evaporate away in a cloud of reddish purple vapor in less than a minute. (Mercury, by the way, also evaporates—one reason it's such an insidious poison.)

As with the other halogens, bromine spends its time almost exclusively in the form of ions, either in an ionic salt or, if it's lucky, lounging in a hot tub. Chlorine is the disinfectant of choice for cool-water swimming pools, but bromine salts are more effective at the higher temperatures of a hot tub.

When not lounging in a hot tub, bromine can sometimes be found sleeping with children. Wait, wait, it's not what you think. Organic bromine compounds, typically tetrabromobisphenol A, are added by law to children's synthetic-fiber pajamas as a flame-retardant. While questions have been raised about the safety of this chemical, the image of burning, molten polyester dripping off the charred body of a child tends to dampen the criticism. (An alternative is snug-fitting cotton pajamas, which are not required to have added flame retardants because cotton doesn't burn as readily, and when clothes fit snugly it is harder for air to get to all sides to sustain combustion.)

The halogens often find themselves in the middle of such dilemmas because they are very active participants in the chemical conversation. Krypton is not.

Citrus-flavored sodas often use brominated vegetable oil as an emulsifier: just enough bromine atoms are added to the oil molecules to increase their density to match that of water, allowing the oil to stay suspended rather than rise to form a separate layer.

The mineral bromargyrite, Ag(Br,I), from the Schöne Aussicht Mine, Dernbach, Germany.

Bromine is liquid at room temperature but evaporates very rapidly into a deep reddish-purple gas.

Sodium bromide tablets used to maintain hot tub water.

Children's pajamas treated with tetrabromobisphenol A.

Elemental

Atomic Weight
79.904
Density
3.120
Atomic Radius
94pm
Crystal Structure

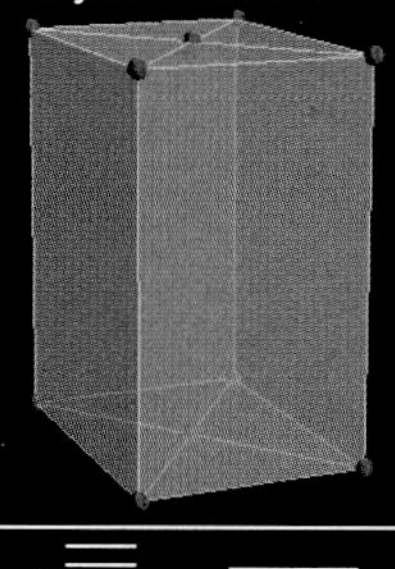

Electron Filling Order

Atomic Emission Spectrum

State of Matter

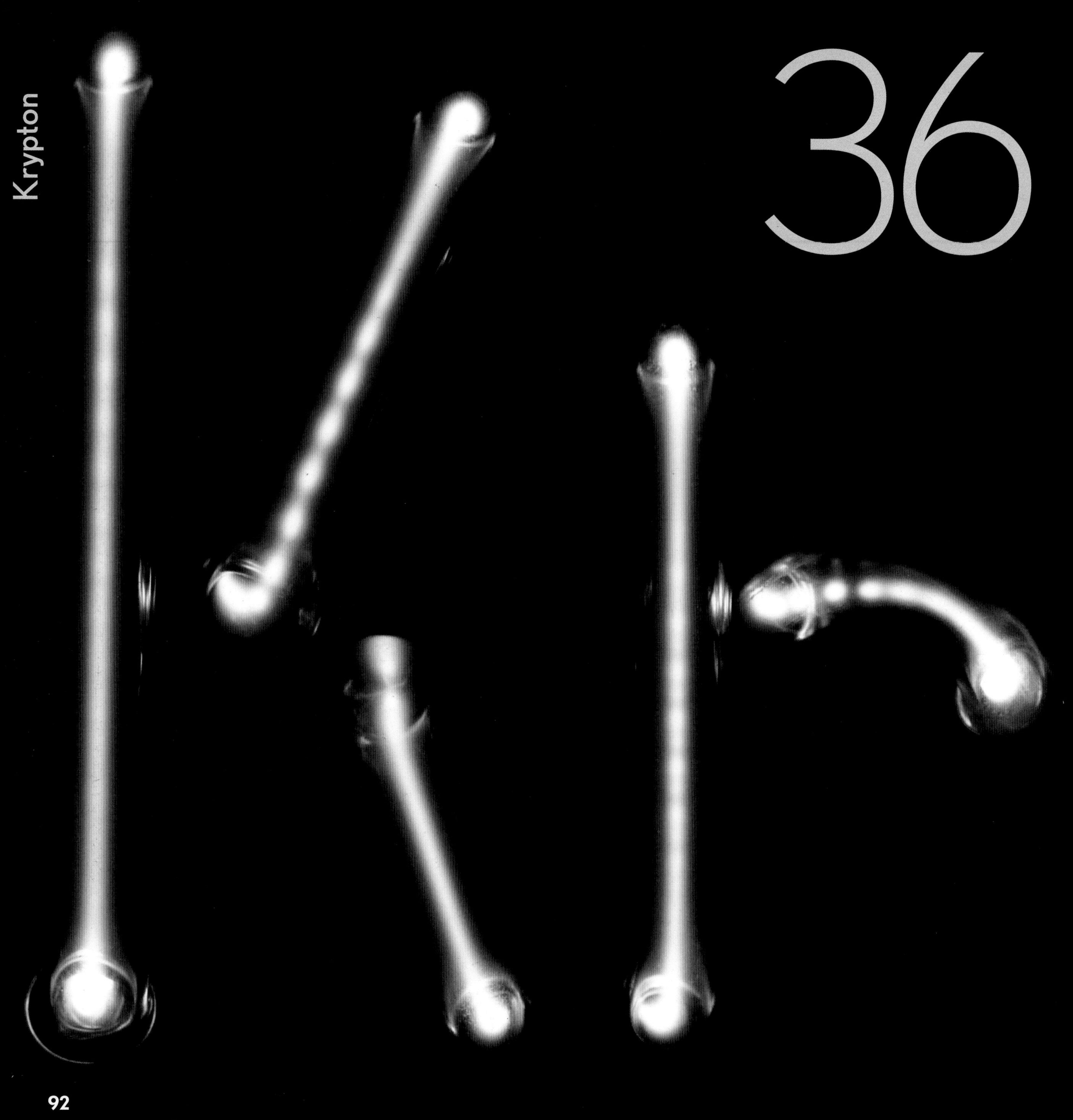
Krypton
36

Krypton

KRYPTON, LIKE THE OTHER noble gases, steadfastly refuses to engage in the primary business of chemistry: bonding. The stubborn inertness of this group renders the noble gases handy when you want to protect something from the rest of the world.

In the case of krypton, this boils down to higher-efficiency lightbulbs. Cheaper incandescent bulbs are commonly filled with argon (18) and/or nitrogen (7), but the higher molecular weight of krypton reduces evaporation of the tungsten (74) filament, allowing it to operate longer at higher temperatures where a larger fraction of the electrical energy goes into visible light instead of heat. (But don't be fooled, even the highest-efficiency incandescent bulb still uses several times as much power as a compact fluorescent bulb generating the same amount of light.)

Krypton, like neon (10), is also exploited for its spectral emission lines when excited by an electric discharge. But while neon glows with its distinctive orange-red color, krypton glows bluish-white, making it useful for photoflashes or for filtering into other colors of the rainbow.

One of krypton's spectral lines had an especially significant role between 1960 and 1983, during which time the meter was officially defined as "1,650,763.73 wavelengths of the orange-red emission line in the electromagnetic spectrum of the krypton-86 atom in a vacuum." (In 1983 this definition was replaced with the one still in force today, "The meter is the length of the path traveled by light in vacuum during a time interval of 1/299 792 458 of a second.")

While length was once defined with krypton, in practice it was almost never actually measured that way. Time, on the other hand, is defined with cesium (55), but more often measured with rubidium.

▲ Before LEDs became a superior alternative, high-end flashlights used krypton bulbs.

◀ Pure krypton is an invisible gas, seen here in a sample ampoule from the days when it was so expensive that this represented a substantial quantity of the gas. Today it is purchased in high-pressure cylinders containing vastly more.

◀ Like all the noble gases, krypton lights up when an electric current flows through it. The colors of these discharges are generally outside the range of colors you can print with standard inks, so this picture is only an approximation of what it looks like in person.

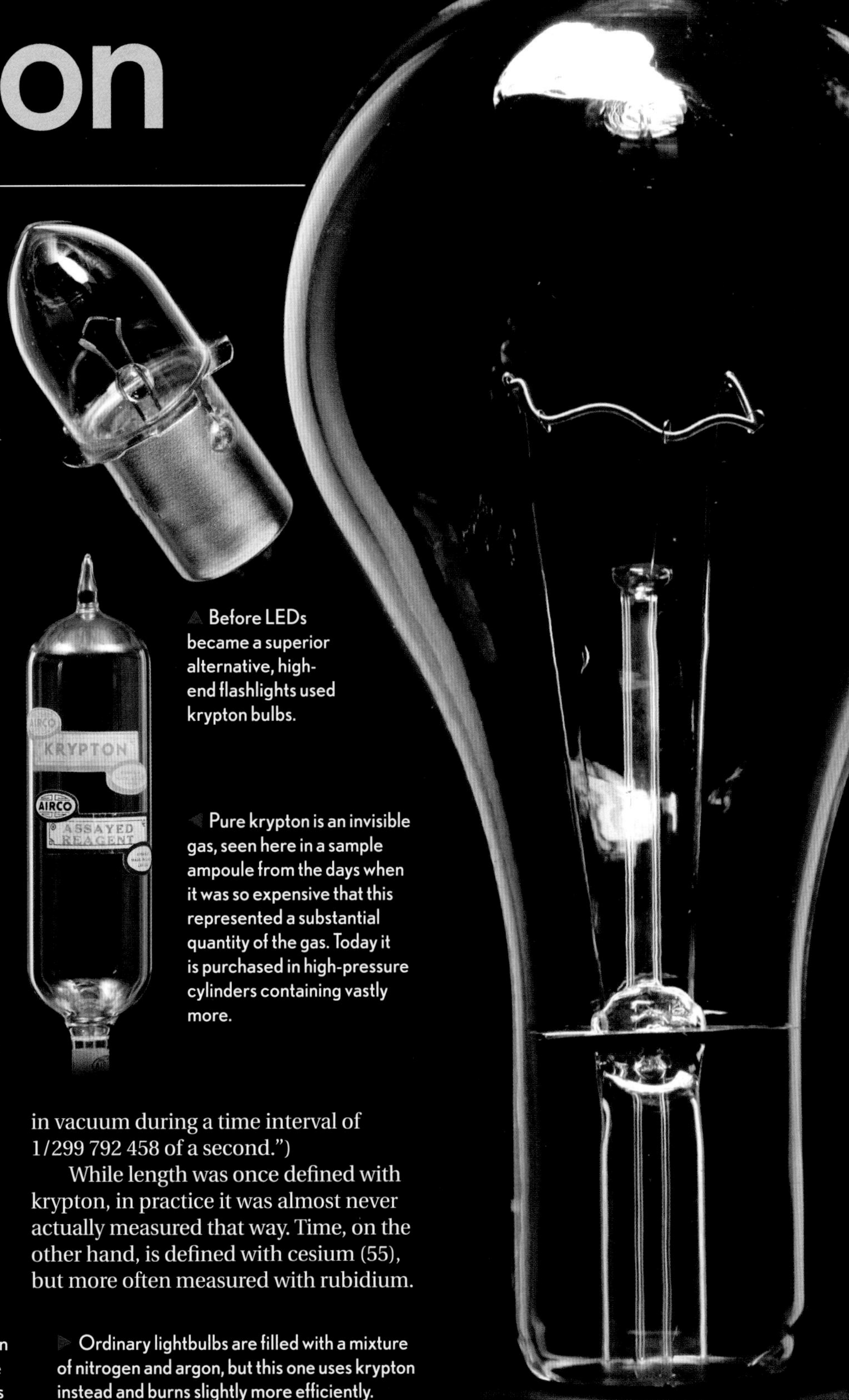

▶ Ordinary lightbulbs are filled with a mixture of nitrogen and argon, but this one uses krypton instead and burns slightly more efficiently.

Elemental

Atomic Weight
83.798
Density
0.00375
Atomic Radius
88pm
Crystal Structure

Electron Filling Order

Atomic Emission Spectrum

State of Matter

Rubidium

Rb

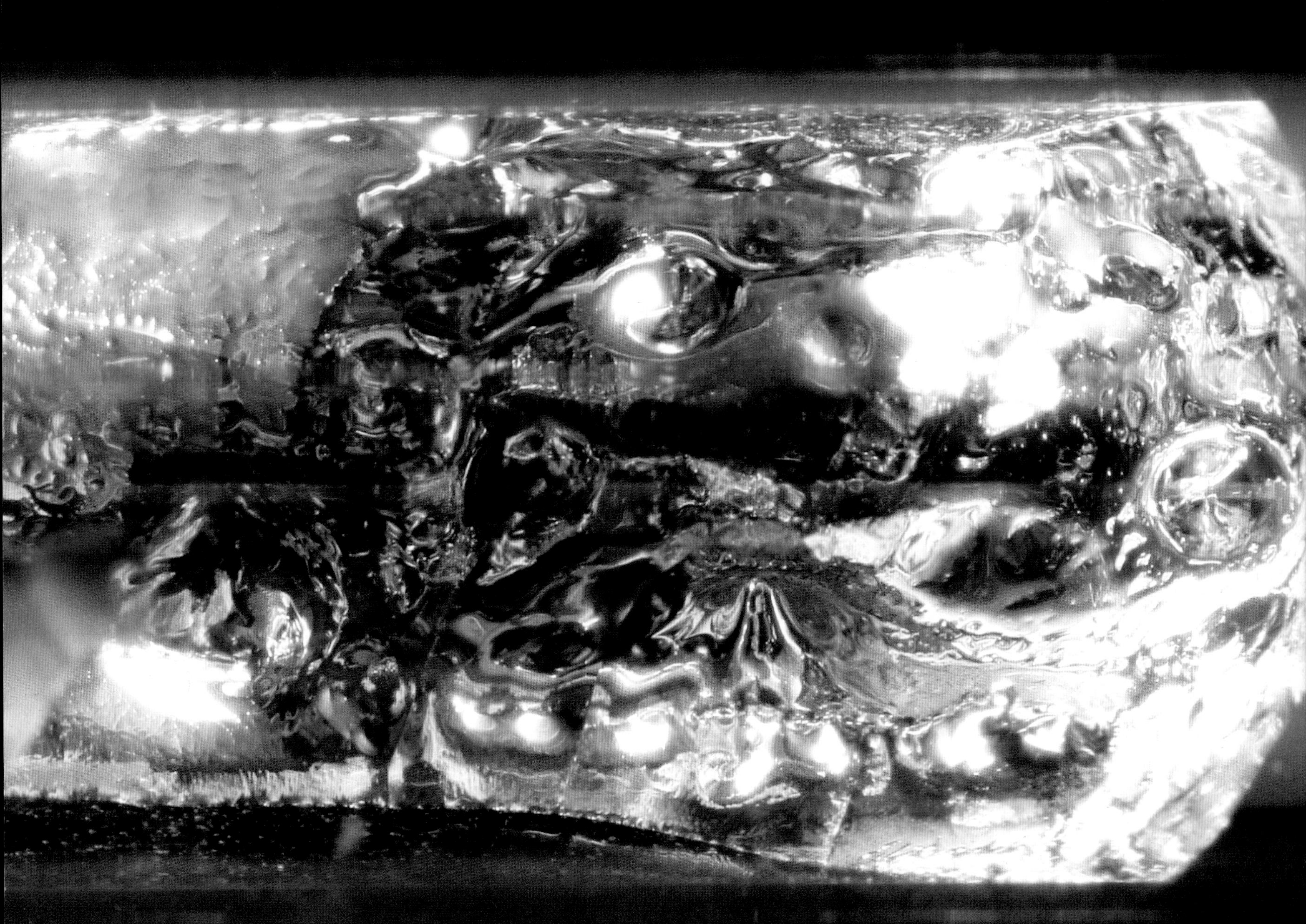

Rubidium

Atomic Weight
85.4678
Density
1.532
Atomic Radius
265pm
Crystal Structure

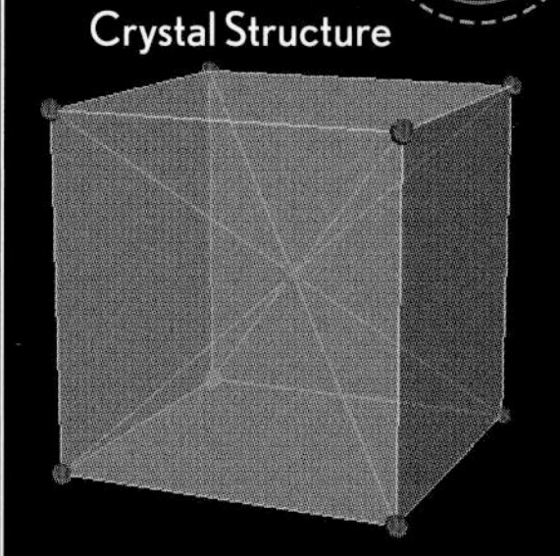

RUBIDIUM IS NOT related to rubies, other than that both derive their name from the Latin for red. The red in rubies comes from chromium (24) impurities, not rubidium, which got its name from the fact that, like so many elements, it was first discovered as an unexplained line in a flame emission spectrum, the line of course being of reddish light. Rubidium itself is not red at all; it's a soft, silvery metal with a very low melting point.

There are precious few actual applications for rubidium. Its namesake spectral line accounts for one, the purple color of some fireworks. Its other applications mostly revolve around the fact that rubidium has a high vapor pressure.

In a rubidium clock, a small (pea-size to fingertip-size) sealed glass ampoule containing a barely visible amount of rubidium is placed inside a combination of heating coils and microwave coils. The heaters vaporize the rubidium and keep it at a stable temperature while the microwave coils are used to measure the precise frequency of a specific hyperfine transition in the main spectral line.

Rubidium atomic clocks are not as accurate as the famous cesium (55) atomic clocks used for decades as the ultimate time standard, but they are still very, very accurate by any reasonable measure. They are also a whole lot smaller and cheaper than cesium clocks, making rubidium time standards commonplace where very accurate time and frequency standards are required.

"Atomic clocks" might sound dangerous, but they are actually more like carefully tuned radios than atomic bombs. Strontium, like rubidium and cobalt (27), is another element that has been unfairly associated with nuclear fallout.

Synthetic rubidium-manganese fluoride crystal ($RbMnF_3$).

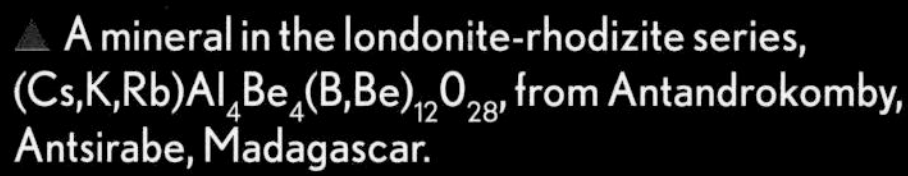

A mineral in the londonite-rhodizite series, $(Cs,K,Rb)Al_4Be_4(B,Be)_{12}O_{28}$, from Antandrokomby, Antsirabe, Madagascar.

A complete rubidium clock cell, less than an inch wide, includes a rubidium vapor cell, heating coils, and transmit and receive antennas.

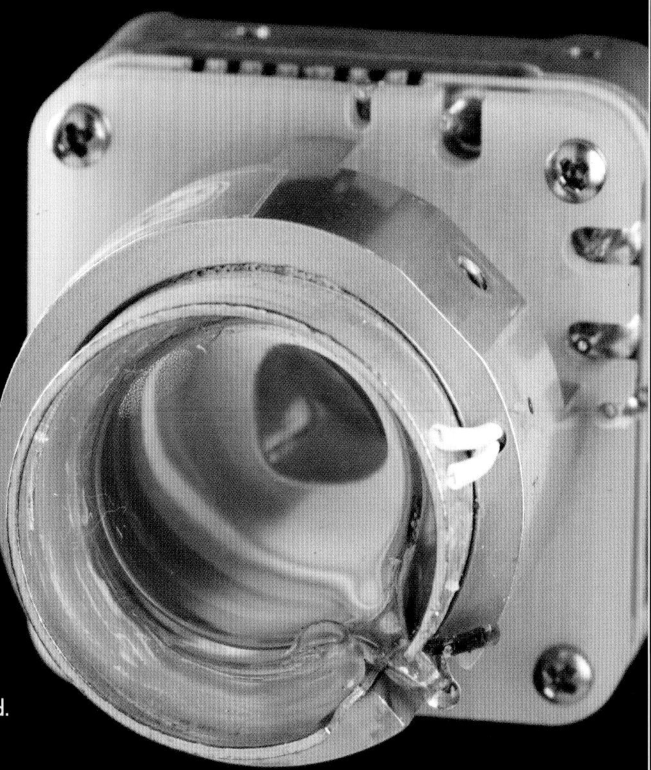

An ampoule containing a gram of highly reactive rubidium metal. If broken open, it would rapidly catch fire.

Rubidium vapor cell from a frequency standard.

Electron Filling Order

Atomic Emission Spectrum

State of Matter

Strontium
Sr
38

Strontium

THE STRONTIUM ISOTOPE ^{90}Sr, a component of nuclear fallout, is the black sheep of the strontium family that has unfairly tainted this element's reputation. Ordinary strontium is not radioactive at all and should not be blamed for any of that atomic bomb unpleasantness.

Perhaps the bomb connection has stuck so firmly in the public mind because strontium has so few other associations. It probably doesn't help that one of those few is with luminous paints, some of which are radioactive. But here, again, strontium is unfairly tainted with guilt by association. Extremely bright modern luminous paints containing strontium aluminate do glow in the dark, but their glow comes not from radioactive decay, as in the case of radium paints, but from efficiently absorbing light from their surroundings and then slowly releasing it over the course of minutes or even hours.

Widely used aluminum-silicon casting alloys have a problem with brittleness, a problem that can be solved by adding a small percentage of strontium. As is often the case, the most convenient way to add a small amount of an exotic element to a batch of alloy is for a specialist manufacturer to make a "master alloy" that contains a much higher percentage of that element. The end user then melts a measured amount of this master alloy into his pot and never has to handle the raw element. The frustrating consequence for element collectors like me is that it's much easier to buy 10 to 20 percent strontium aluminum alloy, by itself a completely useless product, than it is to buy pure strontium.

On an entirely different note, strontium-containing pills are widely sold like vitamins and said to promote bone growth. Due to its chemical similarity to neighboring calcium (20), strontium is indeed bone-seeking (one reason ^{90}Sr fallout is so dangerous). Some strontium compounds seem to show evidence that they may increase bone growth, but whether the form sold in health-food stores has any effect is unclear and unproven.

The benefits claimed for yttrium, on the other hand, are complete and utter nonsense.

▲ Strontium titanate was used as a diamond simulant before the development of cubic zirconia.

▲ An example of the mineral celestine (strontium sulfate).

▶ The active ingredient in this toothpaste is strontium acetate.

◀ The brighter of these luminous powders are europium-doped strontium aluminate, the brightest of modern phosphorescent materials.

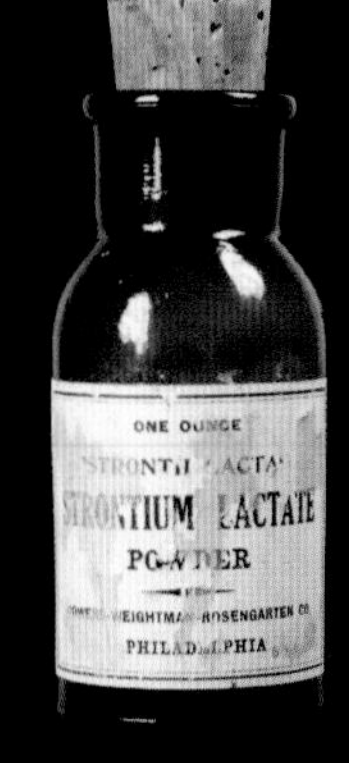

▲ Strontium is bone-seeking because it's in the same column as calcium. Eating it may or may not be healthy.

◀ Pure strontium metal, slightly oxidized despite being stored under mineral oil.

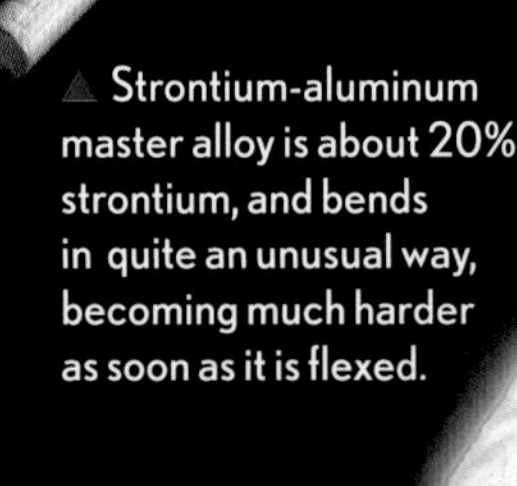

▲ Strontium-aluminum master alloy is about 20% strontium, and bends in quite an unusual way, becoming much harder as soon as it is flexed.

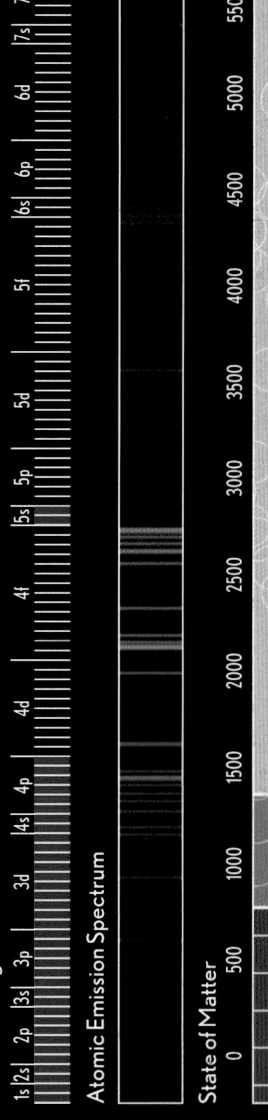

Y

39

Yttrium

Yttrium

YAG (yttrium aluminum garnet) laser crystal boule.

YTTRIUM IS SOMETHING of a hippy element. First, it's named after a village in Sweden, a notably loose country. Second, it is beloved by new age practitioners, who feel that it aids in communication between the spiritual and the practical realms, especially when incorporated into fluorite crystals. (But just to be clear, since this is a book about reality, yttrium really could not care less about our metaphysical states; it's an element, not a transdimensional energy being or whatnot. And by the way, unbeknownst to the new age practitioners who worship them, fluorite crystals actually hate our guts.)

OK, maybe I'm a bit touchy about this, but it really ticks me off when people ascribe magical properties to things in the world that *are* magical but in ways they completely miss.

If you want to see magic, forget about yttrium in fluorite and consider yttrium barium copper oxide (commonly known as YBCO). This material turns into a superconductor when cooled with liquid nitrogen, and superconductors are just plain freaky. For example, if you try to set a magnet on top of a cooled YBCO disk, you can't do it because the magnet will stop about a quarter of an inch above the disk. It just sits there, floating in midair just as plain as day. The only reason this isn't considered black magic of the highest order is the fact that anyone can repeat the trick. (The difference between magic and technology is quite simple: If it works, it's technology; if it doesn't work, we call it magic and get all mushy about it.)

Another slightly magic application is in yttrium aluminum garnet (YAG) crystals, the central component of a powerful class of pulsed lasers, devices able to create beams of light so perfectly collimated that you can bounce them off the moon and see the reflection. (The beam is reflecting not off the moon itself, but off special cat's-eye reflectors put there by the Apollo astronauts specifically for the purpose of bouncing lasers off them.)

While yttrium might have an air of the flaky about it, zirconium is absolutely dead serious.

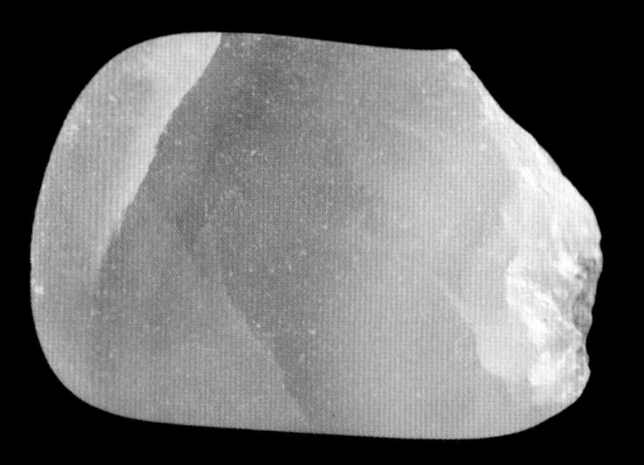

Fluorite crystal claimed to contain traces of yttrium.

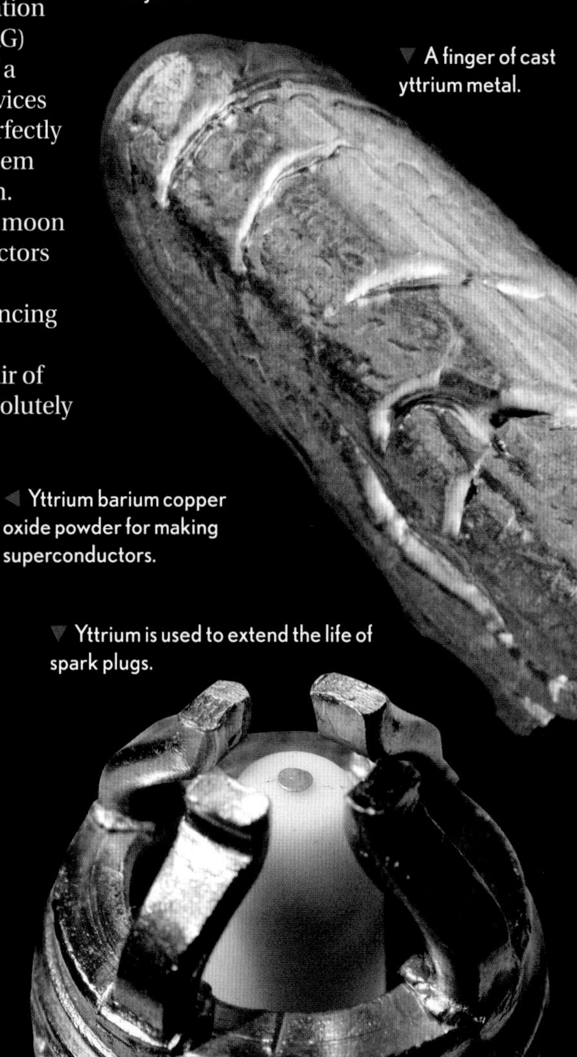

A finger of cast yttrium metal.

Yttrium is used to extend the life of spark plugs.

Yttrium barium copper oxide powder for making superconductors.

A torn ingot of bulk commercial yttrium.

A piece cut from a larger ingot of commercial yttrium metal.

Elemental

Atomic Weight **88.90585**
Density **4.472**
Atomic Radius **212pm**
Crystal Structure

Electron Filling Order: 1s 2s 2p 3s 3p 4s 3d 4p 5s 4d 5p 6s 4f 5d 6p 7s 5f 6d 7p

Atomic Emission Spectrum

State of Matter: 0 500 1000 1500 2000 2500 3000 3500 4000 4500 5000 5500

Zirconium

Zr 40

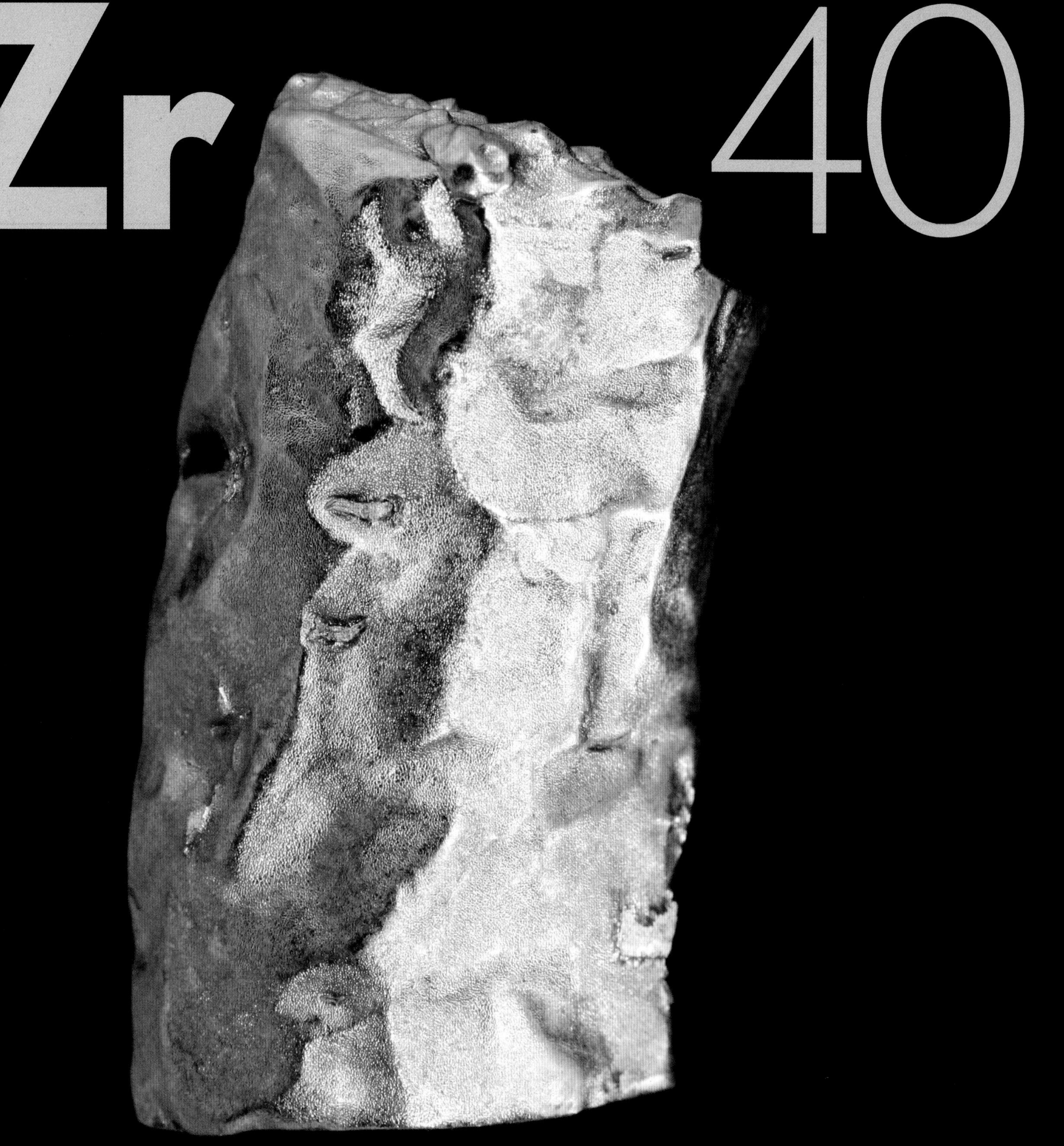

Zirconium

ZIRCONIUM IS A TOUGH, hard metal, and everything to do with it is tough, hard, and abrasive. Tubes made of high-purity zirconium are used to contain the fuel pellets in nuclear reactors, because the metal is both transparent to the neutrons that make the reactor run and able to withstand the hellish environment inside the core of an operating nuclear reactor.

Other manly applications for the element include chemical reaction vessels for highly corrosive substances, incendiary bombs, and tracer rounds. In the form of zirconium oxide, it's used for grinding wheels and specialized types of sandpaper lapping wheels for grinding down welds on oil rigs, giant earth-moving equipment, and dirt bikes.

Ah, but wait—like many gruff and manly things, zirconium has a secret gentle side. Zirconium dioxide in cubic crystal form is what's known as cubic zirconia, or CZ, by far the most common simulated diamond. In malls and cheap jewelry stores the world over, display cases are overflowing with acres of zirconium dioxide. (Even for this application, it's still hard; cubic zirconia is near the top of the hardness scale.)

Really, people should stop thinking of cubic zirconia as fake diamond and start thinking of diamond as overpriced cubic zirconia. There is no real difference in how pretty they are—that's just imagination fueled by having paid way too much for a plain, colorless rock. With an element as down-to-earth as zirconium, one should apply hard-nosed reason even to the selection of engagement rings. (You first.)

While cubic zirconia is a rational choice of jewel, you might want to try something more traditional with the famously jealous Niobe.

Crystal bar of pure zirconium created by thermal decomposition of zirconium iodide.

Zirconium wool in an old single-use flashbulb.

Zirconium laboratory crucible—much cheaper than platinum labware.

A ball bearing made up of super-hard, low-friction zirconia ceramic.

An old Kodak camera from when elements like silver and zirconium were used instead of silicon.

Ceramic knives made of zirconia are incredibly sharp, but chip easily.

Zirconia (ZrO_2) is an important industrial abrasive, as in this flap wheel used by welders.

Elemental

Atomic Weight
91.224
Density
6.511
Atomic Radius
206pm
Crystal Structure

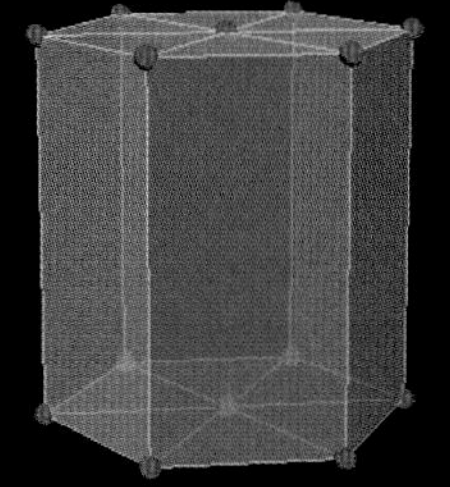

Electron Filling Order

Atomic Emission Spectrum

State of Matter

Niobium Nb 41

Niobium

NIOBE WAS THE DAUGHTER of Tantalus, son of Zeus. Look at a periodic table and you will find her element, niobium, just above tantalum (73). Sadly, the element below tantalum is not named zeusium. It was named dubnium (105) in 1997 after much argument, none of which included a proposal to call it zeusium—evidence that familiarity with the classics is no longer considered essential to a well-rounded education.

Niobe mourned the loss of her children, who were killed by Artemis and Apollo, while I mourn the loss of one my samples of niobium, which was confiscated by the FBI. What I had assumed was an obsolete missile part—a rocket engine with a niobium superalloy nozzle—turned out to be quite up to date and very much missed by the Air Force base from which it had been stolen. (You never can tell what you're going to find on eBay.)

Rocket nozzles are made of niobium alloys because they resist corrosion even at high temperatures. Niobium is also commonly used in jewelry and coins because it can be beautifully anodized, forming a rainbow of colors through the interference of light diffracted through thin transparent layers of oxide on its surface. Corrosion resistance, beautiful colors, and an evocative name combine to make niobium the ideal metal for body piercings. This makes it surprisingly easy to buy pure niobium in shopping malls, if you aren't too embarrassed to go into that kind of store.

If one of those piercings goes wrong, you might find yourself surrounded by a far larger quantity of niobium. Coils of niobium-titanium superconducting wire are used to create the huge magnetic fields inside MRI (magnetic resonance imaging) machines found in hospitals, which among other things can be used to locate objects lost inside the body.

Next in line, molybdenum shares many of the strengths of niobium, but none of its romantic qualities.

Niobium can be anodized to a lovely range of colors.

Niobium-alloy rocket-engine nozzle, confiscated by the F.B.I.

Elemental

Atomic Weight **92.90638**
Density **8.570**
Atomic Radius **198pm**
Crystal Structure

Electron Filling Order · Atomic Emission Spectrum · State of Matter

High-purity crystal niobium bar.

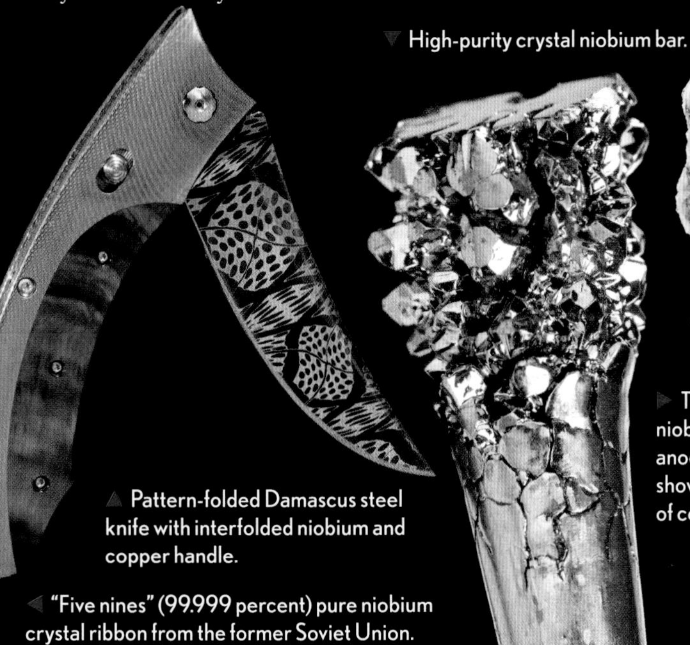

Pattern-folded Damascus steel knife with interfolded niobium and copper handle.

"Five nines" (99.999 percent) pure niobium crystal ribbon from the former Soviet Union.

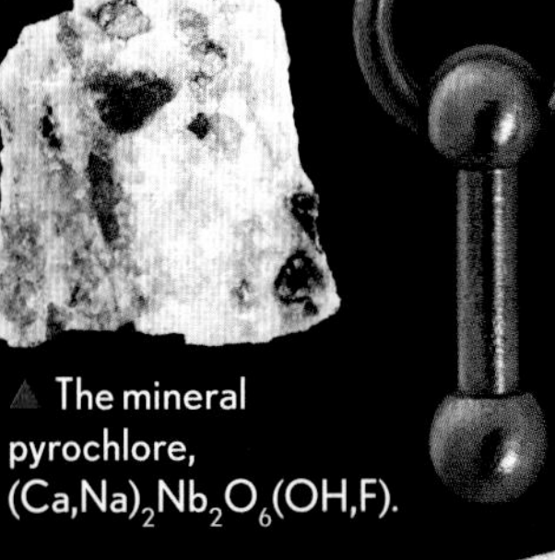

The mineral pyrochlore, $(Ca,Na)_2Nb_2O_6(OH,F)$.

Niobium captive ring barbell for piercing one body part or another.

Thick niobium plate anodized to show a range of colors.

Molybdenum

Molybdenum

The mineral wulfenite ($PbMoO_4$) from the Red Cloud Mine, Arizona.

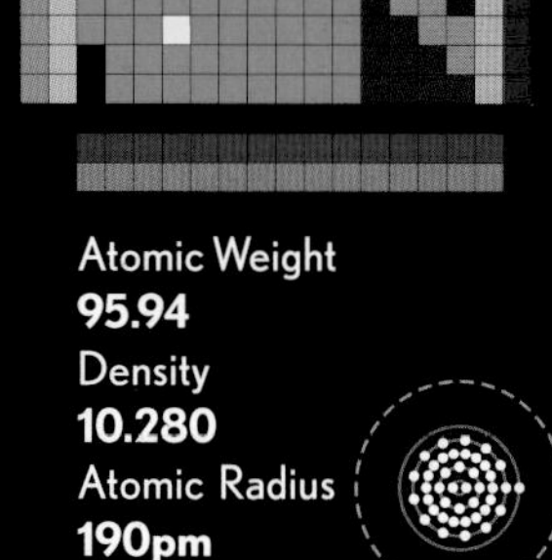

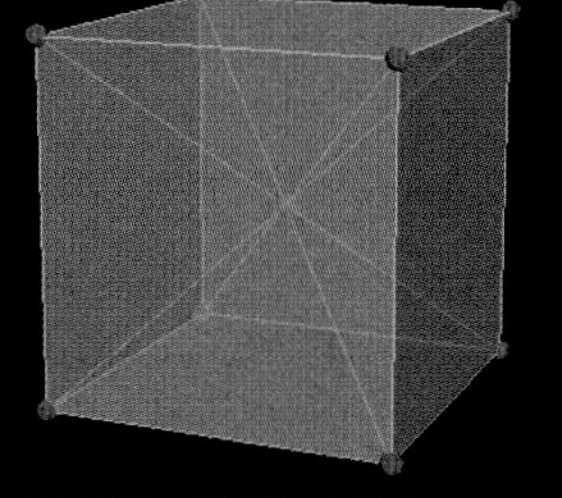

MOLYBDENUM IS A METAL of industry through and through. Its main applications are in steel alloys, where it gives great strength and heat resistance, most notably to M-series high-speed tool steels (M for molybdenum, of course).

Pure molybdenum is much less common but is used in situations where a part must withstand great stress for extended periods at high temperatures, for example in pressure vessels. But while it doesn't lose strength even at fairly high temperatures, molybdenum does oxidize quite rapidly above about 1,000°F (500°C), limiting its use in the most extreme environments.

From resisting movement to encouraging it: molybdenum disulfide is an excellent ultra-high-pressure lubricant. Either as a dry powder or mixed with oil or grease, it is able to withstand crushing pressures and fearsome temperatures without seizing up.

Molybdenum leads to its neighbor, technetium (43), in a very direct way. When the $^{99}Tc_m$ isotope is needed for medical imaging purposes, it has to be created on the spot because its half-life is only six hours. This is done using a device filled with longer-lived ^{99}Mo, which decays into $^{99}Tc_m$, continually replenishing the supply within the device. Because the process of removing the accumulated $^{99}Tc_m$ is called "milking," the device is informally known as a "moly cow." Despite the cute name, it's still a powerfully radioactive object, and the technetium it generates is among the most intense sources of radiation used in medicine, as you can read on the next page.

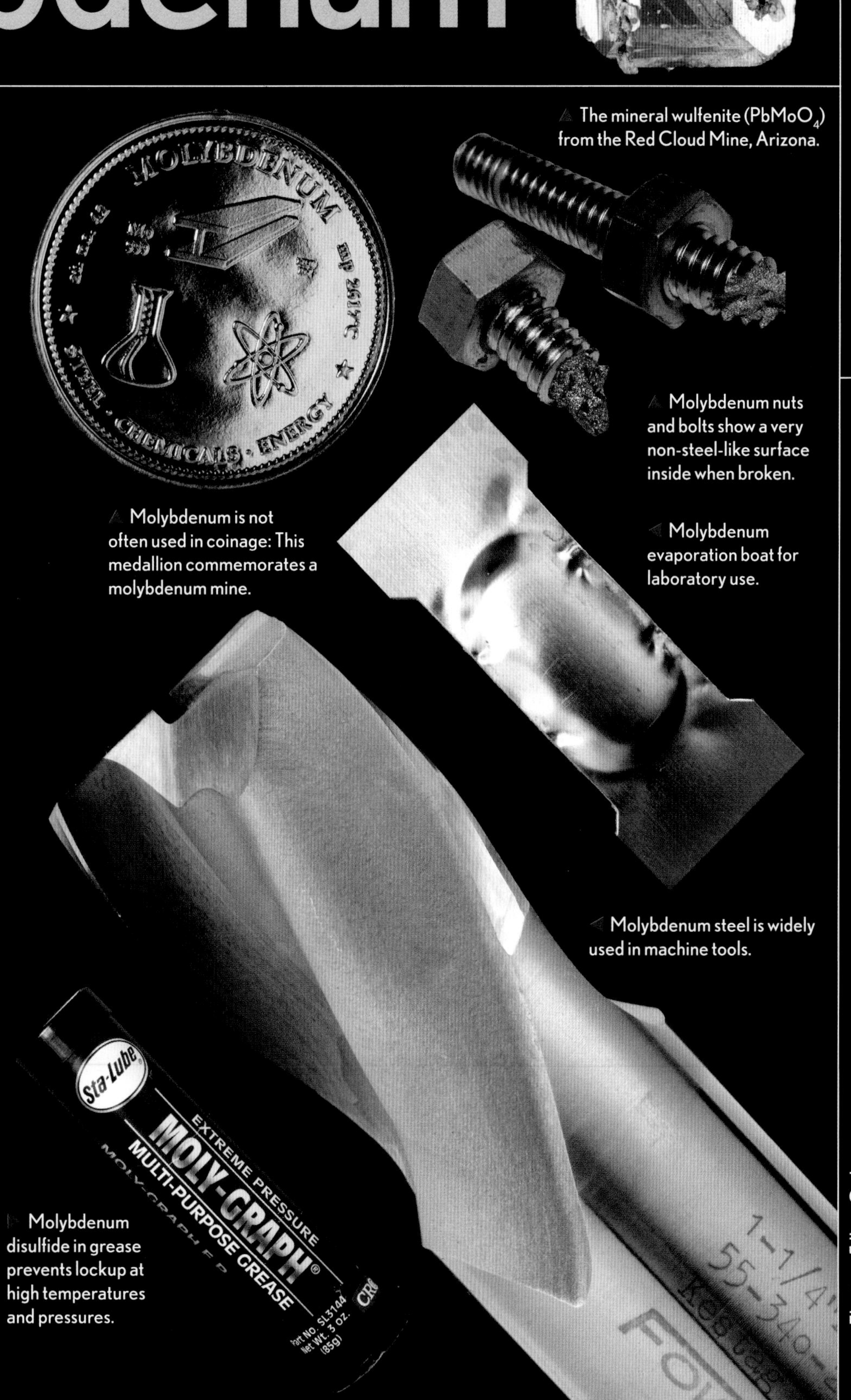

Molybdenum is not often used in coinage: This medallion commemorates a molybdenum mine.

Molybdenum nuts and bolts show a very non-steel-like surface inside when broken.

Molybdenum evaporation boat for laboratory use.

Molybdenum steel is widely used in machine tools.

Molybdenum disulfide in grease prevents lockup at high temperatures and pressures.

Molybdenum steel is a common high-strength alloy, but large pure moly bars, like this one, are unusual.

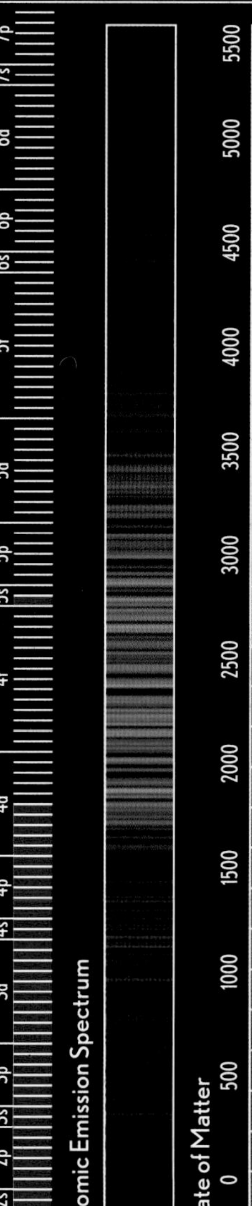

Technetium

Tc

43

Technetium

TECHNETIUM IS A BIZARRE ANOMALY, a radioactive element sitting right smack in the middle of the most stable, workmanlike range of the periodic table, the fifth period transition metals. There's a clear line between radioactive and stable elements: everything below bismuth (83) is stable; everything above it is radioactive—except technetium and promethium (61), which stick out like sore thumbs.

And if you have a sufficiently sore thumb and think it might be bone cancer, you might find yourself being injected with a particularly short-half-life version, $^{99}Tc_m$, which is bone-seeking and can thus be used to create images of where bone growth is occurring, using a gamma-ray camera.

So fiercely radioactive is $^{99}Tc_m$ that when medics transport it, they use little carts loaded with lead (82) or tungsten (74) "pigs," heavily shielded containers surrounding syringes of technetium ready to be injected. Since significant radiation escapes even though the pigs, the carts have surprisingly long handles.

It must be fairly intimidating to see a doctor enter your room with a device designed to keep the thing she proposes to inject into you as far away from her as possible. But the fact is that while you may get one such injection in a lifetime, medical personnel are exposed to the material day in and day out, and must be extremely careful to avoid accumulating a dangerous dose over time.

Technetium got its name because it was the first nonnaturally occurring element to be created: it exists only through technology (except for vanishingly small amounts in certain pitchblende ores, a fact not discovered until 1962). With ruthenium we return to stability—there are seventeen more elements before we reach the next radioactive one.

Cart filled with lead "pigs" for holding $^{99}Tc_m$ doses.

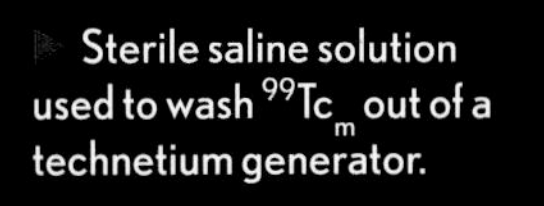

Sterile saline solution used to wash $^{99}Tc_m$ out of a technetium generator.

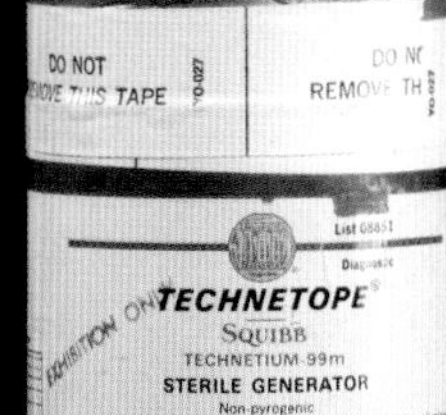

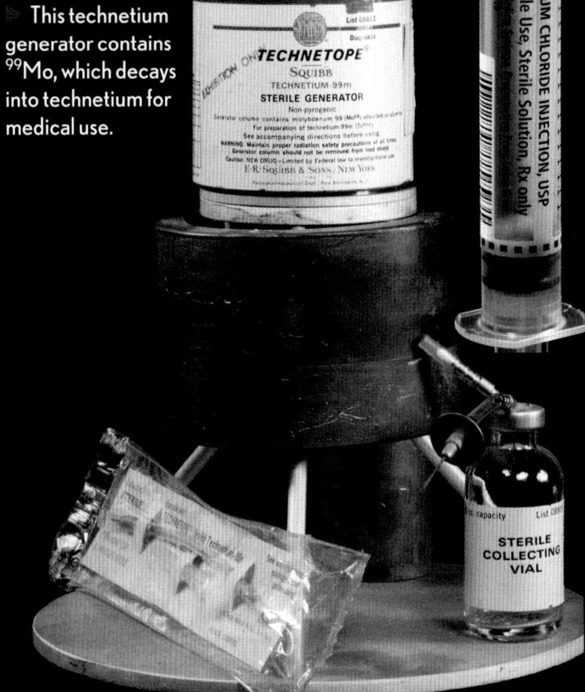

This technetium generator contains ^{99}Mo, which decays into technetium for medical use.

Gamma ray image formed when $^{99}Tc_m$ injected into a patient gathered around areas of bone growth.

Technetium is not considered naturally occurring, but in 1962 traces of it were found in samples of pitchblende (uranium oxide) from Africa.

A thin layer of pure technetium electroplated onto a copper substrate.

Elemental

Atomic Weight
[98]
Density
11.5
Atomic Radius
183pm
Crystal Structure

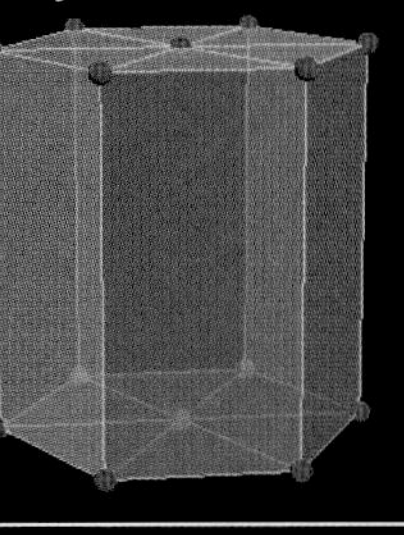

Electron Filling Order: 1s 2s 2p 3s 3p 4s 3d 4p 5s 4d 5p 6s 4f 5d 6p 7s 5f 6d 7p

Atomic Emission Spectrum

State of Matter: 0 500 1000 1500 2000 2500 3000 3500 4000 4500 5000 5500

Ruthenium

Ru

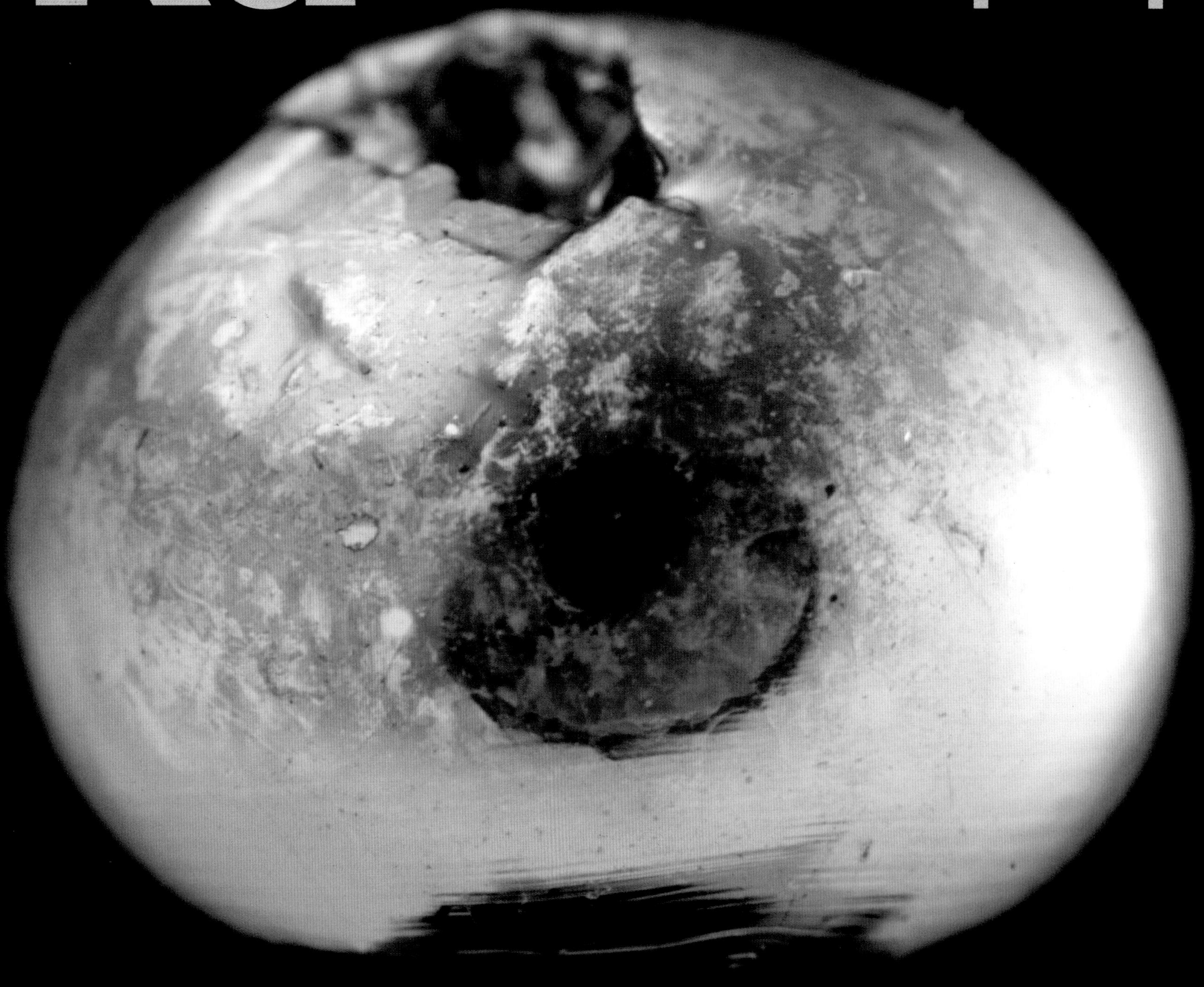

Ruthenium

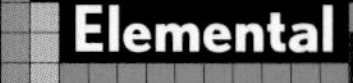

Atomic Weight
101.07
Density
12.370
Atomic Radius
178pm
Crystal Structure

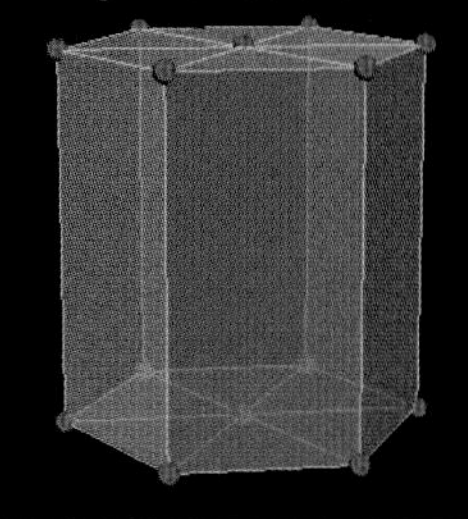

IN THE MIDDLE AGES, a region that includes modern-day Russia, Ukraine, and Belarus was known as Ruthenia. Since ruthenium was discovered and named for this region before germanium was named, a case can be made that it is the first element named for its discoverer's country. But the region known as Rus in discoverer Karl Klaus's day does not correspond to any modern nation-state, so I don't count it.

Ruthenium is the first of the precious metals, one of the minor platinum group metals that occur with platinum (78) in ores and share many of platinum's desirable properties. True to its classification as a precious metal, in daily life you're most likely to encounter ruthenium in jewelry, as a thin plating that has a darkish gray, pewter-like shine to it. Because of its high resistance to corrosion, plating an incredibly thin layer of very expensive ruthenium onto a much less expensive base metal is more economical than using a solid mass of moderately expensive pewter.

But as with most of the platinum group metals, ruthenium's main applications are as a catalyst and an alloying agent. Ruthenium appears in a particularly exotic example: single-crystal superalloys used in high-performance turbine blades when mere cost is of little concern.

Where ruthenium plating gives jewelry a dark shine, its neighbor, rhodium, is known for producing a brilliant luster.

▲ Ruthenium chloride is a vivid red.

▼ An experimental ruthenium-based solar cell.

◀ This button of ruthenium was created by melting powder in an argon-arc furnace, the easiest method that works.

▶ Cheap costume jewelry is often plated with ruthenium when a darker finish is desired.

Electron Filling Order
1s 2s 2p 3s 3p 3d 4s 4p 4d 4f 5s 5p 5d 5f 6s 6p 6d 7s 7p

Atomic Emission Spectrum

State of Matter
0 500 1000 1500 2000 2500 3000 3500 4000 4500 5000 5500

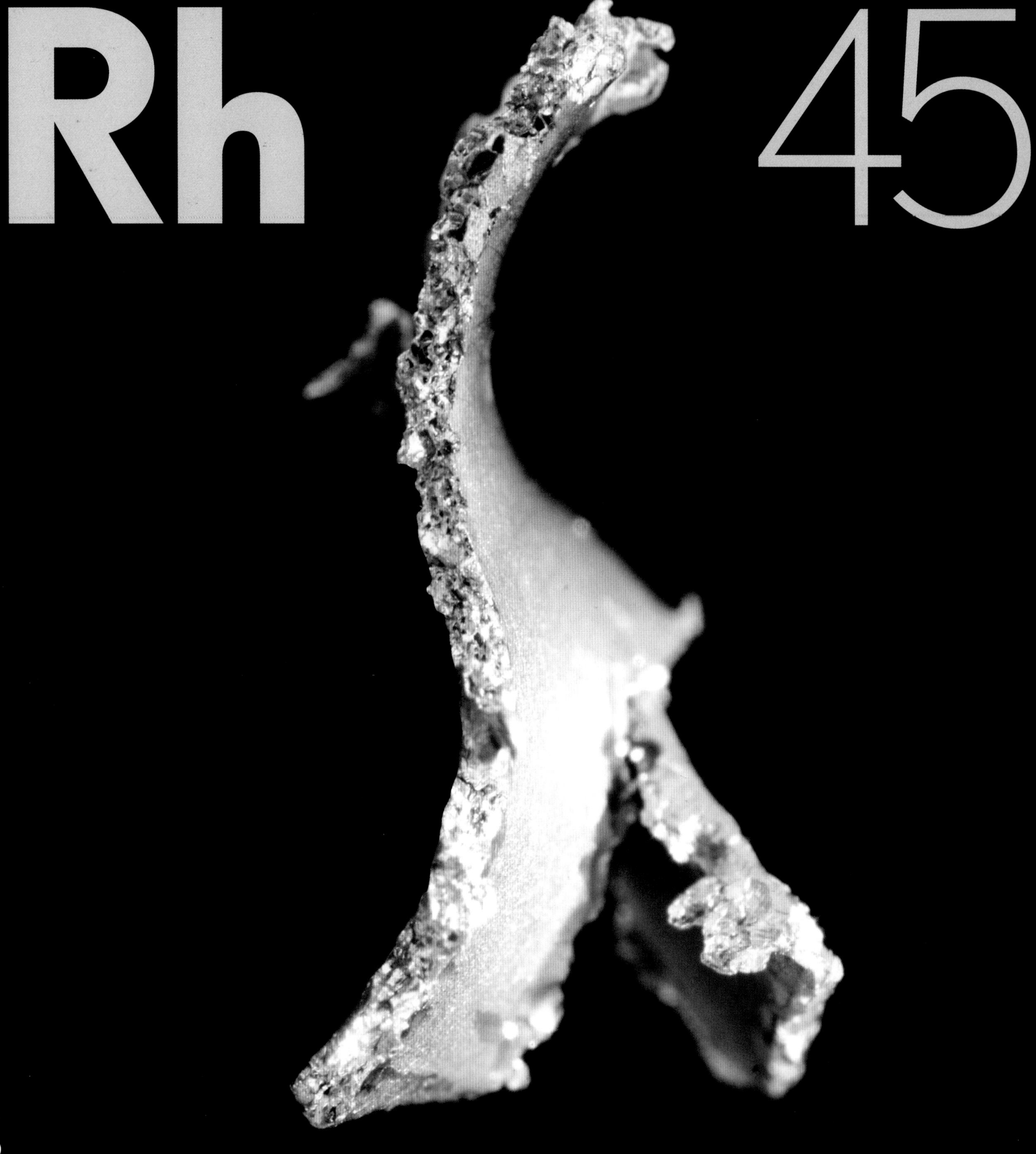

Rhodium

Rh

45

Rhodium

Atomic Weight
102.90550
Density
12.450
Atomic Radius
173pm
Crystal Structure

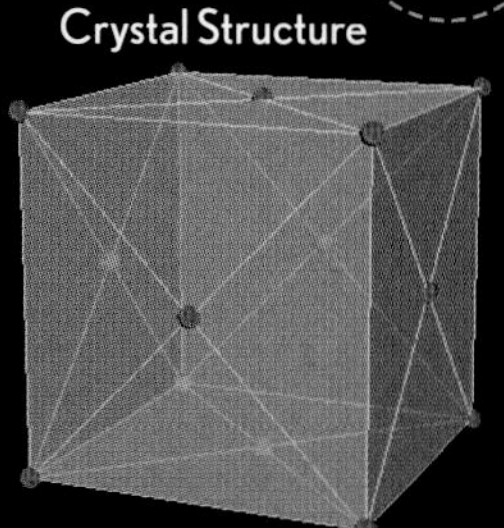

RHODIUM IS FAMOUS for its extraordinary price swings. If you bought a pound of rhodium in January 2004 and sold it in June 2008, you would have multiplied your investment by a factor of 22, turning $5,000 into $110,000 in four years. (And if you bought $5,000 worth in June 2008 and sold it five months later, it would have been worth $380. Rhodium bites, so beware.)

These huge price swings are partly due to speculation and partly to the fact that, as with all the minor platinum group metals, the supply of rhodium depends mainly on how much platinum (78) is being mined. Rhodium occurs as a minor constituent of platinum ore, so the more platinum you mine, the more rhodium residue you get. But if the demand for rhodium goes up, the supply can't respond unless the price of platinum also goes up—it is not economical to mine more platinum just for its rhodium impurity.

The other thing rhodium is famous for is being shiny. Cheap costume jewelry designed to look like silver or platinum is often rhodium plated, because a micron-thick film of rhodium is shinier than all the platinum in the world. (In fact, an expert eye can tell rhodium plating because it's *too* shiny.) This shininess also comes in handy for mirror coatings, for example in searchlights.

But, sadly, the biggest use for rhodium by far is not shiny at all: it's as a catalyst in automobile catalytic converters, the ultimate fate of a distressingly large fraction of the world's precious metals. The only thing that reflects light better than rhodium is silver (47), but it's not that great for jewelry, since it tarnishes in air so easily, and is used only for scientific mirrors that must have the absolutely highest possible reflectivity. In jewelry applications, rhodium plating is a fine substitute for silver, as is a thin layer of our next element, palladium.

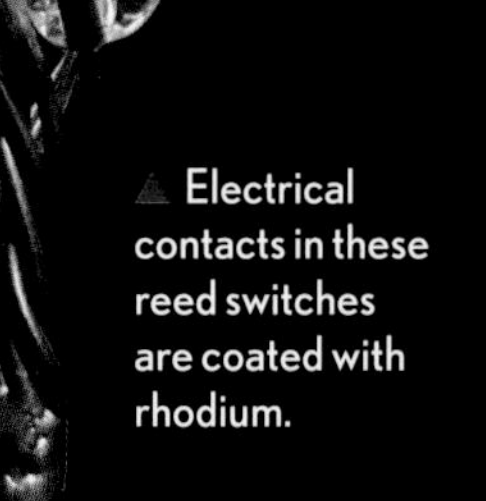

▲ Electrical contacts in these reed switches are coated with rhodium.

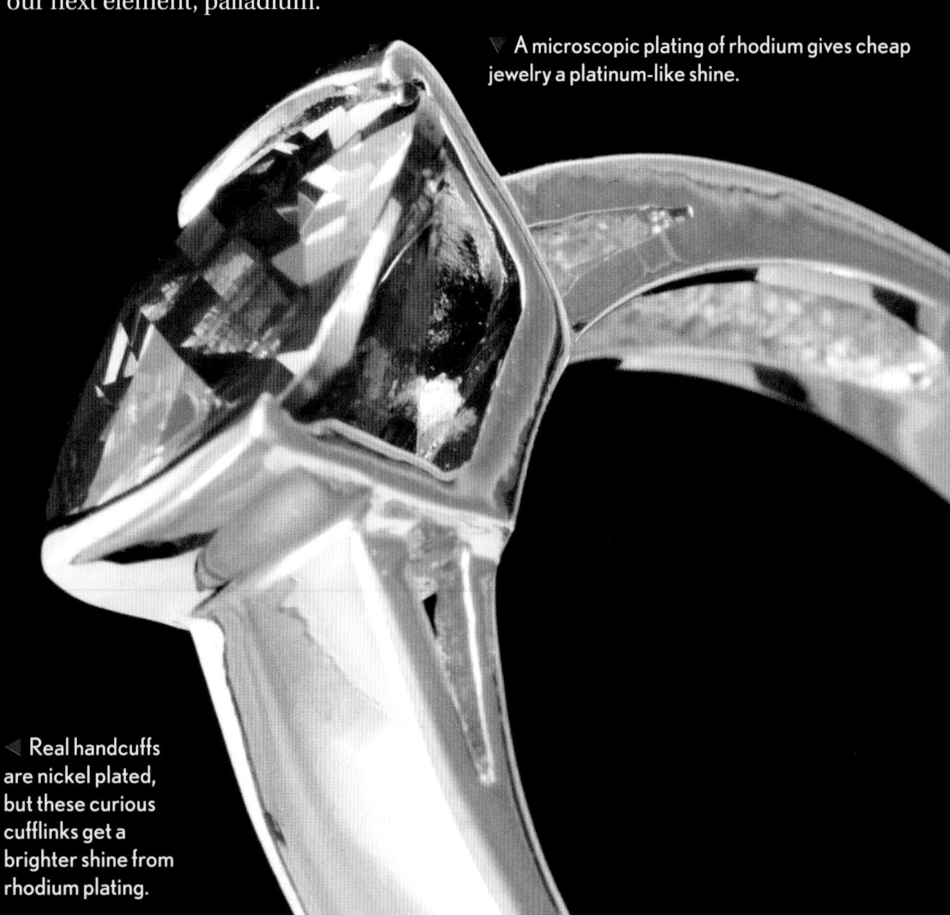

▼ A microscopic plating of rhodium gives cheap jewelry a platinum-like shine.

◀ The torn edge of a piece of rhodium foil shows its internal grain structure.

◀ Real handcuffs are nickel plated, but these curious cufflinks get a brighter shine from rhodium plating.

Electron Filling Order
Atomic Emission Spectrum
State of Matter

Palladium

Pd 46

Palladium

YOU'VE HEARD OF GOLD LEAF, gossamer-thin sheets of gold (79) used since antiquity to cover, or gild, objects. Palladium, too, can be beaten into sheets nearly as microscopically thin as gold leaf, and in that form it is used to imitate silver (47). Which is ironic, since it costs about twenty times as much. Unlike palladium leaf, actual silver leaf would tarnish, and since leaf of this sort can be less than a thousand atoms thick, any attempt to clean off the tarnish would quickly wear off all the silver.

Some solid palladium is used in jewelry, but as with rhodium (45), its main application is in automobile catalytic converters. These devices reduce smog in cities by burning up the residues of unburned fuel in the exhaust. Tiny particles of palladium (often mixed with other minor platinum group metals) are embedded in a ceramic honeycomb through which the hot engine exhaust is piped. On the surface of these catalyst particles, unburned fuel is able to combine with oxygen from the air at a much lower temperature than is normally required, converting the unburned fuel to carbon dioxide and water.

While burning without a flame is a neat trick, the most amazing thing about palladium is its absolutely astonishing ability to absorb hydrogen gas. Without applying any external pressure, a solid chunk of palladium will absorb 900 times its volume in hydrogen gas. The hydrogen just disappears into the solid metal. Where does it go? Hydrogen is able to sneak in between palladium atoms in their crystal lattice. If it weren't so expensive, tanks filled with palladium mesh would be an obvious way to store huge amounts of hydrogen with no need for high pressures. Not surprisingly there is ongoing research attempting to find rare earth alloys that could work nearly as well as palladium but at much lower cost. Palladium may be used to imitate silver, but for better or worse, silver will always be the real thing.

▲ Palladium bullion round, a form in which precious metals are traded.

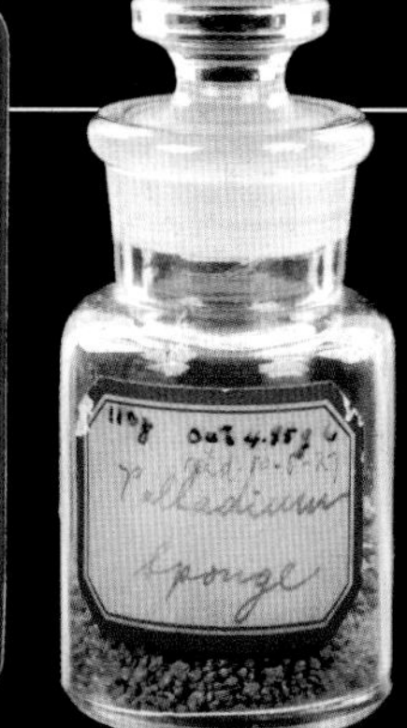

▲ Antique palladium "sponge," a term that means a fine powder.

▶ Native, or naturally occurring, palladium metal.

▼ Peculiar palladium-foil object halfway between a coin and a stamp.

▶ Automobile catalytic converter.

◀ Lovely torn piece of pure palladium.

Elemental

Atomic Weight **106.42**
Density **12.023**
Atomic Radius **169pm**
Crystal Structure

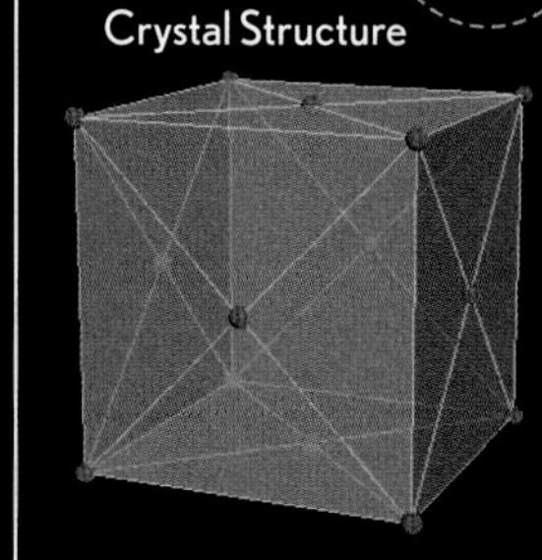

Electron Filling Order
Atomic Emission Spectrum
State of Matter

Silver

Ag 47

Silver

THE MAIN PROBLEM with silver is that it tarnishes, which would seem like an immediate disqualification for a metal of kings. But despite its shortcomings, silver is the first element we've encountered that has been associated since antiquity with glory and riches. I guess if you can afford a staff of silver polishers, tarnish is not such a big deal.

While silver and gold (79) go hand in hand, silver is definitely the junior partner. Historically it has typically sold for about a twentieth of the price of gold, though in the last century the ratio has reached as much as a hundred to one. This difference makes silver affordable as a coinage metal: Gold is just *too* valuable for everyday use. Silver has been used in commonly circulating coins for almost 3,000 years, whereas even the smallest gold coin may be more than most people want to carry around in their pockets.

But silver is not always second fiddle. Gold is not the very best at any one thing—it's not the most corrosion-resistant element, it's not the hardest, it's not the most valuable, it's not the most *anything*. But silver can claim two prizes: Of all the elements, it is the best electrical conductor and the most reflective.

For mirrors that need to be absolutely as reflective as they possibly can be, silver is without equal, though of course it has to be protected from tarnishing. While silver is used in some electrical applications, copper (29) is only about 10 percent less conductive at a small fraction of the cost. (Gold is widely used in electrical contacts not because it's the best conductor, but because it's a quite good conductor that absolutely does not tarnish or oxidize, meaning its good conductivity doesn't deteriorate.)

From the precious heights of silver we descend to cadmium, a distinctly low-class, peasant element.

Silver is not worth that much, so larger ingots—including 10oz and 100oz—are common.

Shorts made of silver thread would protect against electromagnetic fields, if that were actually a problem against which protection was needed.

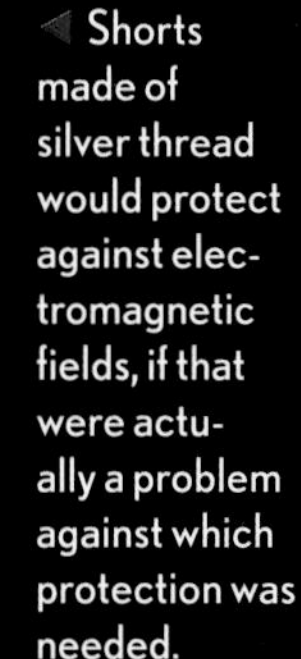

Silver improves the heat conductivity of this heat-sink compound.

Clockwise, a silver lab part, pendant, irradiated dime, and scientific front-surface mirror.

Silver tracheostomy tubes.

Tetradrachm showing the name Alexander (the Great). Minted in 261 BC, these coins are almost unfathomably old, yet ones like it are easily available—no one ever throws away a coin.

Elemental

Atomic Weight
107.8682
Density
10.490
Atomic Radius
165pm
Crystal Structure

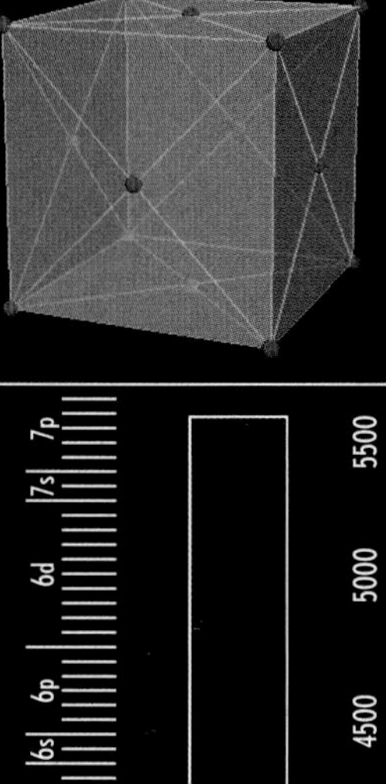

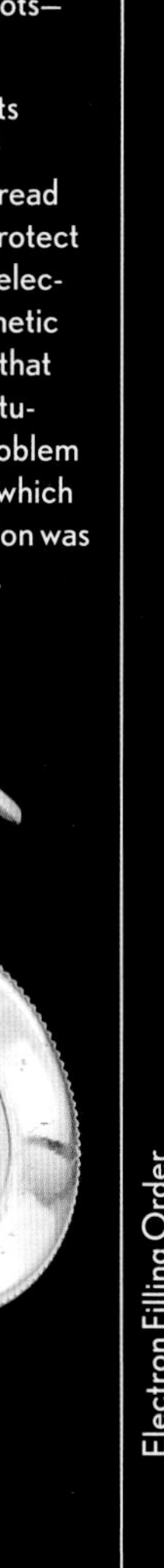

Cd 48

Cadmium

Cadmium

CADMIUM IS PERHAPS best known for its use in nickel-cadmium (ni-cad) batteries, though these days they are being replaced in many applications by nickel metal hydride and lithium-ion batteries that are lighter, more powerful, and less toxic. Cadmium, unfortunately, is a lot like lead (82) and mercury (80) in that it accumulates both in the environment and in the body, causing long-term damage to life everywhere it's found. (This is why ni-cad batteries should be taken to stores or collection points that recycle them, not thrown in the garbage where their cadmium will eventually get into the environment.)

The other place you will find large amounts of cadmium is in cadmium-plated fasteners, used these days mainly on aircraft. While common zinc (30) plating is good enough for domestic applications, cadmium offers an unbeatable combination of properties when it's really, *really* important that a given bolt not rust or cause corrosion of the parts it's touching (for example, if it's the bolt holding together the landing gear of the airplane you're in).

One bright spot with cadmium is cadmium yellow, an extremely intense pigment favored by the impressionists. Claude Monet, asked about the colors he used, replied, "In short, I use white lead, cadmium yellow, vermilion, madder, cobalt blue, chrome green. That's all." Four elements in one quote—not bad for a painter! Since vermilion is mercuric sulfide, Monet's favorites constitute practically a full house of toxic pigments, missing only Paris green (see arsenic, element 33, for a discussion of that nasty stuff).

Fortunately, the next element is considerably more benign.

◄ A fish cast by the author out of solid cadmium, for no reason whatsoever.

◄ Cadmium-plated brake rotor.

► Treatment with dichromate gives these cadmium-plated castle nuts a gold color.

▼ Common nickel-cadmium rechargeable battery.

◄ Cadmium foil intended for radiation shielding.

► The classic cadmium yellow pigment is cadmium sulfide.

◄ The mineral greenockite is natural crystalline cadmium sulfide.

Elemental

Atomic Weight
112.411
Density
8.650
Atomic Radius
161pm
Crystal Structure

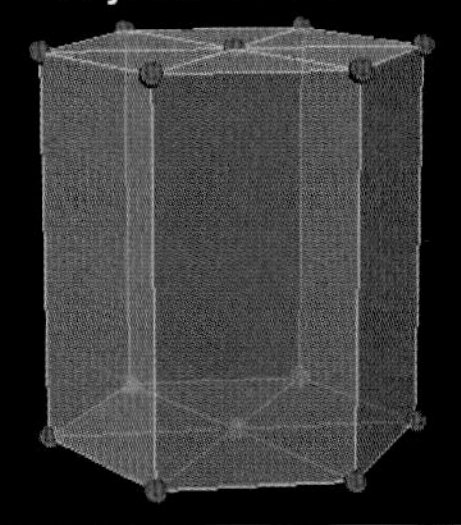

Electron Filling Order

Atomic Emission Spectrum

State of Matter

Indium

In

49

Indium

INDIUM IS NOT NAMED for India or Indiana or any other piece of geography. It is named for the strong indigo-blue spectral emission line that was the first evidence of its existence. It's said that until 1924, only a gram of it had been isolated in the whole world, but these days hundreds of tons a year go into the production of LCD televisions and computer monitors.

In that application it is in the form of indium tin (50) oxide, a transparent conductor of electricity that allows signals to be communicated to the individual pixels in a display without blocking the light from all the other pixels.

The pure element itself is also a good conductor but not at all transparent, being a soft, silvery, and quite fun metal. In pure form, it's so soft you can easily dent it with your fingernails, or even shave off slices of it with your pocketknife. So far as is known, indium is not toxic, always a nice bonus for an element that is fun to play with.

Because indium is one of the very few metals that wet glass (rather than being repelled by its surface), it can be used as a gasket material in high-vacuum applications where any sort of rubber gasket would be hopelessly porous by the standards of the vacuum you are trying to achieve.

Indium shares an interesting property with its neighbor, tin: When bars or rods of either metal are bent, they "cry," producing a cracking sound as the internal crystals break and rearrange. While quite a few people have heard the tin cry, the indium cry is a more exclusive experience.

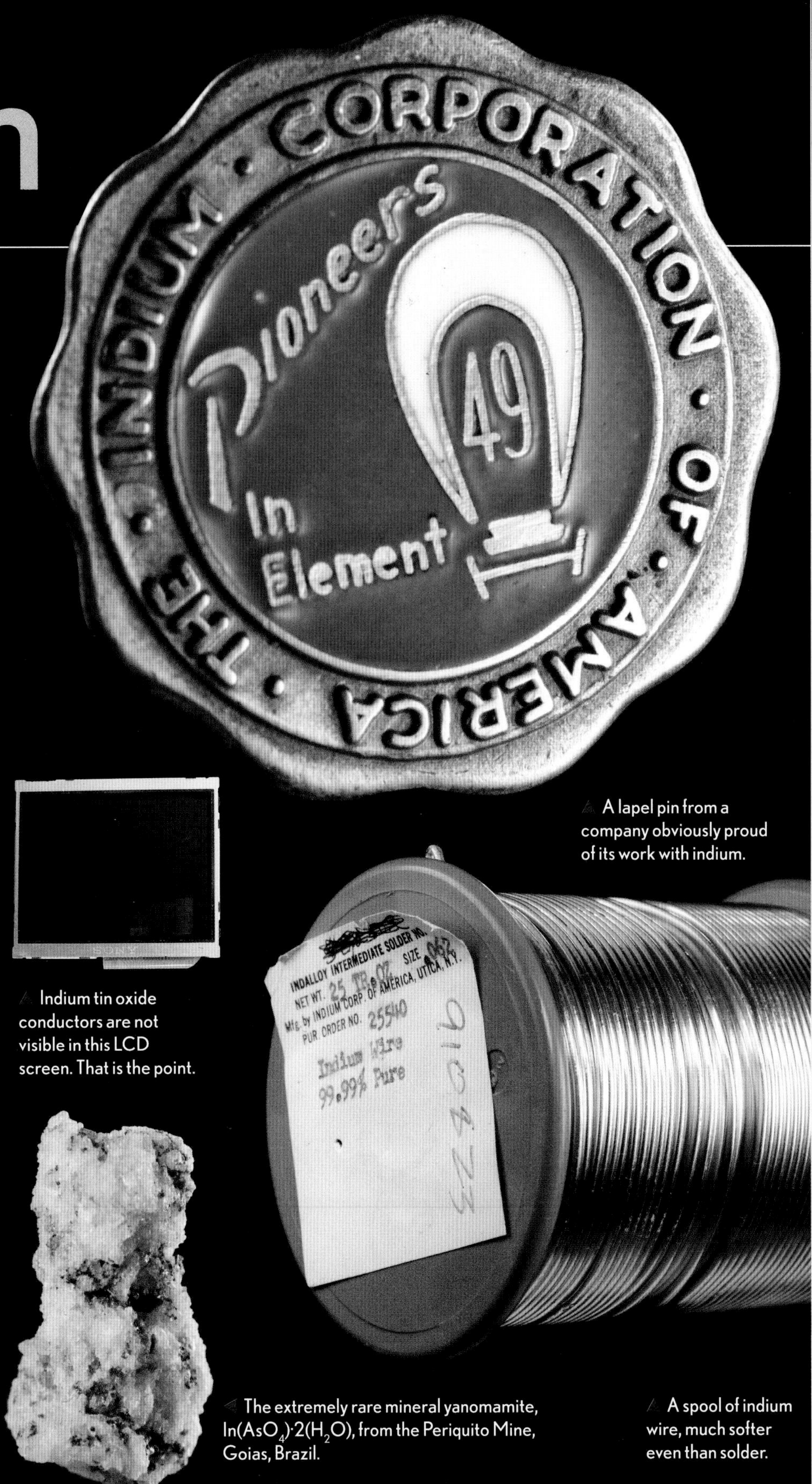

A lapel pin from a company obviously proud of its work with indium.

Indium tin oxide conductors are not visible in this LCD screen. That is the point.

Pure indium is almost always sold in 1kg bars, of which this is about half. It is so soft that a bar like this can be cut in half with a knife (though it takes some effort).

The extremely rare mineral yanomamite, $In(AsO_4)\cdot 2(H_2O)$, from the Periquito Mine, Goias, Brazil.

A spool of indium wire, much softer even than solder.

Elemental

Atomic Weight
114.818
Density
7.310
Atomic Radius
156pm
Crystal Structure

Electron Filling Order: 1s 2s 2p 3s 3p 3d 4s 4p 4d 4f 5s 5p 5d 5f 6s 6p 6d 7s 7p

Atomic Emission Spectrum

State of Matter: 0 500 1000 1500 2000 2500 3000 3500 4000 4500 5000 5500

Tin

Sn 50

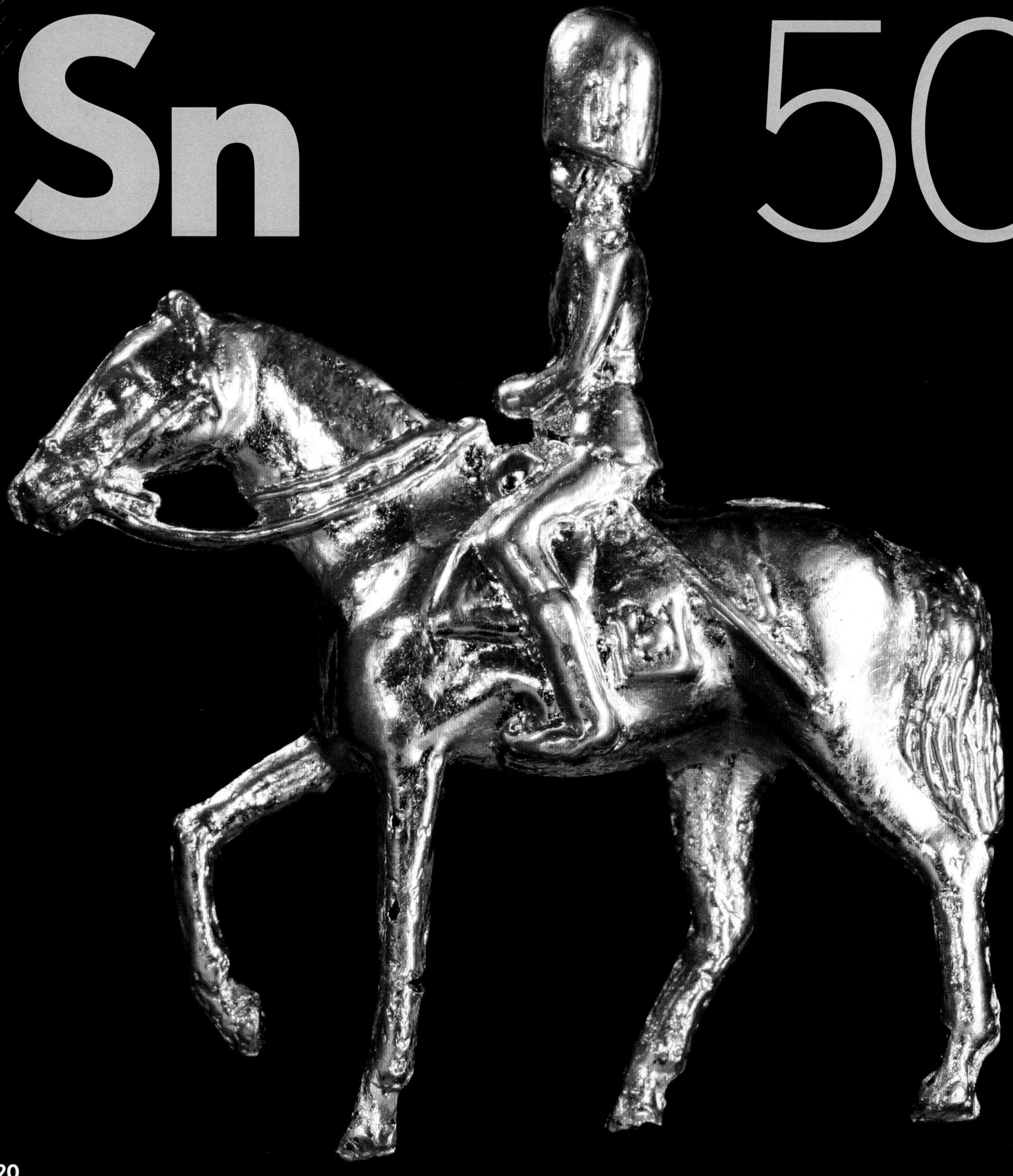

Tin

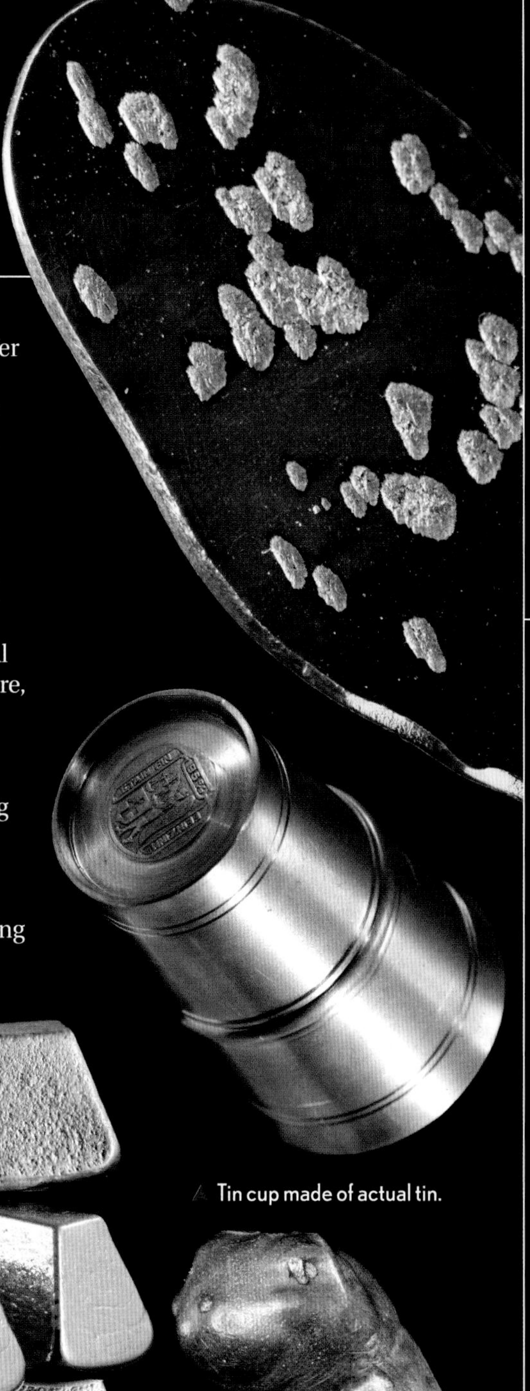

Gray tin allotrope growing on a piece of metallic tin.

AH, TIN, what a lovely, lovely element. Pretty much completely nontoxic, stays shiny forever, easy to melt and cast into minutely detailed shapes, not horribly expensive—really, what more could you ask for?

Tin soldiers were almost never made of pure tin. Lead (82) is cheaper and easier to melt, so lead-antimony or tin-lead alloys were a more frequent choice. These days, of course, nearly all toy soldiers are made of plastic, which is a definite improvement from a product-safety point of view. (Today the idea of using lead as a main ingredient in children's toys would go over about as well as a lead balloon.)

In fact, many things called tin—including tin cans, tinfoil, and tin roofs—are not and never were made of actual tin. The word has come to mean any sort of thin sheet metal, to the point that if you visit a scrap-metal yard they will refer to sheets of metal being lifted by giant electromagnets as "tin," oblivious to the fact that tin is completely nonmagnetic.

A few coins have been minted in tin, but an odd property limits this application: At wintry temperatures, tin starts to convert—very slowly over the course of months—from a silvery metal into a dark gray powder. It's not rusting or oxidizing or undergoing any chemical change at all. Instead the crystal structure, or allotropic form, is changing from its normal metallic form to a cubic crystal structure known as gray tin. When this happened to tin organ pipes during long European winters, the phenomenon came to be called the tin pest.

If you had your money turn to gray powder, you might think you were dealing with the next element, antimony.

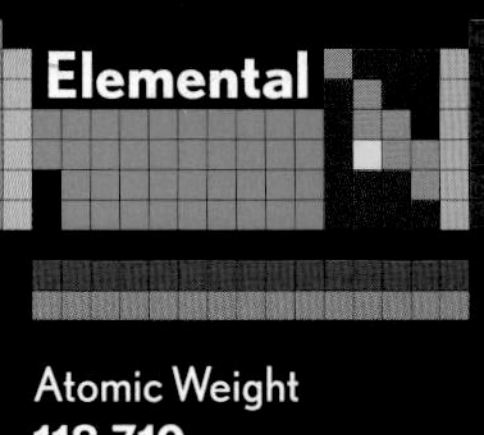

Atomic Weight
118.710
Density
7.310
Atomic Radius
145pm

Crystal Structure

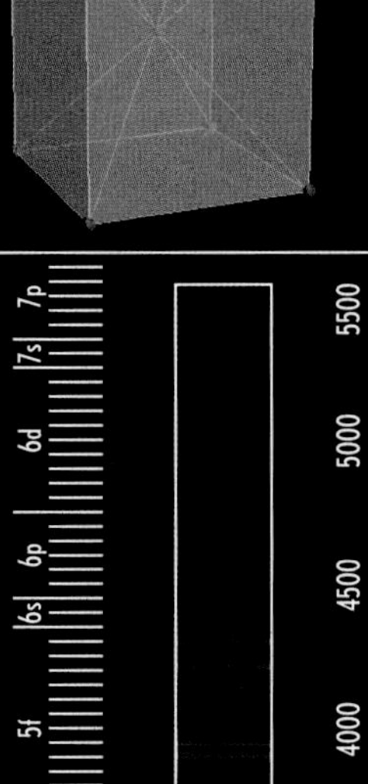

Electron Filling Order

Atomic Emission Spectrum

State of Matter

Lead-free solder is mostly tin.

Tin cup made of actual tin.

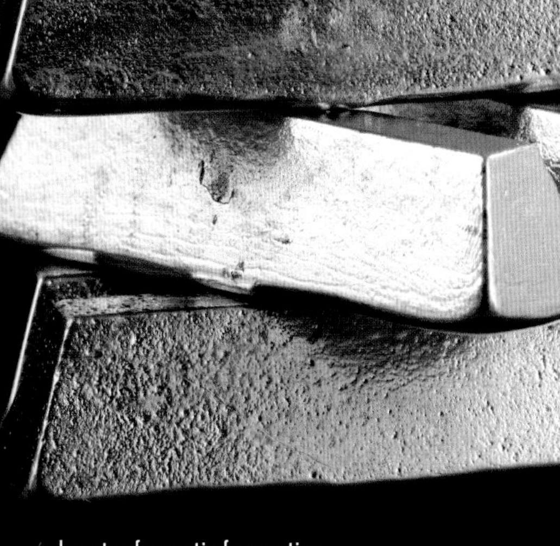

Ingots of pure tin for casting.

Cute tin caterpillar.

The mineral cassiterite (tin oxide) from the Viloco Mine, La Paz, Bolivia.

The classic tin soldier was more often made of a tin-lead alloy, but this one is 99.99 percent pure tin.

Antimony

Sb 51

Antimony

NO, ANTIMONY is not anti-money, it just doesn't want you to lose sight of your other priorities. When not passing judgment on your lifestyle, antimony is a typical metalloid, clearly metallic in appearance, yet brittle and more crystalline than ordinary metals.

Adding antimony to lead (82) makes the lead much harder, and just the right mixture of lead, tin (50), and antimony has the wonderful property that it expands just a tiny bit when it solidifies from a molten state. By pouring this alloy into hand-carved master molds, Johann Gutenberg was able to create crisp, hard, reusable letterforms for printing, a little invention he called movable type. After a brief run of 550 years, printing with movable type is now pretty much history, replaced by computer and photolithographic processes whose development would have been impossible without the widespread literacy fostered across the globe by Gutenberg's clever use of antimony.

Linotype machines may be fading from memory, but another application of lead-antimony alloys remains as popular as ever: bullets.

Bullets are famously made of lead, but pure lead is too soft, so antimony is added to create harder bullet lead alloys. Lead-acid car batteries use lead electrode plates that are similarly hardened with antimony.

A lovely property of antimony that I have not seen reported anywhere else is that blocks of it emit beautifully melodic pinging noises after being cast. No doubt crystals inside are breaking and slipping as the blocks cool down, in effect tapping the bars from the inside and causing them to ring like Tibetan chimes. Other metals click and crackle sometimes when cooling, but I have never heard such musical cooling-off sounds as from recently cast bars of antimony.

While antimony makes music while cooling, the very name *tellurium* itself is musical.

Pure antimony sputtering target that looks not unlike the antimony stars favored by alchemists.

Wine steeped in antimony goblets like these induces vomiting, one of the ways in which antimony was used in medicine.

Neither tin toys nor antimony toys are pure; both have historically been made of alloys with some of both metals, plus lead.

Beautiful lumps of broken crystal like this are how bulk antimony is sold.

An ingot of cast antimony broken in half to show the internal crystals, whose formation created beautiful music while the ingot was cooling.

The antimony Foo Lion incense burner—an excellent justification for the existence of eBay.

Elemental

Atomic Weight
121.760
Density
6.697
Atomic Radius
133pm
Crystal Structure

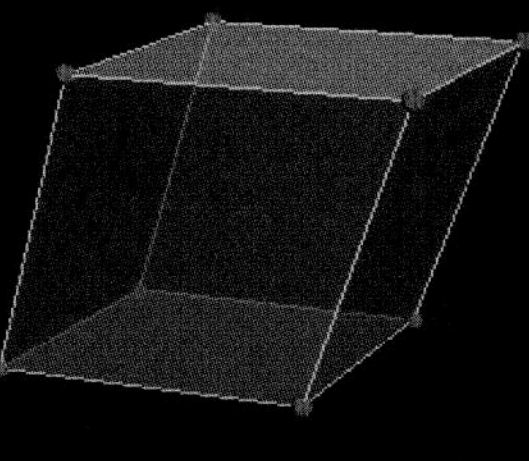

Electron Filling Order

Atomic Emission Spectrum

State of Matter

Tellurium

Te

52

Tellurium

TELLURIUM IS the most beautiful element name. From the Latin for earth, it has a poetry about it unmatched by any other. (I have a special fondness for the name: When a company was planning to publish a computer game called Wolfram, creating a potential trademark problem for my software company, Wolfram Research, I defused the situation by convincing them that Tellurium was a much better name for their game.)

It may have a pretty name and a pretty crystal structure, but this element's properties make it anything but pretty when you meet it. Exposure to very small concentrations can cause you to smell of rotten garlic for weeks, a fact that for some reason limited early interest in researching the substance.

Despite this problem, and despite the fact that tellurium is one of the rarest of all elements—eighth or ninth least abundant in the earth's crust—it has found a number of important applications. You might well have some in your house right now in the form of tellurium suboxide, the magic layer in DVD-RW and Blu-ray rewritable discs whose reflectivity can be switched back and forth between two states by laser heating.

Combine extreme rarity with use in popular disc formats, solar cells, and experimental memory chips, and some people predict an explosion in tellurium prices. On the other hand, DVD and Blu-ray are losing out to online movies, other types of solar cells don't use tellurium, and who knows if it will be tellurium phase change memory, or carbon (6) nanotube memory, or something not yet invented that will really take off? A collapse in tellurium prices strikes me as just as likely as a run on the stuff.

I'm afraid I don't have any similarly useful investment advice about the next element, iodine.

▲ Bismuth telluride is used in thermoelectric coolers, like this one from a one-can-of-soda refrigerator.

▶ The mineral calaverite (gold tellurite).

▼ Beautiful crystals form on the surface of a disk of molten tellurium as it hardens.

▲ CD-RW and DVD-RW disks use tellurium suboxide in their rewritable data layer.

◀ Tellurium is hardly ever used in pure form, but these beautiful slender crystals are how it is distributed commercially.

I

Iodine

53

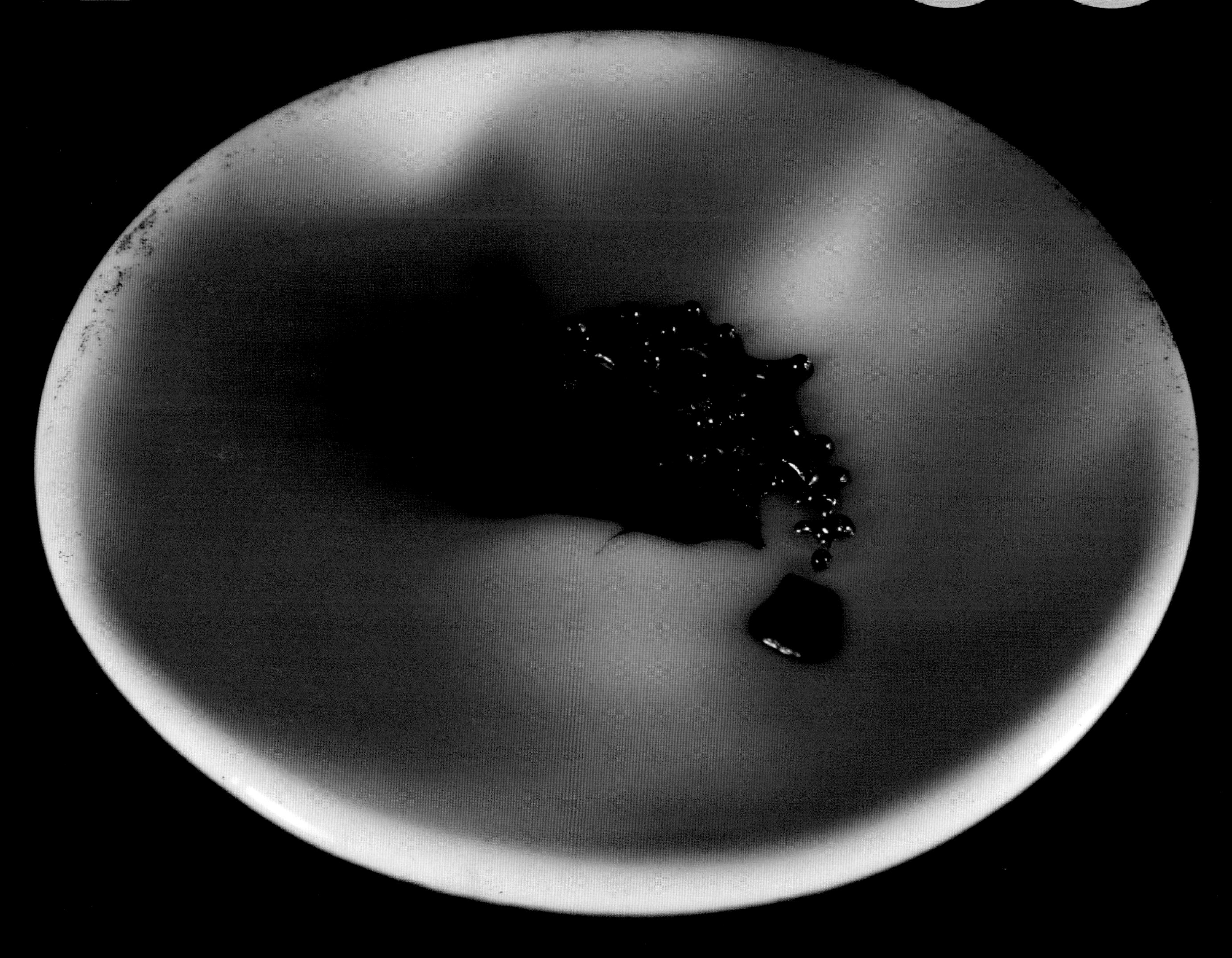

Iodine

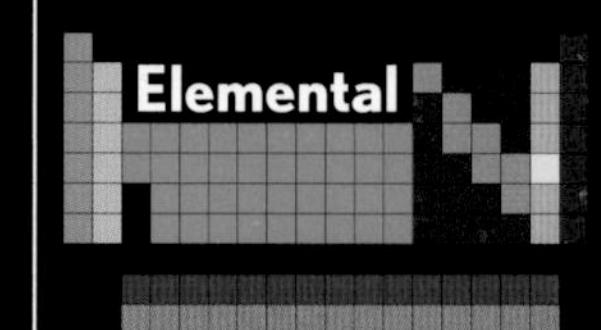

Atomic Weight
126.90447
Density
4.940
Atomic Radius
115pm
Crystal Structure

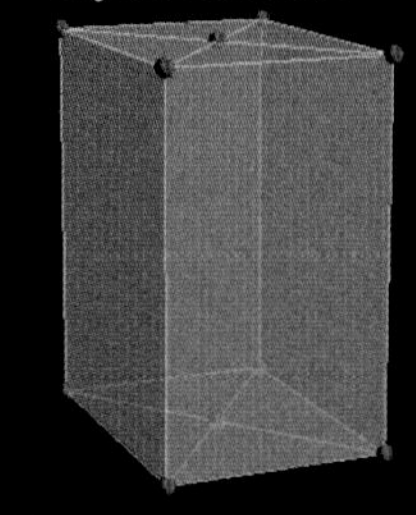

AS YOU MOVE DOWN the halogen column, the elements mellow a bit, from vicious fluorine (9) to deadly chlorine (17) to barely-liquid bromine (35)—until you reach iodine, an element so comparatively benign that it is used to cure hoof fungus in horses.

Iodine is a solid at room temperature, but like bromine it's just barely hanging on. Gentle heating will melt it, at which point it immediately evaporates into a thick, dense, beautiful violet vapor.

Iodine taught me the difference between smoke and vapor. You can photograph smoke against a black background by lighting it from the side, because smoke is made up of tiny particles that reflect light. It is *impossible* to photograph a vapor, even a colored vapor, against a black background because no matter how much light you shine in from the side, there are no particles to reflect light, only individual molecules. The only way to see a vapor is by how it absorbs light coming toward you from a bright background. This fact wasted considerable time when I attempted to take a nice picture of iodine vapor for my black-background poster.

Iodine, usually in the form of a few percent dissolved in alcohol (which causes the stinging, not the iodine), was once widely used as a disinfectant, and to a limited extent it still is. Like chlorine and bromine above it in the periodic table, iodine disinfects by brute chemical attack on microbes, something they are not able to develop resistance to. If the fancy topical antibiotics we use today ever lose their effectiveness, the halogens will always be there to save us, or at least our hooves.

After every halogen comes a noble gas, but here it's the least noble of the lot, xenon.

Iodine evaporates into a beautiful violet vapor when heated. There's a torch under the plate in this photograph.

IODINE SWAB 1½ CCM. Mild Tincture Iodine U.S.P.

Iodine deficiency causes goiter, but goiter is uncommon today because of iodized salt, making iodized gum unnecessary.

Iodine dissolved in alcohol has been used as a disinfectant for generations. The sting is from the alcohol, not the iodine.

Resublimated iodine for veterinary use as a disinfectant.

U. S. P. XIII TINCTURE OF IODINE CONTENTS ½ OZ. CAUTION POISON ALCOHOL 44 50% Certified PHARMACAL COMPANY INCORPORATED NEW YORK, N. Y. DISTRIBUTORS

Pretty, old tincture-of-iodine bottles are common collectors' items.

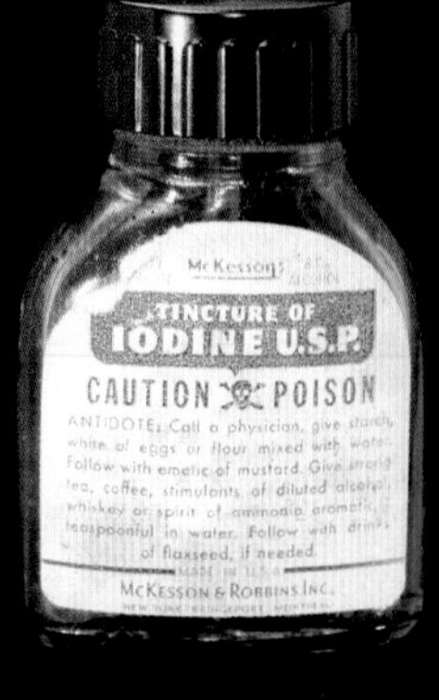

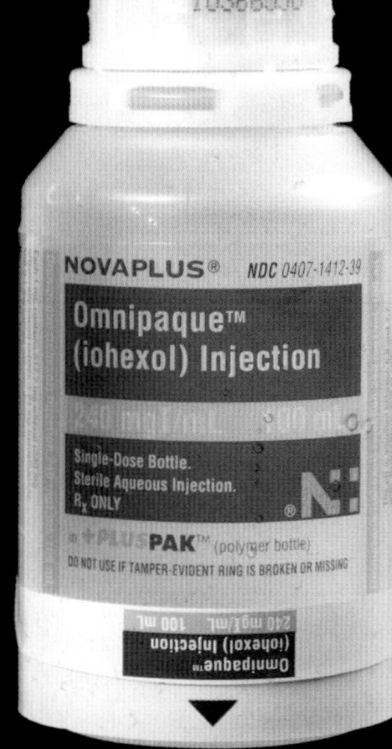

An iodine-containing contrast agent used in imaging the heart in CT scans.

Electron Filling Order
Atomic Emission Spectrum
State of Matter

54

Xenon

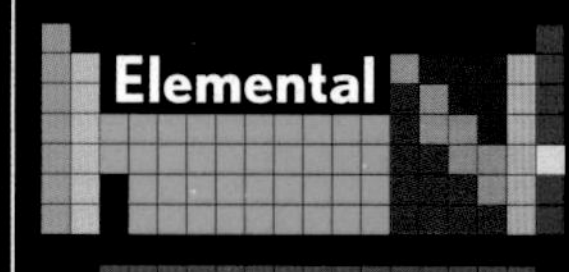

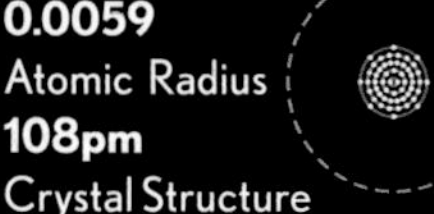

Atomic Weight
131.293
Density
0.0059
Atomic Radius
108pm
Crystal Structure

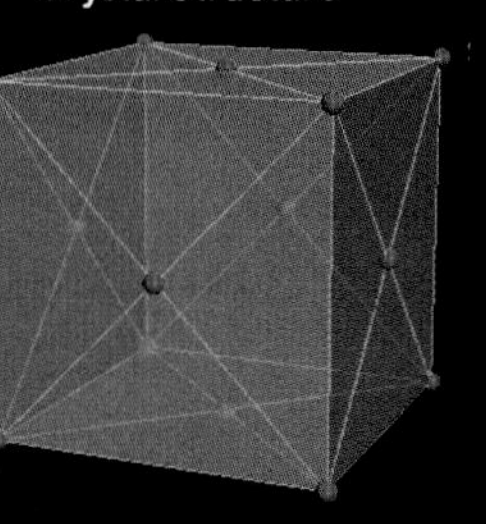

▲ Blue-tinted film colors the light from a xenon-filled incandescent bulb in order to make it look like an expensive xenon metal halide headlamp.

FOR MOST PRACTICAL PURPOSES, xenon is noble: inert and nonreactive, just like the other gases in this column of the periodic table. It's even the most expensive. But in what can only be described as an egregious case of slumming, in 1962 xenon was caught in the act of forming compounds with common elements.

Since then dozens of xenon compounds, usually involving fluorine (9), have been discovered and prepared. Xenon difluoride, for example, is commercially available from any laboratory [illegible]og. It comes in a bottle just as pla[illegible]ay. This is shocking, just shocking—it simply isn't something that noble gases are supposed to *do*.

Indiscretions aside, most applications of xenon do still make use of its typically noble inertness. Incandescent lightbulbs filled with xenon can burn hotter and brighter because of the low thermal conductivity of the gas. But arc lighting is where xenon really shines.

The central problem in cinema projectors and spotlights is creating a parallel beam of light, which is done by using light from a tiny, intense source and bouncing it off a parabolic focusing mirror. The more compact the source of light at the mirror's focal point, the better the beam. Imax projectors use fantastically bright 15-kilowatt xenon short-arc lamps to create their huge projected images. The bulbs are filled with xenon at such a high pressure that they must be stored and handled in special protective enclosures and clothing due to the risk of explosion.

On a much smaller scale, xenon metal halide lamps are in those really annoying new headlights that dazzle you when you encounter certain brands of overpriced cars on the street at night.

As naturally as noble gas follows halogen, alkali metal follows noble gas, and next in line is the most reactive of that tribe.

◄ Inhaled radioactive ^{133}Xe is used to study lung function.

► High-power xenon flashtube used by studio photographers.

► Xenon short-arc projector lamp.

◄ The xenon gas in this tube is being excited by a high-voltage discharge, creating a lovely pale-violet glow.

► Genuine xenon metal halide headlamp.

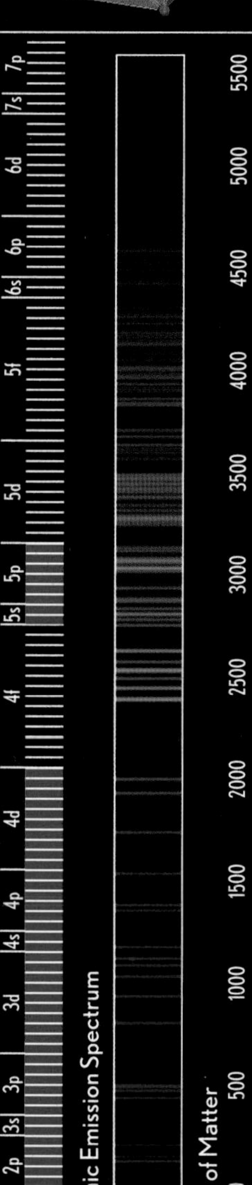

Cesium

Cs 55

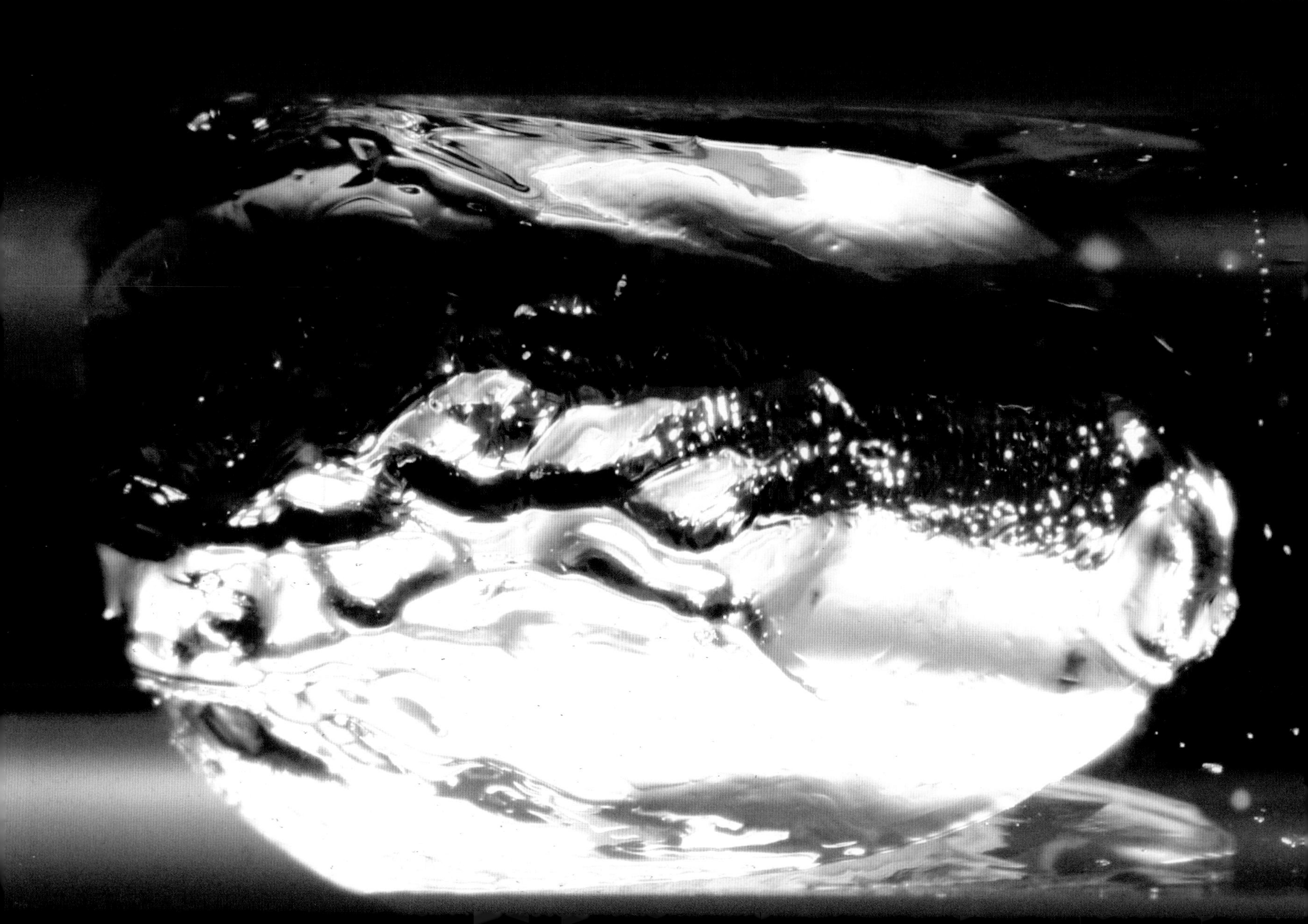

Cesium

CESIUM IS WIDELY listed as the most reactive of all the alkali metals, and technically it is. When you drop a piece in a bowl of water it *instantly* explodes, sending water flying in all directions. But that doesn't mean it makes the biggest bang of the alkali metals. Sodium (11) takes longer to explode when tossed in water, but the whole time you're waiting, a plume of hydrogen (1) gas is building up, and when all that hydrogen ignites, the explosion is much bigger than anything you can get with cesium. I know this because I spent a few days filming the whole series of alkali metal water explosions, debunking a certain British television show that used a stick of dynamite to "improve" their cesium explosion to match what they expected from its technical reactivity. That was a fun couple of days.

But the main business of cesium is not explosions, it's time. The current official definition of the second reads as follows: "The second is the duration of 9,192,631,770 periods of the radiation corresponding to the transition between the two hyperfine levels of the ground state of the cesium 133 atom." To realize this standard in practice, you beam a signal of approximately that frequency through a collection of cesium atoms and watch how much of the signal is absorbed by them as you slowly adjust the frequency around the target value. When the maximum amount is being absorbed, indicating that your signal is dead on the transition's energy level, the frequency of your signal is *by definition* exactly 9.19263177000000 . . . gigahertz—if, that is, your cesium atoms are in perfect isolation and there are no stray electric, magnetic, or gravitational fields influencing them.

International Atomic Time, which underlies the more commonly used Coordinated Universal Time, is kept by synchronizing the operation of 300 cesium atomic clocks around the world. The most accurate ones are cesium fountain clocks in which lasers throw bundles of a few million isolated cesium atoms upward in a vacuum chamber and measure them as they are in free fall, almost completely devoid of outside influences. If the NIST-F1 cesium fountain clock in Boulder, Colorado, had been built by dinosaurs 70 million years ago, it would now be off by less than a second.

From floating cesium atoms, we move on to an element whose very name means heavy.

◀ The cesium in this ampoule melts if you hold it in your hand for a minute, yielding the prettiest gold liquid. If the ampoule were to break in your hand, the resulting fire would be extremely unpleasant.

Ultra-miniaturized cesium atomic clock created by NIST.

A cesium getter—when activated by heat, it would clear the last traces of oxygen and water from a vacuum chamber.

Powdered cesium formate used in oil-well drilling.

A piece of solid magnesium metal floats on concentrated cesium formate solution, used in oil-well drilling to clear rock chips.

Vacuum chamber at the base of the cesium fountain clock at the National Physical Laboratory in England.

Elemental

Atomic Weight
132.90545
Density
1.879
Atomic Radius
298pm
Crystal Structure

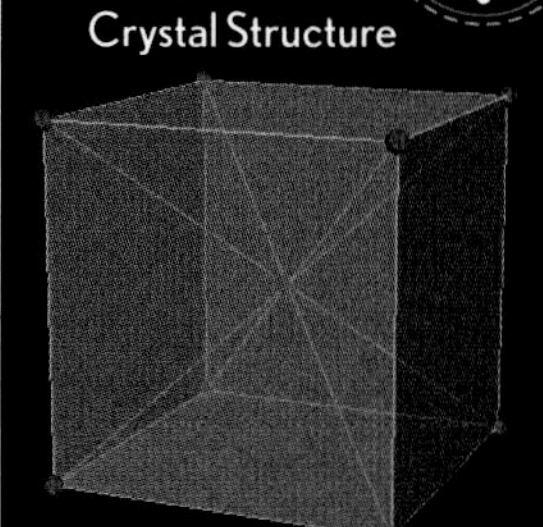

Electron Filling Order

Atomic Emission Spectrum

State of Matter

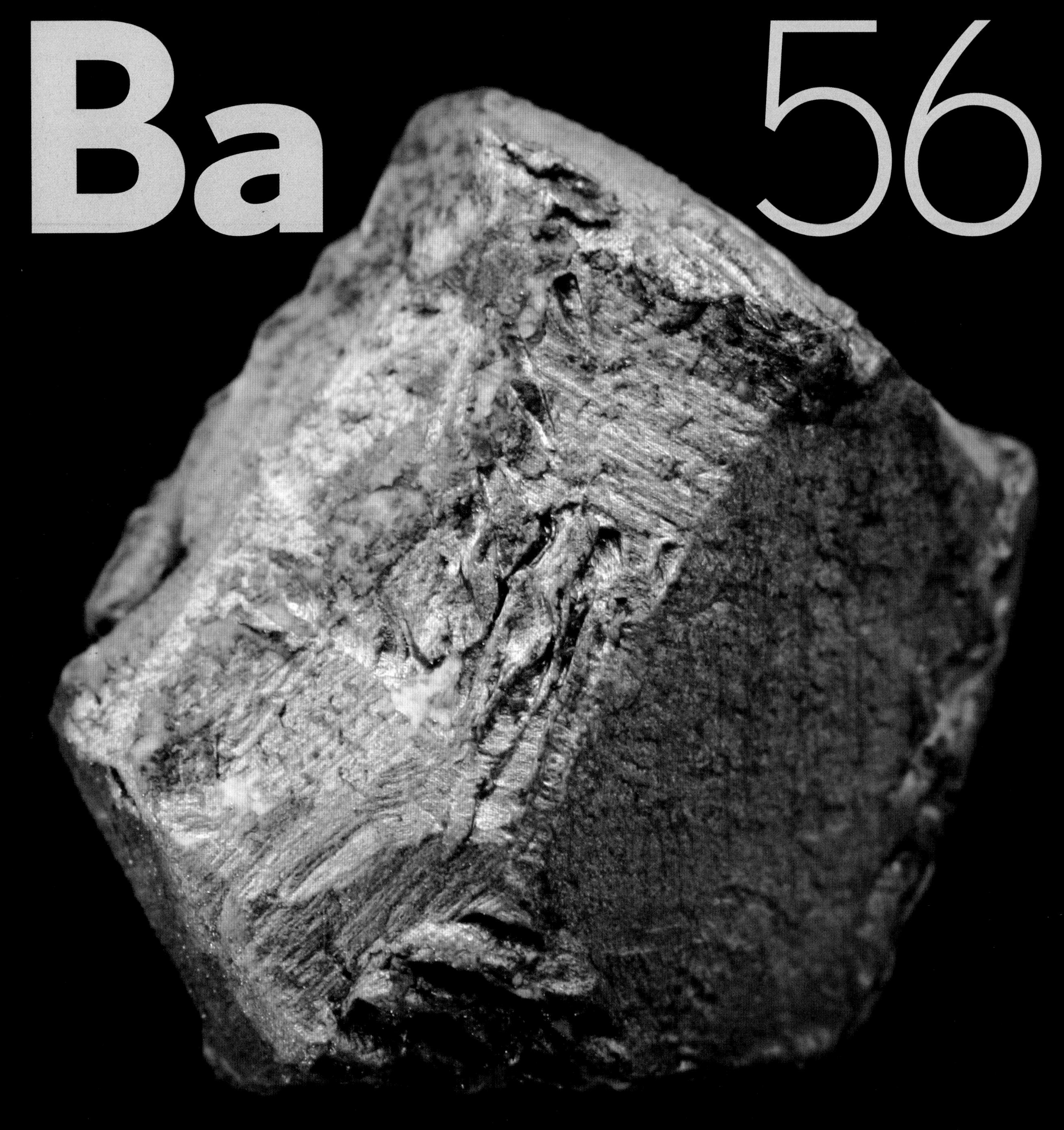
Barium
Ba
56

Barium

BARIUM, FROM the Greek for heavy, is not particularly heavy. It's actually less dense than titanium (22), a metal known for being light. But while barium is not heavy in pure form, many of its compounds are, and many of its applications take advantage of the density of slurries of barium compounds.

One such application is oil-well drilling, where barium sulfate "mud" is pumped down into the hole as it is being drilled. The density of the solution helps f[illegible]k chips up and out of the hole. B[illegible]sulfate solutions are also used to venture somewhere else the sun never shines, in the form of barium enemas. Barium sulfate is opaque to x-rays, so depending on which part of the digestive system you want to image, you either swallow it or introduce it through the other end and then take an x-ray that will show in detail all the twists and turns of the digestive tract.

Pure barium reacts rapidly with oxygen (8), a property that makes the metal form useless for most applications but turns out to be very useful when you want to get rid of this gas. Old-style vacuum tubes usually have a patch of silvery barium metal evaporated onto the inside of their glass envelope. The barium reacts with any stray oxygen, water vapor, carbon dioxide, or nitrogen (7) that was left over during manufacture or makes its way over time through the tube's glass and seals. Similar barium "getters" are used in many kinds of lamps and vacuum systems to remove the last traces of oxygen or moisture.

In the post-vacuum-tube era, the neatest use of barium is in YBCO superconductors, described under yttrium (39). From superconducting magnetic levitation we turn next to the rare earths, a group known for their varied magnetic properties.

Almost all common vacuum tubes contain a barium getter of some sort. This one has a large patch of the metal flashed onto the inside of the glass envelope.

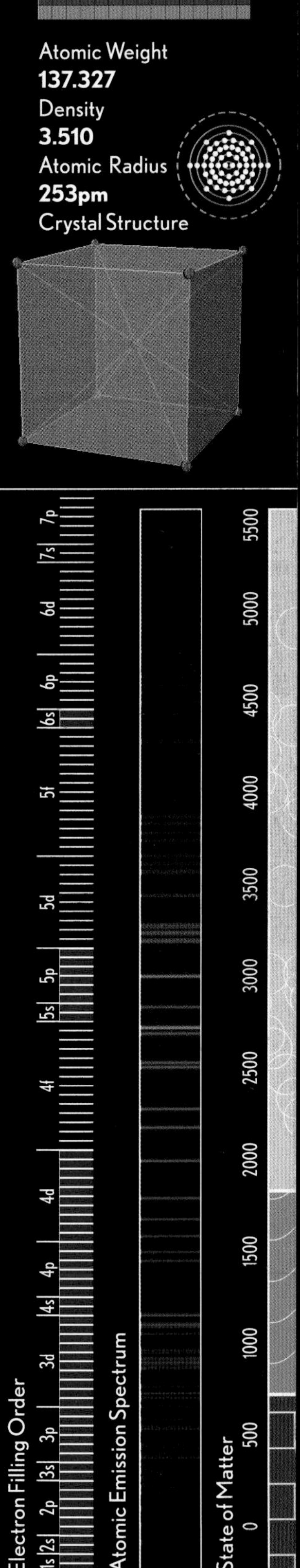

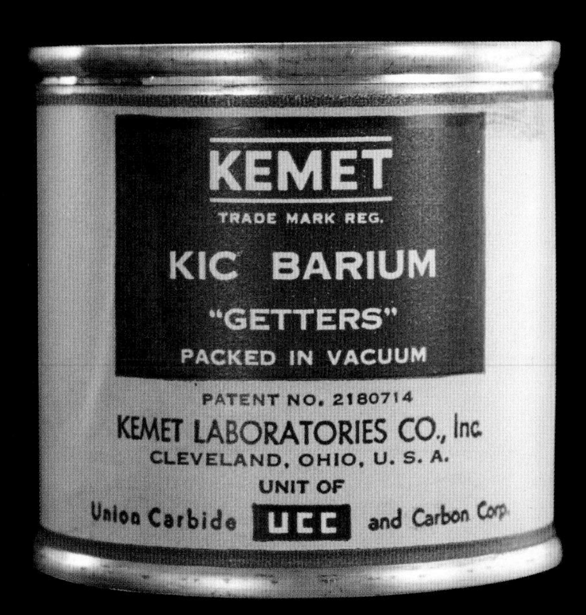

Barium getters must be stored in sealed cans for shipping, lest they try to clear the whole planet of oxygen.

Pure barium is a shiny metal, like so many other elements.

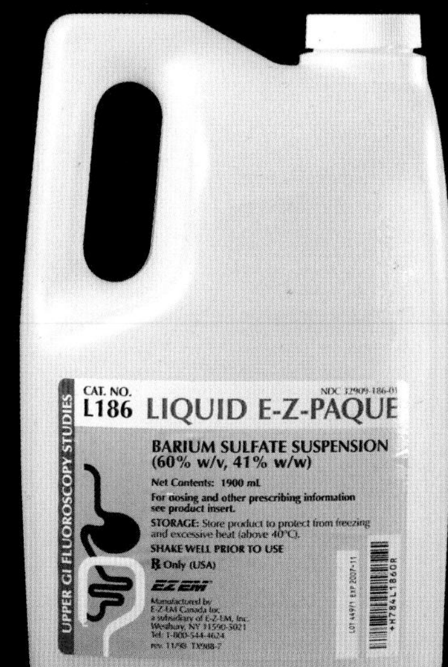

Barium sulfate is commonly used to image the digestive tract from both ends.

The mineral barite (barium sulfate) from the Julcani Mine, Huancavelica, Peru.

Lanthanum

La 57

Lanthanum

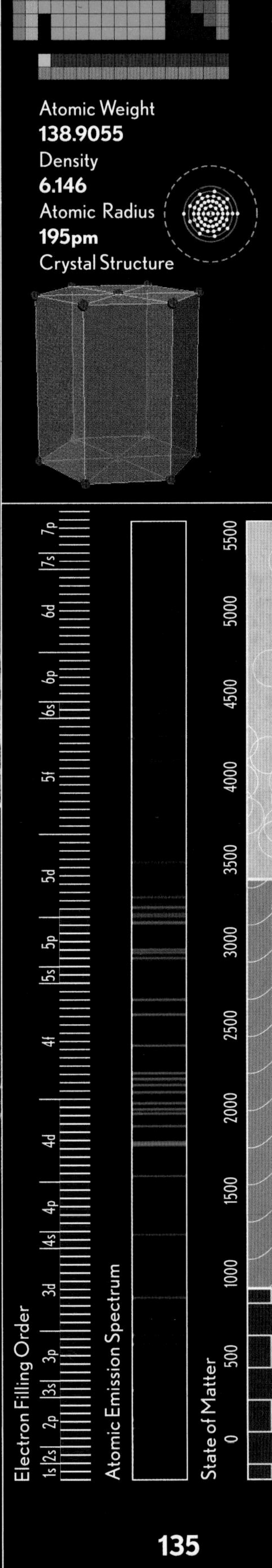

LANTHANUM IS THE FIRST of the rare earth series known as the lanthanides, the uppermost of the two rows of elements commonly shown separately below the main periodic table. All of the lanthanides are chemically almost identical, and they are all found together in the same ores. In some cases it took years to realize that what chemists had thought was one element was actually a mixture of several different rare earths.

The main differences are in magnetic properties; some rare earths such as neodymium (60) make the strongest magnets, while others such as terbium (65) are used in alloys that change shape in magnetic fields.

As for lanthanum itself, it is one of the most abundant of the rare earths (which really aren't that rare), and is used in many applications where it doesn't matter much which one you're using. An example is lighter "flints," which are actually alloys of iron and "mischmetal," German for "mixed metal" and referring to a mixture of lanthanum and cerium (58) with smaller amounts of praseodymium (59) and neodymium. (Mischmetal isn't a precise alloy, it's basically just whatever mixture came out of the mine that day. In many applications the rare earths are interchangeable, so there's no point going to the expense of separating them.)

Rare earth oxides are heat-resistant and glow brightly when hot, making them useful in lantern mantles, which are basically incandescent lamps heated with gas instead of electricity.

Giving lie to the term "rare" earth, lanthanum is more than three times as abundant in the earth's crust as lead (82) and cerium is nearly twice as abundant as lanthanum.

Blocks of mischmetal, primarily lanthanum and cerium, are used for spark effects in movies.

The mineral bastnasite, (La,Ce)(F,CO3).

A large torn ingot of pure lanthanum metal.

Lanthanum oxide glows brightly in camping-lantern mantles.

A dramatic shower of sparks created by holding a mischmetal block to a grinding wheel.

Cerium

Ce

58

Cerium

ON EARTH CERIUM is almost as abundant as copper (29) and is quite inexpensive, especially in the form of cerium oxide, widely used as an abrasive powder for polishing glass.

Cerium metal is pyrophoric, meaning that it can catch fire when you scratch, file, or grind it. In practice this doesn't mean the whole lump catches fire, but rather that the shavings burn as they are formed, making the material very sparky. Not surprisingly this makes cerium useful in lighter flints, where its highly pyrophoric nature is moderated by alloying it with iron (26). Large lumps of undiluted mischmetal, lanthanum-cerium mixtures described under lanthanum (57), are used in movie special effects to create huge trails of sparks, for example behind a car that's dragging on concrete.

One of my favorite rare earth samples is a campfire starter that is basically just a huge lighter flint mounted in a plastic handle. When you scrape it hard with the back of a knife, it puts out an exuberant shower of sparks that should have no trouble igniting a pile of dry tinder. Not that I've ever used it to start a fire, I just like seeing the sparks.

Other applications include small amounts in aluminum (13) and magnesium (12) alloys and tungsten (74) welding electrodes.

What praseodymium lacks in applications it more than makes up for in width.

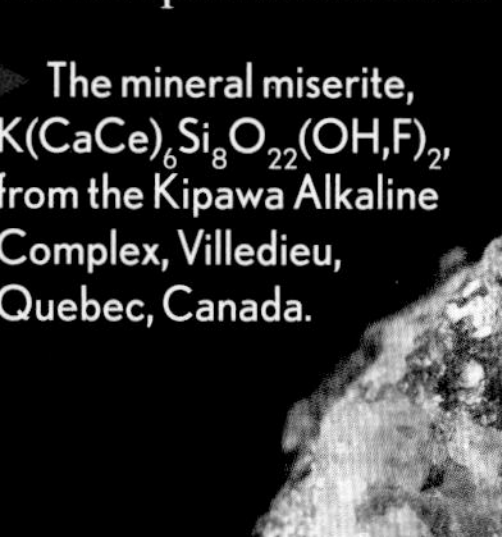

▸ The mineral miserite, $K(CaCe)_6Si_8O_{22}(OH,F)_2$, from the Kipawa Alkaline Complex, Villedieu, Quebec, Canada.

▴ Cerium oxide powder is a commonly used abrasive for grinding and polishing glass.

◂ A cut ingot of pure cerium, one of the least-expensive rare earths.

▸ A huge half-inch diameter rod of cerium-lanthanum-iron alloy, basically a giant lighter flint, creates a shower of sparks when scraped with a steel blade.

Elemental

Atomic Weight
140.116
Density
6.689
Atomic Radius
158pm
Crystal Structure

Electron Filling Order
1s 2s 2p 3s 3p 4s 3d 4p 5s 4d 5p 6s 4f 5d 6p 7s 5f 6d 7p

Atomic Emission Spectrum

State of Matter
0 500 1000 1500 2000 2500 3000 3500 4000 4500 5000 5500

Praseodymium Pr 59

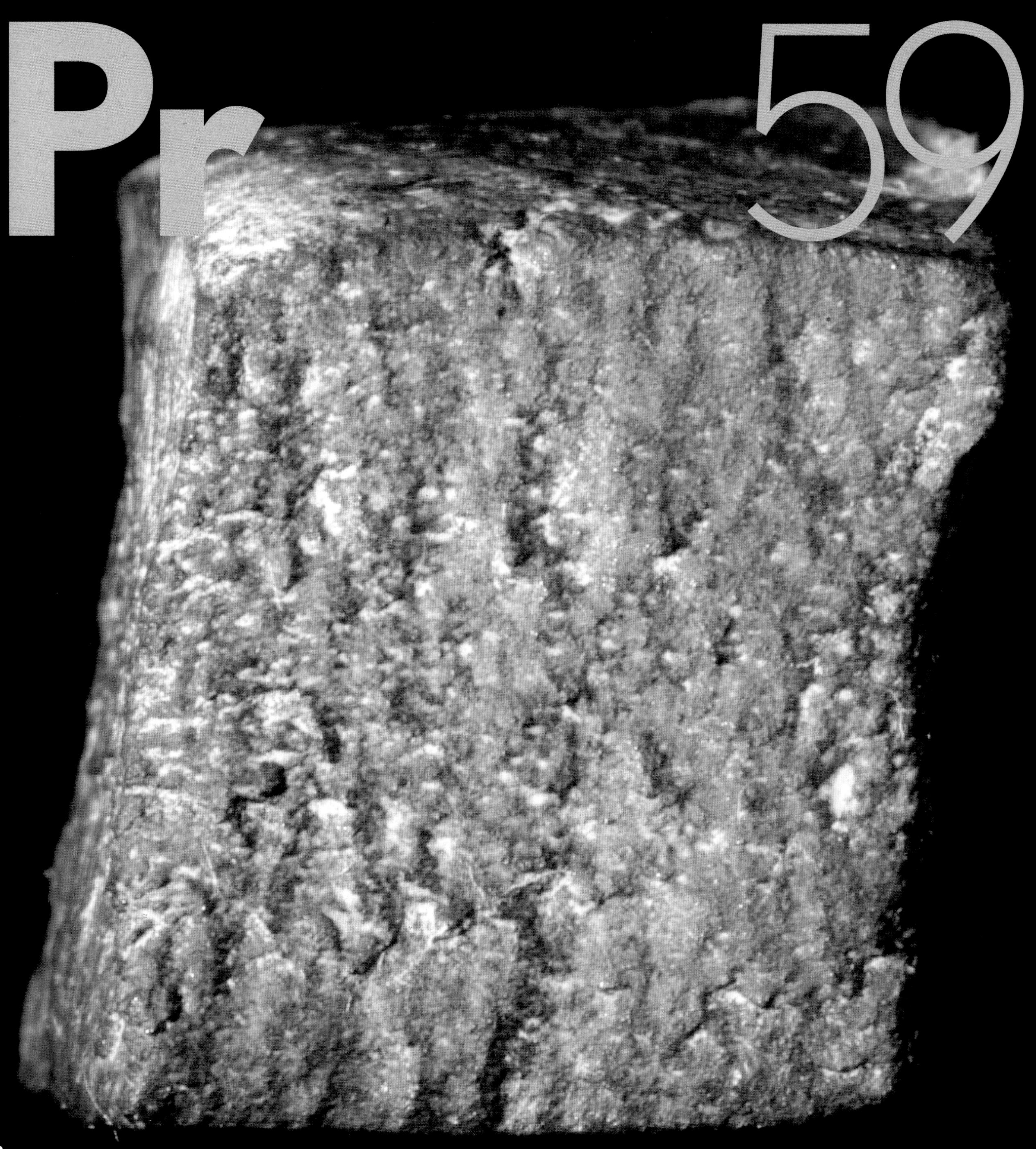

Praseodymium

IF YOU'RE ENGRAVING element names on a set of tiles, praseodymium is the one to watch out for: It's the widest element nam[illegible] (that is, in a proportionally spaced font—[illegible]utherfordium, element 104, has mor[illegible]etters, but only one *m*). This is a f[illegible]rn when you're planning to buil[illegible] did some years ago, a wooden tabl[illegible]n the shape of the periodic table—a periodic table table, if you will, with e[illegible]ved wooden tiles for each element. [illegible]e not planning such a project, [illegible]y not find this information of much use.

Many rare earths are not actually very rare—they got their name mainly because they are difficult to isolate. The solvent extraction method used today to separate rare earths relies on the slightly different solubilities of rare earth compounds between two liquids that do not dissolve in each other (in essence oil and water). Even though the differences in solubility are small, it's possible to arrange a countercurrent system in which many, many extraction steps are carried out in a continuous stream, progressively increasing the degree of separation until the substance in one phase is nearly pure.

C[illegible]ntercurrent solvent extraction completely revolutionized the availability of all the rare earths and dramatically lowered their cost in pure form. The sudden availability of large quantities of rare earths at reasonable prices set off a search for useful things that might be don[illegible] with them. This search was more successful in some cases than others.

Praseodymium, for example, found use in "didymium" eyeglasses, which have one very specific application—glassblowers view their work through them. A mixture of praseodymium and neodymium (60) gives the lenses a seemingly weak blue tint that is actually the result of very strong absorption of a specific wavelength of yellow light. This wavelength corresponds to the bright sodium yellow emission lines that give hot soda-lime glass its intense color. It's quite remarkable: When you look through didymium lenses, you can stare directly at a torch that is heating glass to the melting point and see nothing but the dull blue glow of the torch flame and a faint ruddy orange glow from the hot glass. Take the glasses off and the blinding yellow light forces you to look away.

What a difference one proton makes: From the obscurity of praseodymium to an element you almost certainly have some of around the house, neodymium.

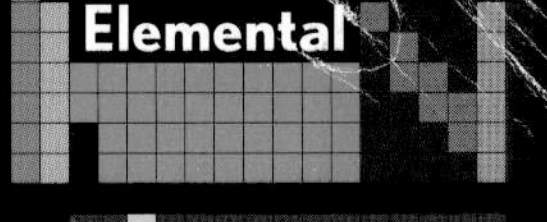

Atomic Weight
140.90765
Density
6.640
Atomic Radius
247pm
Crystal Structure

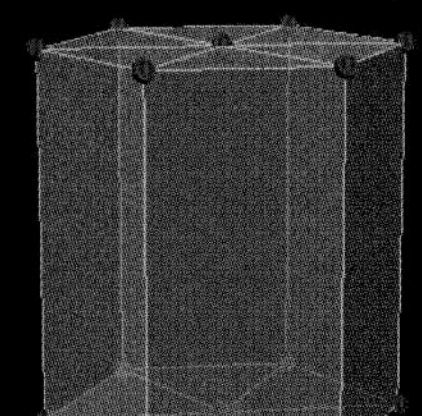

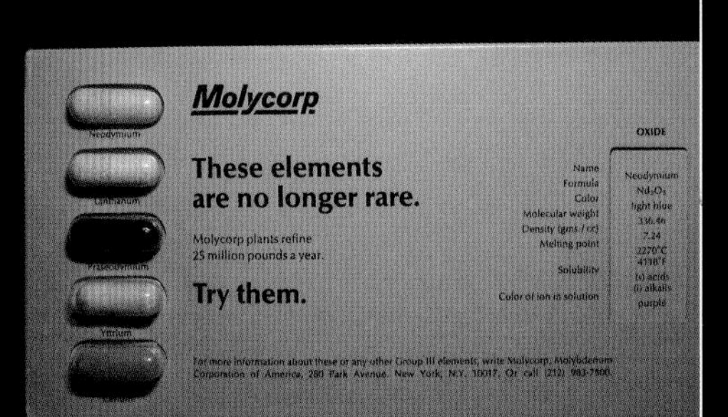

Sample set given away to advertise the fact that the rare earths were suddenly available at reasonable prices.

Lens from a pair of didymium glassblowers' glasses.

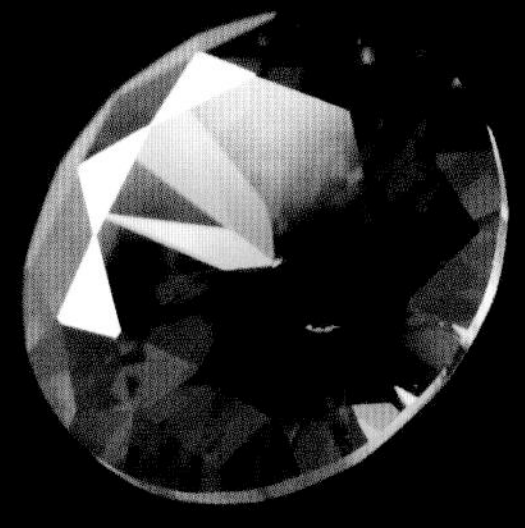

Praseodymium creates the color in fake cubic zirconia–based peridot.

Carbon arc light rods with praseodymium-doped cores to create daylight-white light for motion-picture filming.

A blue filter containing praseodymium turns an inefficient yellowish incandescent bulb into an even less efficient daylight-spectrum bulb.

A block of pure praseodymium, slightly oxidized.

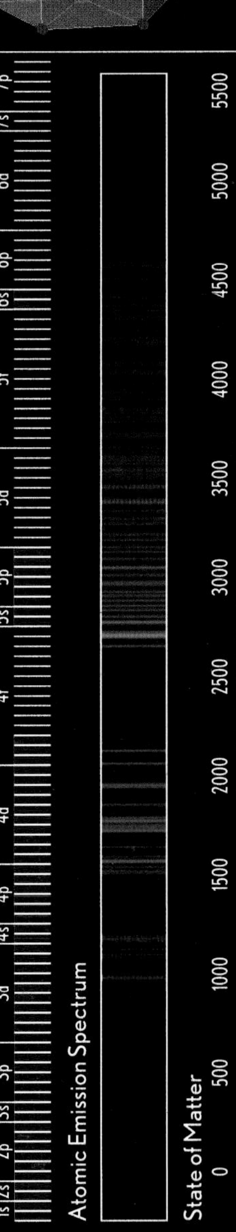

Neodymium

Nd

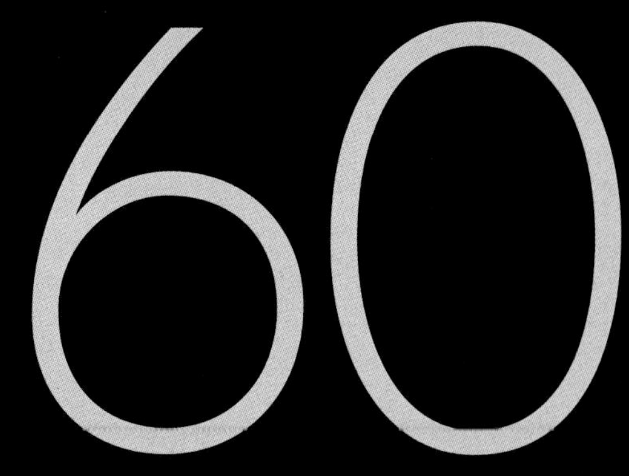

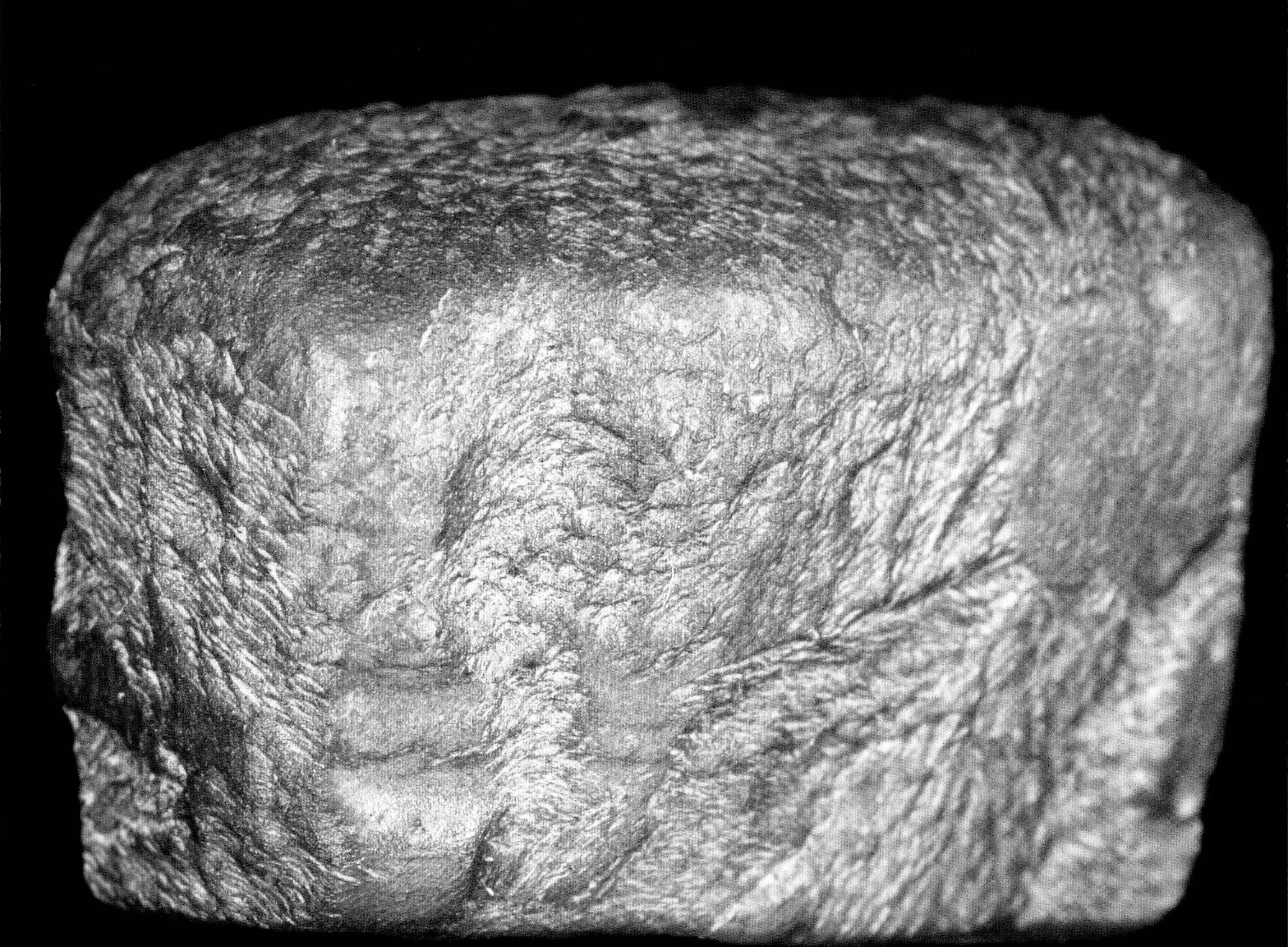

Neodymium

NEODYMIUM IS THE BEST known of the lanthanide series of rare earths because of neodymium magnets (which are actually made of a neodymium-iron-boron alloy). These are by far the strongest readily available permanent magnets, so strong that they are genuinely dangerous to be around, especially if you have more than one.

They can jump toward each other from a foot or more away. Heaven help you if you're holding one of them when that happens. Even very small ones can give you a blister; large ones, a few inches on edge, can destroy a finger or an entire hand. Swallowing one small neodymium magnet is no big deal: you just wait for it to come out the other end. Swallowing two a few hours apart is an immediate medical crisis—they find each other through different loops of the intestine and stick, causing perforations that can be life threatening.

This flesh-pinching property is used to hold earrings and other fake piercing–type jewelry on people who aren't ready for the commitment of an actual piercing.

Neodymium has optical properties when mixed with glass. Really stupid incandescent bulbs have neodymium in their glass to filter out some of the yellow incandescent light, yielding a whiter light closer to daylight. Stupid because this filtering makes them even less efficient than already horribly inefficient ordinary yellowish incandescent bulbs. A better alternative is to use daylight-spectrum compact fluorescents, which are several times more efficient and use europium (63) phosphors to help them emit a more pleasant spectrum, instead of neodymium to absorb part of a less-pleasant spectrum.

Neodymium glass is also a laser material able to amplify light pulses after being pumped full of energy with flashbulbs. The next element, promethium, needs less help to glow.

Elemental

Atomic Weight
144.24
Density
7.010
Atomic Radius
206pm
Crystal Structure

▲ Neodymium magnets make this tiny motor surprisingly powerful.

▲ Neodymium magnets are crucial in lightweight hi-fi headphones.

▶ Strong magnets attached to an oil filter trap metal fragments as they pass by.

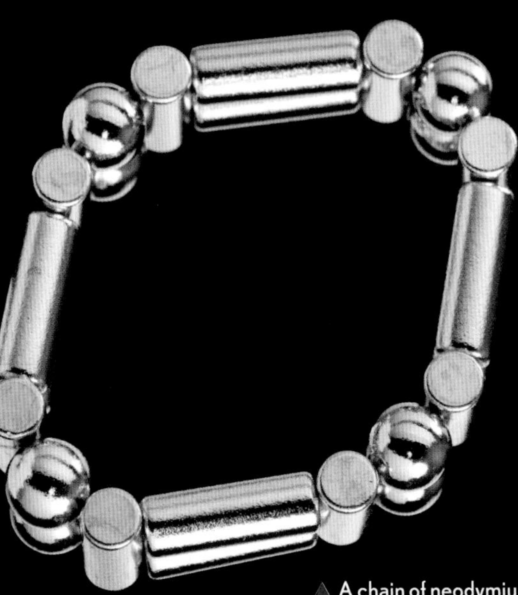

◀ Pure neodymium metal.

▲ A chain of neodymium magnets is strong enough to wear as a bracelet without a string connecting any of the beads.

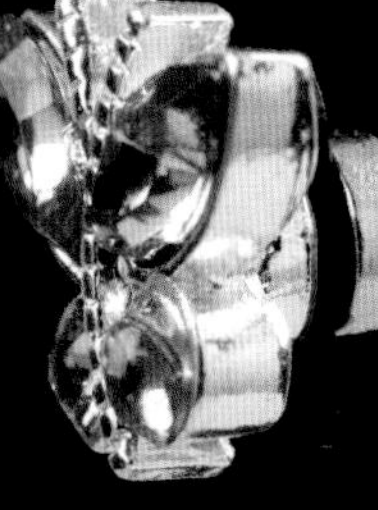

▲ Tiny neodymium magnets are strong enough to hold on earrings without a piercing.

▶ A ring of neodymium magnets arranged as you would find them in a miniature motor.

Promethium
Pm
61

Promethium

PROMETHIUM AND technetium (43) are the two exceptions to the general rule that elements below bismuth (83) are stable. Due to a combination of factors in the way the protons and neutrons in these two elements pack into shells in their nuclei, they just can't find a satisfactory stable arrangement, meaning there are no stable isotopes of either element.

Technetium has interesting applications in medicine, but promethium's applications are meager. There was a brief shining moment, after people stopped using radium (88) and before tritium became available, that promethium was used to create luminous dials and markings by mixing it with zinc (30) sulfide phosphorescent material. Few examples of these devices survive, and none of them work anymore because the half-life of the ^{147}Pm isotope used is only 2.6 years.

Promethium was replaced by tritium, an isotope of hydrogen (1), because tritium is much safer. The radiation from tritium does not penetrate the glass tubes it is kept in, and if the tubes break, the tritium escapes and rapidly floats high up and away from people, because tritium, like hydrogen and helium (2), is far lighter than air. (Promethium and radium paint, on the other hand, are sticky substances that can flake off and get into everything, making for a messy and expensive cleanup.)

After promethium, it's back to stable elements for the next 21, starting with samarium.

A tiny amount of promethium in this compact fluorescent bulb's glow switch keeps the gas inside ionized.

This promethium luminous button was produced using leftover stock kept for making diving watches.

A compact fluorescent bulb that uses a promethium glow switch (most do not).

Promethium luminous paint, as seen on this compass dial, was used briefly after radium was discontinued and before tritium took over.

Elemental

Atomic Weight
[145]
Density
7.264
Atomic Radius
205pm
Crystal Structure

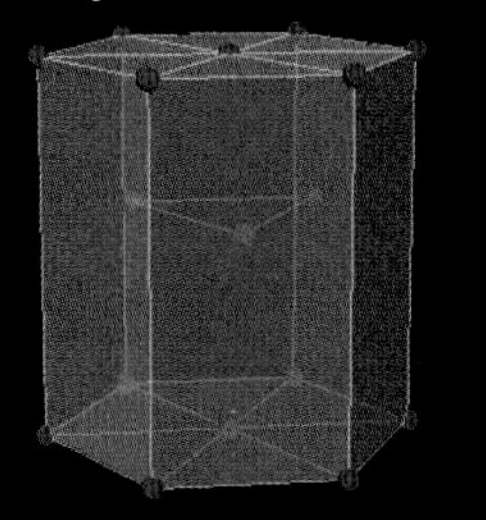

Electron Filling Order
1s 2s 2p 3s 3p 4s 3d 4p 5s 4d 5p 6s 4f 5d 6p 7s 5f 6d 7p

Atomic Emission Spectrum

State of Matter
0 500 1000 1500 2000 2500 3000 3500 4000 4500 5000 5500

Samarium

Sm

Samarium

Elemental

Atomic Weight
150.36
Density
7.353
Atomic Radius
238pm
Crystal Structure

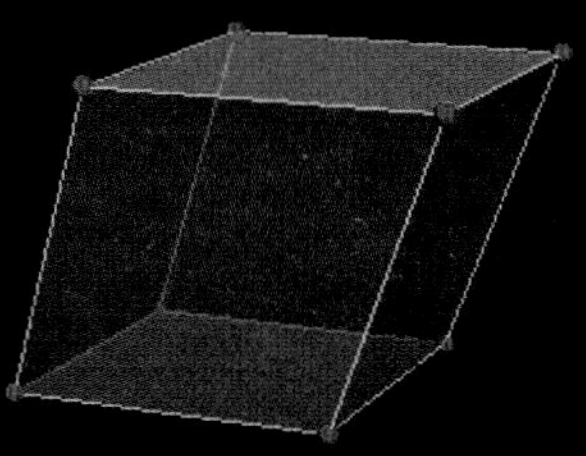

SAMARIUM IS NOT named for Samaria, the ancient city, but rather for the mineral samarskite, which in turn is named for its discoverer, the Russian Vasili Samarsky-Bykhovets (though his name, for all I know, may go back to Samaria if you look far enough). Because Samarsky was not yet dead when the element was named, a case can be made that samarium is a second example, predating seaborgium (106), of an element named after a living person. Unlike seaborgium, however, samarium was not named specifically to honor the person. Indirect naming through a previously named mineral doesn't count in my book (and this, you will note, is my book).

Neodymium-iron-boron magnets are the strongest available today, but samarium-cobalt magnets can operate at higher temperatures where neodymium magnets would lose their magnetism. For some reason, they are preferred in fancy electric-guitar pickups, though unless you plan to set your guitar on fire, I have no idea how they could possibly make a difference in that low-temperature application.

Other than in magnets, samarium's applications are scattered; as with pretty much any element, you can find examples of chemical reagents, medicines (a radioactive isotope of samarium, in this case), and various uses in research—for example, in research on possible uses of samarium. And if I say it has no other important applications, someone is going to complain that, no, such and such is terribly important. But you know what I mean.

The situation with europium is more illuminating.

A coin stamped in pure samarium, one of a series created from nearly every practical element.

Samarium-cobalt magnets are not as strong as the neodymium variety, but they can operate at higher temperatures.

An electric-guitar pickup with samarium-cobalt magnets.

Dendritic crystals of pure samarium metal.

The mineral monazite contains some of nearly every rare earth.

Electron Filling Order

Atomic Emission Spectrum

State of Matter

Europium

Eu

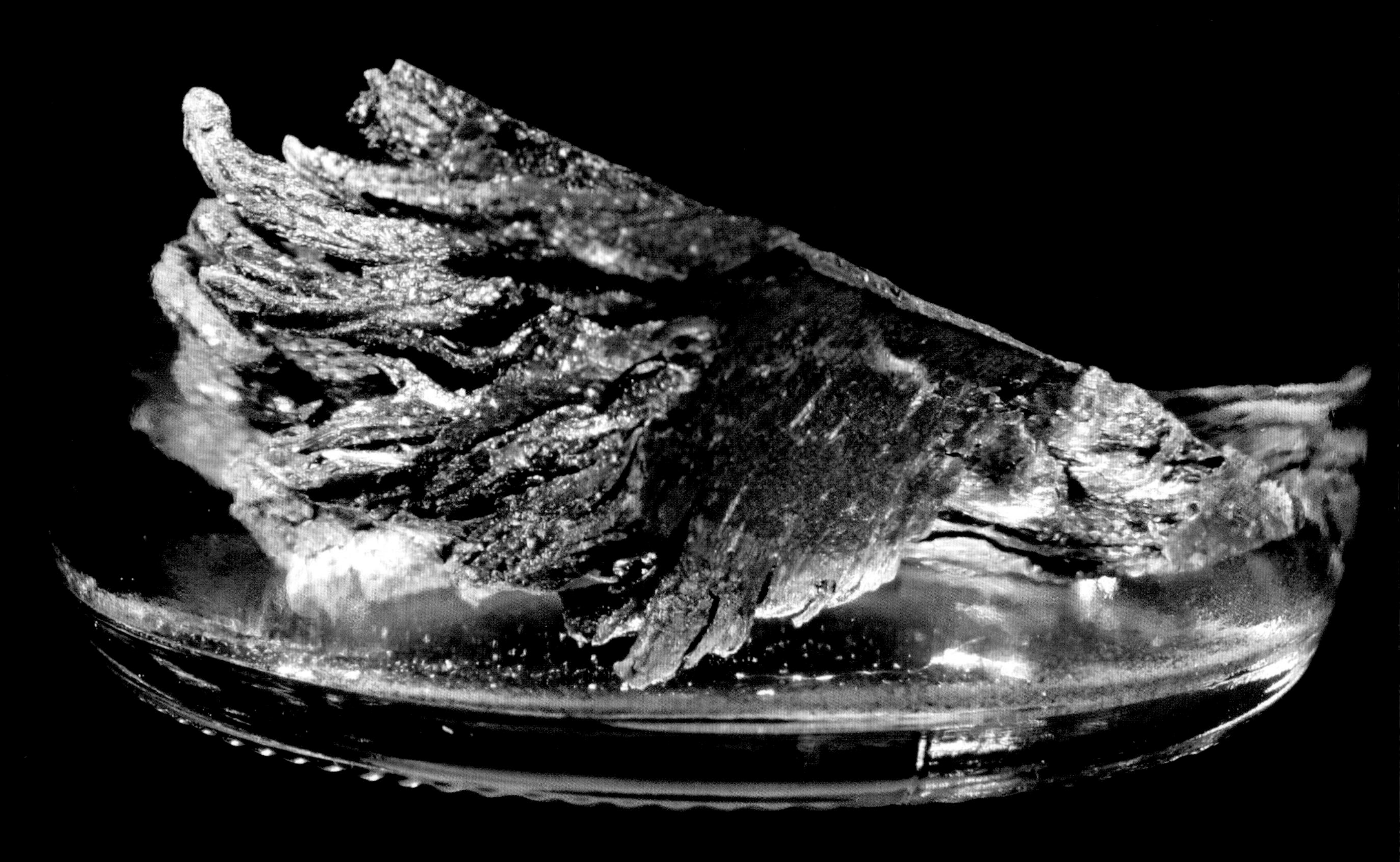

Europium

EUROPIUM IS NAMED for the continent of Europe. As with ruthenium (44) this is sort of like being named for a country, but not the same, so I don't count it among the four that really are—germanium (32), polonium (84), francium (87), and americium (95).

Europium's applications, somewhat unusual for a rare earth, center not on magnetism but on luminosity. It's used in phosphorescent paint, including some amazing modern varieties that can glow brightly for many minutes, or dimly for many hours, after being exposed just briefly to a strong light source.

Europium is also used in the phosphors in those increasingly scarce CRT (cathode-ray tube) monitors and color television sets. Soon to be historical curiosities, these devices are giant vacuum tubes in which a focused beam of electrons is accelerated by thousands of volts toward tiny dots of red, green, and blue phosphor on the inside of the front wall, the screen. The color of light emitted by each dot is determined by the elements and compounds in that patch. Red was a problem in early color television sets because no good, bright red phosphor was known, and the other two colors had to be intentionally dimmed to maintain the correct color balance. With the invention of europium-based red phosphors, color television could suddenly become bright and vibrant, thus contributing even more effectively to the rotting of children's minds the world over.

Compact fluorescent lightbulbs, those wonderful devices that have liberated us from Edison's horribly inefficient incandescent lamps, also use europium in the mix of phosphors they employ to create a pleasant spectrum of light. I am now so accustomed to the bright, beautiful daylight spectrum light from my compact fluorescents that I find the dingy, old yellow light of incandescent bulbs to be quite depressing.

With gadolinium we return to rare earths with magnetic applications, though of quite a different sort.

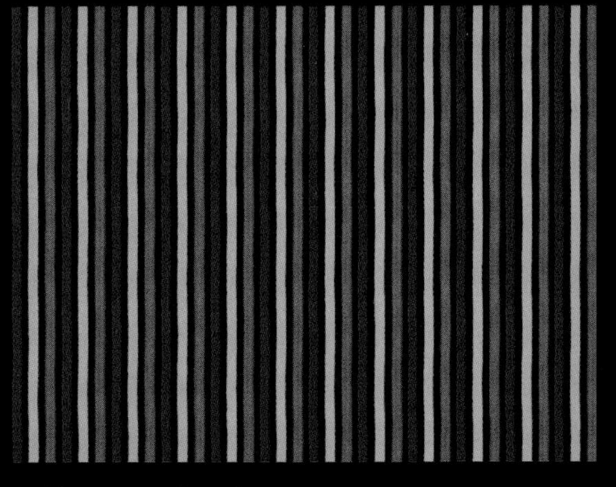

Europium phosphors contribute a brilliant red to CRT televisions.

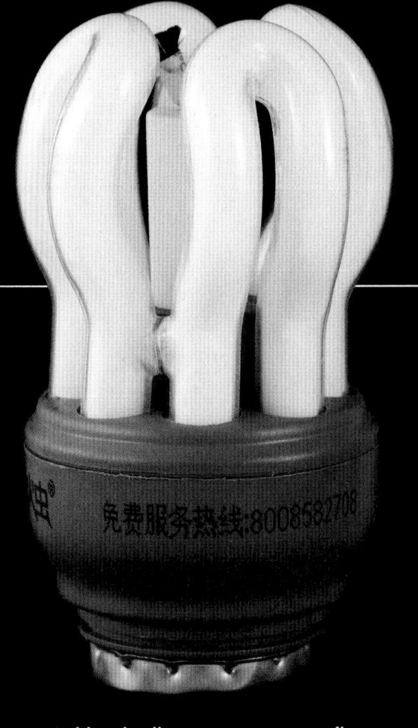

Nearly all common compact fluorescent bulbs use europium phosphors to create a pleasing spectrum of light.

Monazite sand contains nearly all the rare earths.

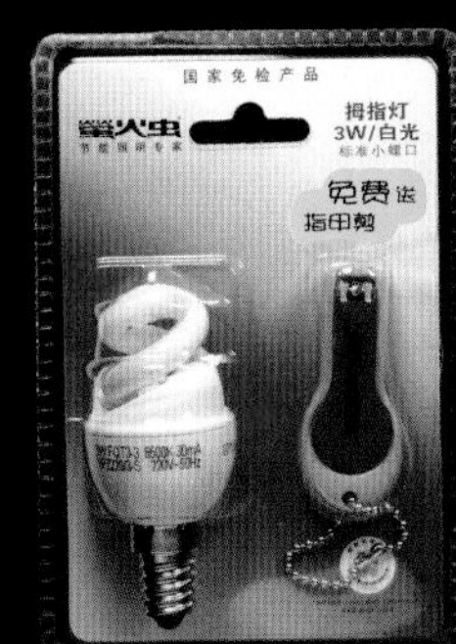

Compact fluorescent bulb sold with a pair of nail clippers. Wacky enough for Japan, but actually spotted in China.

This tiny two-watt compact fluorescent bulb produces a bit of light for nearly no power.

Pure europium oxidizes over time, even when stored under oil.

Elemental

Atomic Weight
151.964
Density
5.244
Atomic Radius
231pm
Crystal Structure

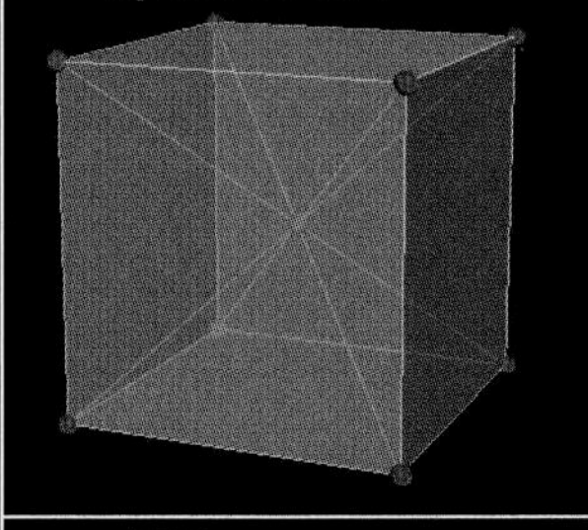

Electron Filling Order

Atomic Emission Spectrum

State of Matter

Gadolinium Gd 64

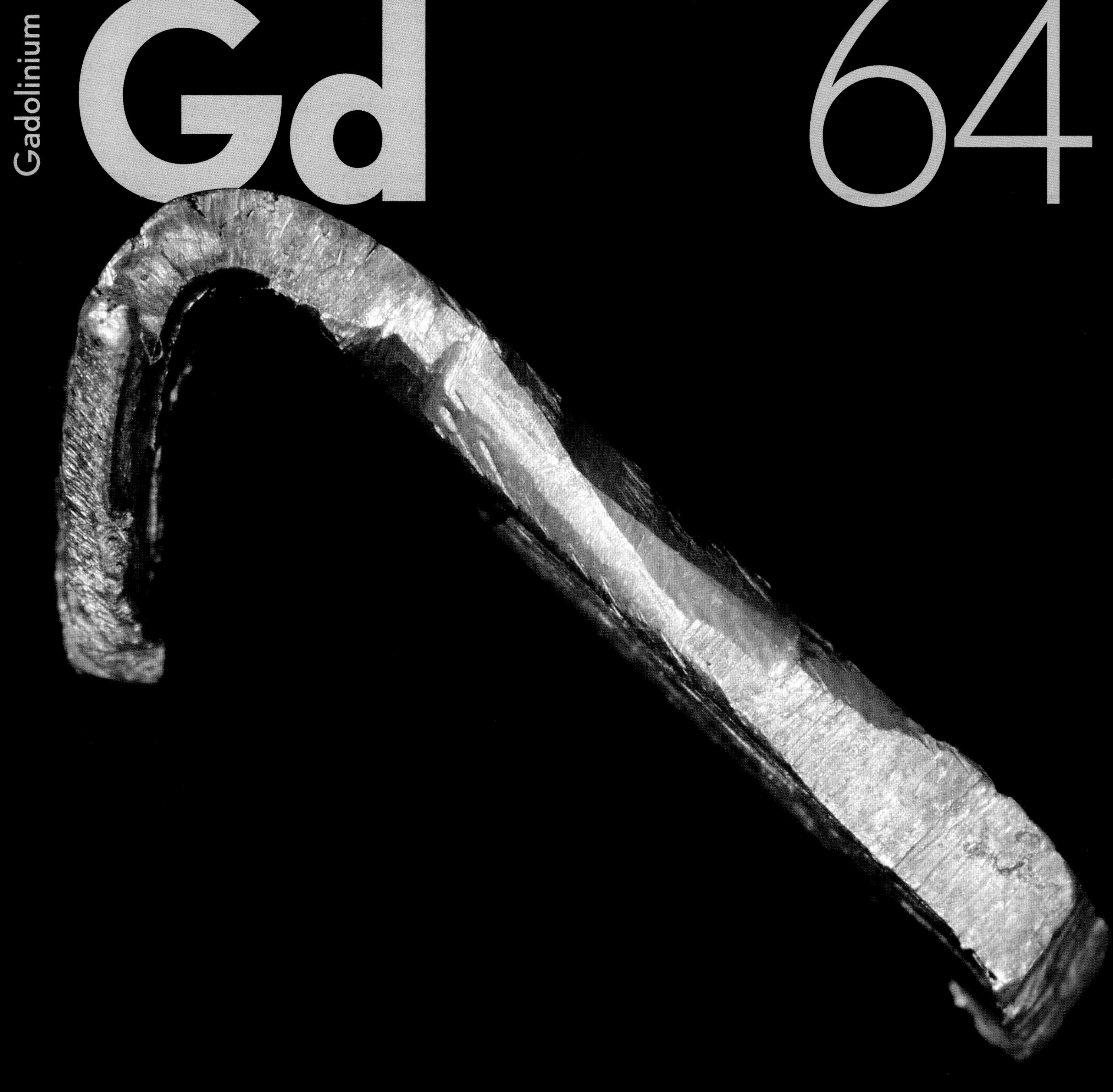

Gadolinium

GADOLINIUM COMPOUNDS are highly paramagnetic, a property that causes them to be injected into humans. One of gadolinium's main applications is as a contrast agent for MRI scans. The idea is similar to the way barium (56) sulfate is used as a contrast medium for x-rays of the gastrointestinal tract.

Soft tissue is quite transparent to x-rays, but a coating of barium sulfate, opaque to x-rays, will reveal details of the inside surface of the digestive tract. Similarly, gadolinium responds strongly to the magnetic fields in an MRI machine, so if you inject it into the bloodstream (in the form of gadopentetate dimeglumine), an MRI will show just where the blood is—and isn't. An MRI can pinpoint the precise location of internal bleeding by visualizing in three dimensions exactly where blood is leaking from a vessel, or can locate a constriction or blockage, showing clearly where the bloodstream is narrowed or stops.

Though no commercial applications have yet been found for this next phenomenon, gadolinium has a Curie point at just about room temperature (19°C, 66°F), which makes it very convenient for showing someone just what the heck the Curie point is. In case you're wondering, this is the temperature at which a material goes from being ferromagnetic (attracted to a magnet) to paramagnetic (not attracted to a magnet). If you cool down a lump of gadolinium in ice water, it will stick to a magnet, but when it warms up it will fall off.

Curie point transitions are just one of many odd magnetic properties of the rare earths, and not really as odd as the tendency of terbium to change shape when it's placed in a magnetic field.

Coin struck in pure gadolinium, for no reason other than that it can be done.

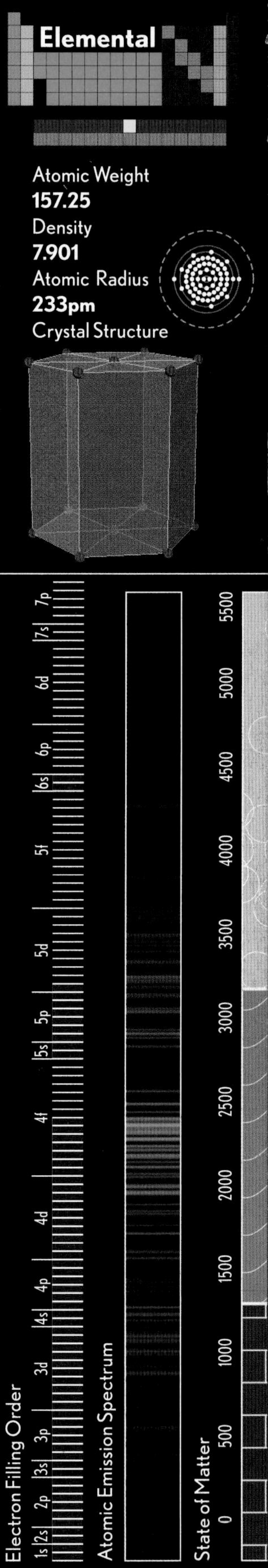

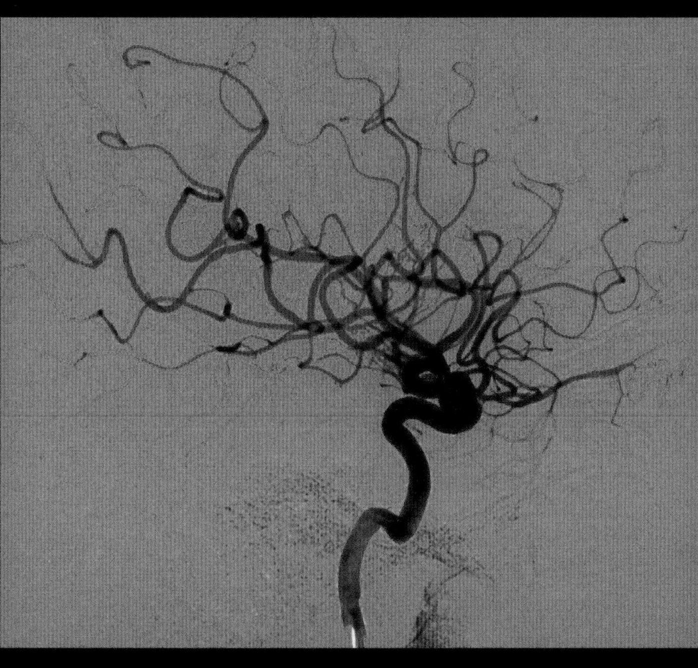

A gadolinium contrast agents allows a leaking blood vessel to be seen in this MRI image.

A hook shape of pure gadolinium gives at least some semblance of variety to the appearance of the rare earths. Of course, it's still just another gray metal.

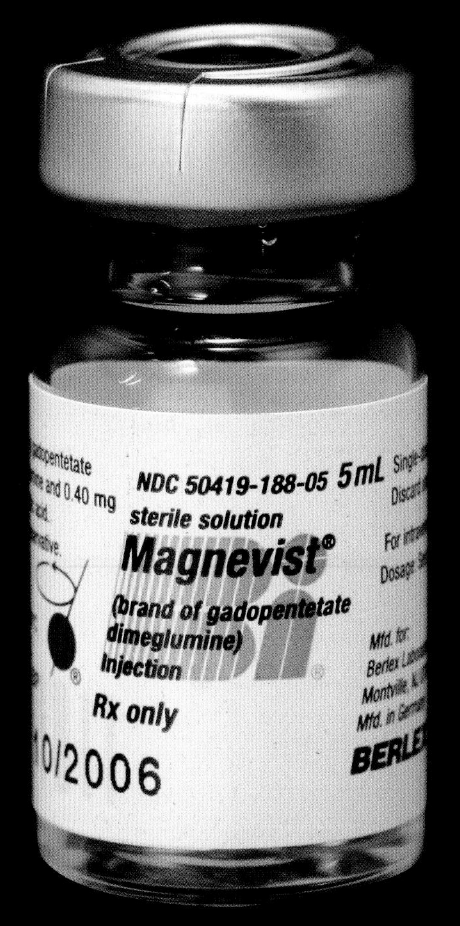

A vial of gadolinium MRI contrast agent.

Terbium

Tb 65

Terbium

TERBIUM ITSELF, and to an even greater degree a particular alloy of terbium called terfenol, has the unusual property of changing shape when placed in a magnetic field. A rod of the stuff will instantaneously grow longer or shorter by a small amount depending on the strength and alignment of the field it's in. This property might not seem very useful, but you can use it to turn any solid surface into a loudspeaker.

If you press the end of a terfenol rod down on a wooden tabletop, then apply a magnetic field whose strength follows an audio signal (by wrapping the terfenol rod in what amounts to a speaker coil), the rod will shake the whole table, turning its entire surface into a huge sound-radiating surface that plays the same role as a speaker cone.

Why not just press an ordinary loudspeaker onto the table and get the same effect? Because that will just muffle the speaker. This is an example of the problem of impedance matching. Ordinary speakers exert a small force to move a lightweight speaker cone over a relatively large distance. In order to move the large mass of a solid wood tabletop, a large force must be exerted over a short distance, which an ordinary speaker magnet and coil just can't do. A terfenol rod is one of the very few ways to make this type of loudspeaker, and you can actually buy fairly inexpensive terfenol devices designed for exactly this purpose!

If only dysprosium had such a wide spectrum of uses.

Terbium-doped red glass decorative teardrop.

This SoundBug speaker contains the terfenol solid surface audio driver shown below.

A rod of terfenol alloy inside a copper coil creates a solid-surface audio driver.

Cut slice of pure solid terbium.

Bumpy rod of very-high-purity terbium.

Elemental

Atomic Weight
158.92534
Density
8.219
Atomic Radius
225pm
Crystal Structure

Electron Filling Order

Atomic Emission Spectrum

State of Matter

Dysprosium

Dy

66

Dysprosium

IT'S NOT THAT there isn't *anything* useful to do with dysprosium. It's a minor component in the terfenol alloy discussed under terbium (65). It can optionally be a minor component in the neodymium-iron-boron magnets discussed under neodymium (60). It's a minor component in several other applications. But in terms of trying to find really interesting applications unique to dysprosium, the element lives up to its name: from the Greek *dysprositos,* meaning "hard to get at."

Usually if you Google an element name, you'll find lots of companies posting information about how their products use that element, or scientific papers exploring its interesting properties. Look up dysprosium, and you have to go to the fourth page of results before finding anything that isn't a periodic table website's entry for dysprosium—usually an obligatory "it's an element so we have to have a page about it" sort of page.

But that doesn't mean there are no important applications for dysprosium! It just means that the people who know about them don't feel the need to talk in public. Beyond the world of what you find on the internet, or even in books and scientific papers, there is a whole other realm of private knowledge kept inside companies as trade secret information. Dysprosium, for example, is widely used in the form of dysprosium iodide and dysprosium bromide salts to impart valuable spectral lines to the red color range in high intensity discharge lighting. You have no doubt spent many hours under the glow of commercial lighting systems containing dysprosium, but unless you happen to know someone who works in the industry, you are not likely to run into this information. (Not even, as of this writing, in Wikipedia, which seems to know just about everything about every element—but in fact does not.)

The next two rare earths, holmium and erbium, are a bright spot on the way to thulium.

Himalayan sea salt is sold with claims that it's healthier to eat than regular salt, in part because it contains a laundry list of elements, including dysprosium (which is not healthy to eat, making the claims more than a little suspect). It's also sold in huge solid chunks like this one, which has been turned into a lamp by hollowing out a cavity inside.

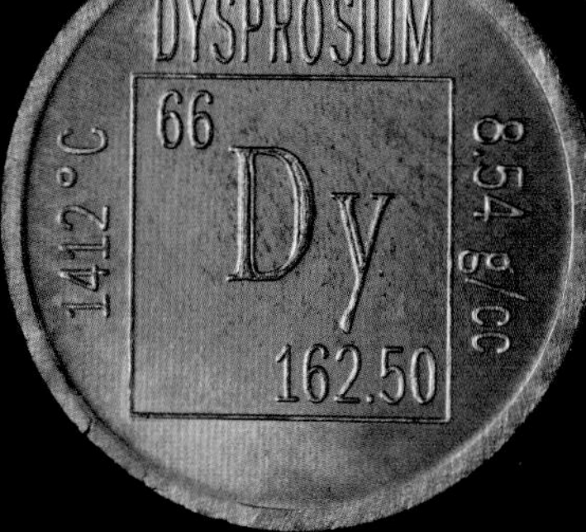

A coin struck in pure dysprosium. Yes, we live in a strange world.

Hollow cathode lamps, which create the characteristic spectrum of the element they contain, are available for almost every element, making them handy when you have few other photographs to represent a particularly obscure rare earth.

Pure dendritic crystals of dysprosium.

Elemental

Atomic Weight
162.5
Density
8.551
Atomic Radius
228pm

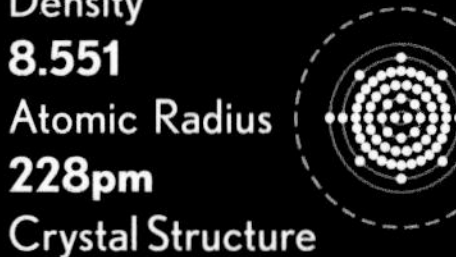

Crystal Structure

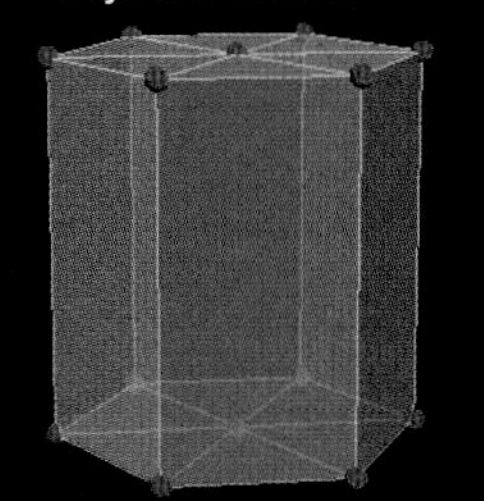

Electron Filling Order: 1s 2s 2p 3s 3p 4s 3d 4p 5s 4d 5p 6s 4f 5d 6p 7s 5f 6d 7p

Atomic Emission Spectrum

State of Matter: 0 500 1000 1500 2000 2500 3000 3500 4000 4500 5000 5500

Holmium

Ho

Holmium

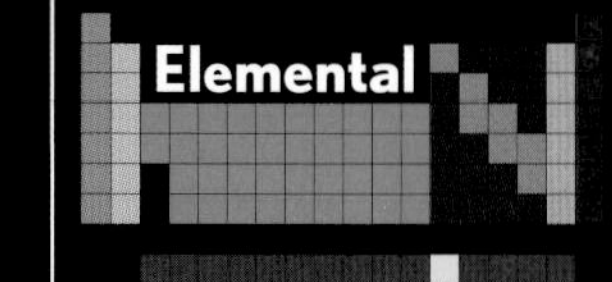

Atomic Weight
164.93032
Density
8.795
Atomic Radius
226pm

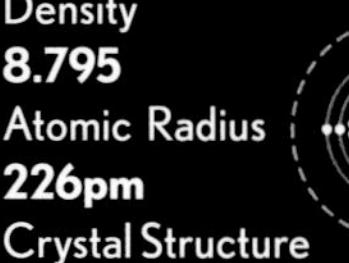

Crystal Structure

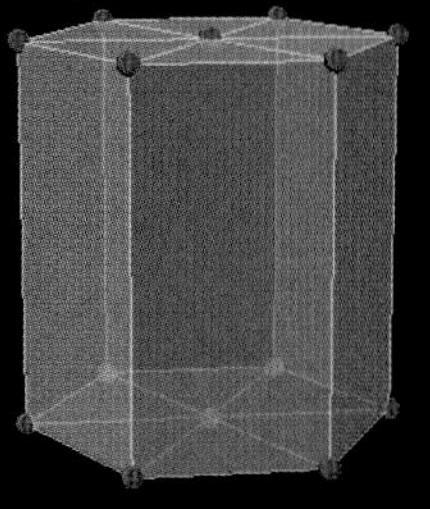

Electron Filling Order

Atomic Emission Spectrum

State of Matter

HOLMIUM ACHIEVES the pinnacle of aspirations for a rare earth element. All of them have some kind of interesting magnetic property, but holmium has the highest value for a particularly important one, the magnetic moment.

What this means is that when holmium is placed in a magnetic field, the holmium atoms line up with the field and concentrate it, bringing the magnetic lines of force closer together, thus raising the field's local intensity. If you put a slug made of holmium—called a pole piece—at the end of a magnet, you get a stronger magnet.

Holmium pole pieces are used in MRI machines, in which extremely intense magnetic fields line up the atoms in the body so their nuclear spins can be measured. These magnets are so strong that elaborate precautions are taken to make sure that no metal objects ever get near them. True story: I had an MRI done once and the technician insisted on x-raying my eyes first, which baffled me until I learned that it was because I had answered yes on the admission form to whether I had done any welding or metalworking recently. It turns out they worry in such cases that tiny fragments of metal might be lodged under the patient's eyelids, and in the tremendous magnetic field of the MRI they might come loose and scramble an eyeball. (This is the kind of question you *know* they ask only because that exact thing must have happened to someone.)

Sticking with medical applications, the lasers used for laser surgery are often holmium-doped YAG (yttrium aluminum garnet) solid-state lasers: As with other rare earths, holmium impurities in glass and crystal materials create color centers that can store optical energy and release it in the form of a laser pulse.

While holmium wins the prize for rare earth magnetic properties, the optical properties of the rare earths reaches a pinnacle with erbium.

Holmium chloride is the form in which this element is introduced into high-intensity discharge lights, where holmium's spectrum is useful.

MRI machines use holmium pole pieces to concentrate their magnetic field.

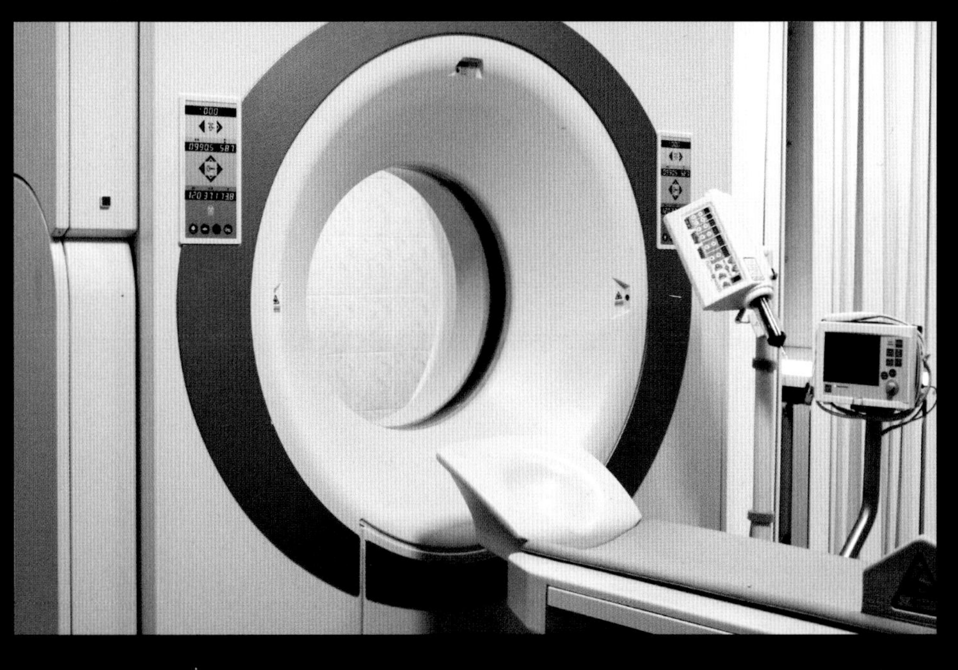

Pure holmium coin.

Polycrystalline surface of pure holmium metal.

Erbium

Er

68

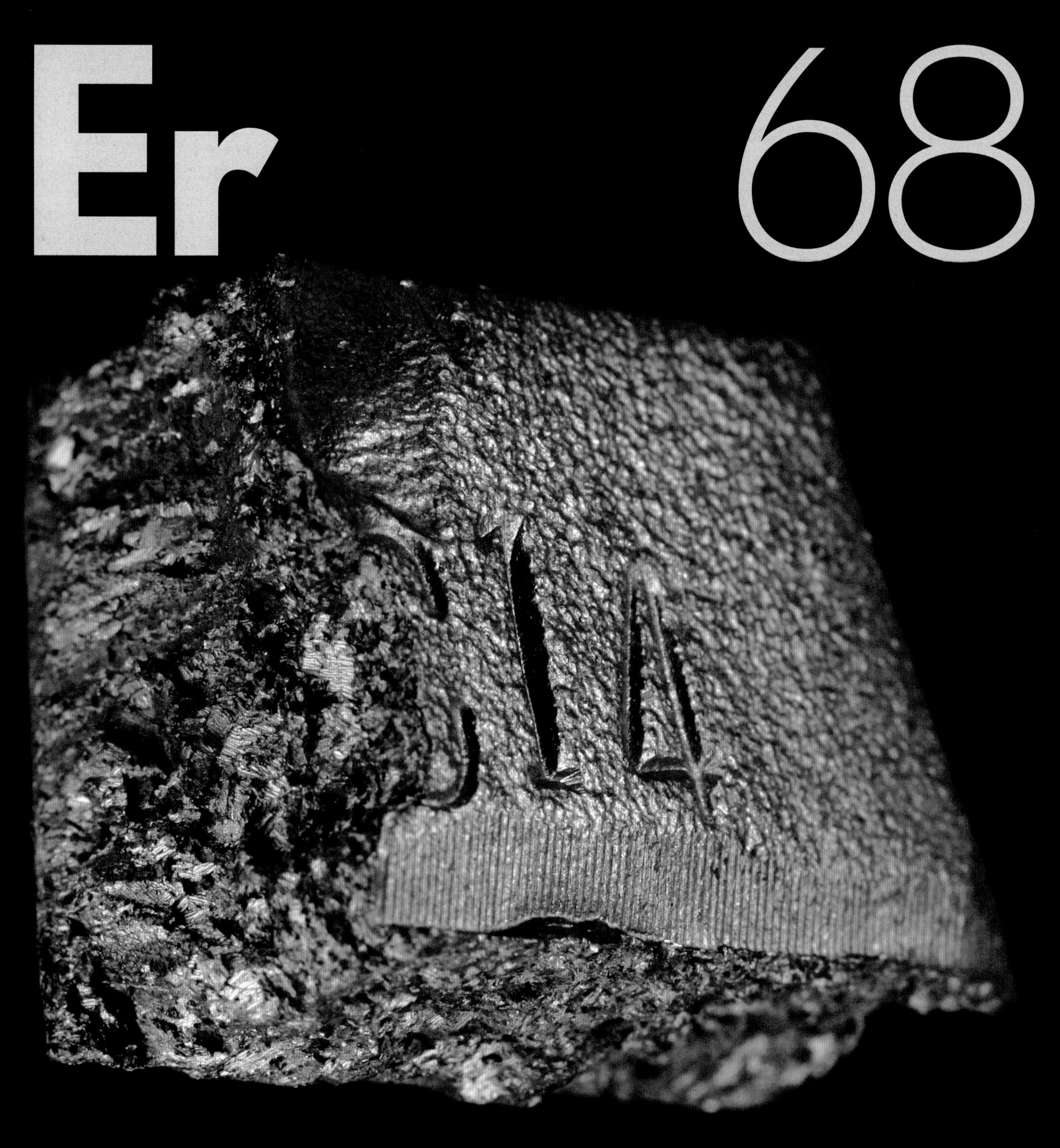

Erbium

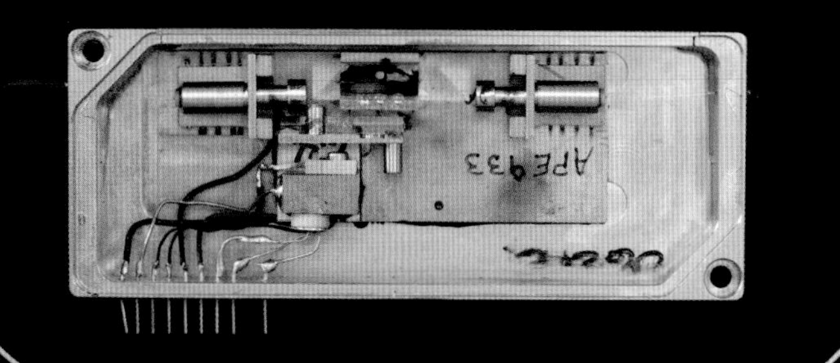

Elemental

Atomic Weight
167.259
Density
9.066
Atomic Radius
226pm
Crystal Structure

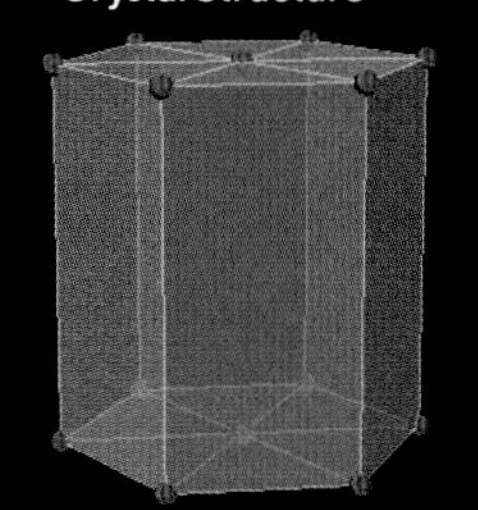

Electron Filling Order

Atomic Emission Spectrum

State of Matter

ERBIUM PLAYS a crucial role in modern communication systems because it allows a pulse of light to be amplified in a fiber-optic cable without ever having to convert it into an electrical signal. A weak pulse of light coming down an optical fiber enters a section that has had a small impurity of erbium introduced into the glass of the fiber. The pulse exits this erbium-doped section much brighter than when it came in—amplification happens entirely inside the fiber; there is no intercepting of the pulse, it just comes out stronger than it went in.

Of course, whenever you end up with more energy than you started with, the extra energy has to come from *somewhere*. (Anyone who tells you otherwise is probably trying to sell you something, and whatever it is, don't buy it.)

To operate this device, known as an erbium-doped fiber amplifier, you first have to pump energy into the doped fiber using a laser. The energy is stored around the erbium atoms in the form of electrons promoted to higher-energy excited states. The energy stays trapped there until a pulse of just the right wavelength of light comes through and triggers the electrons to decay back to their ground state and release their stored energy as light.

The process is called stimulated emission, and is also the way lasers work (the word "laser" is an acronym for light amplification by stimulated emission of radiation). Crucially, the light emitted this way always travels in the same direction as the light that stimulated its emission, so the added light lines up with the incoming pulse and goes out the front end, not backward toward where the pulse came from.

Lasers and related optical devices are among the most ubiquitous and useful devices ever invented, which makes it all the more disappointing to finally reach thulium.

A high-power, erbium-doped, laser-pumped waveguide amplifier.

Erbium impurities create the pink color in these pretty glass rods.

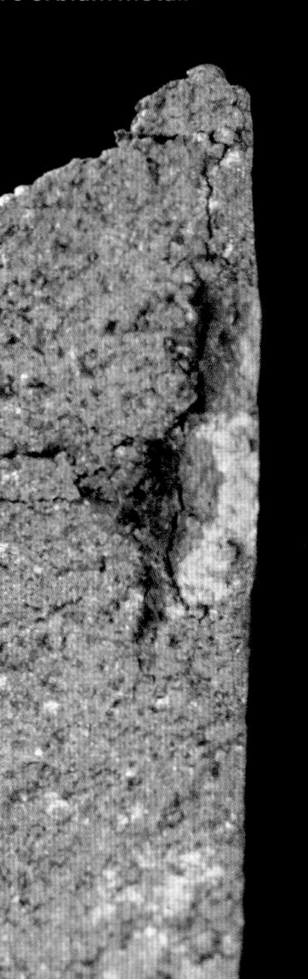

Solid pure erbium metal.

Exotic bismuth-tellurium-erbium alloy created for research purposes.

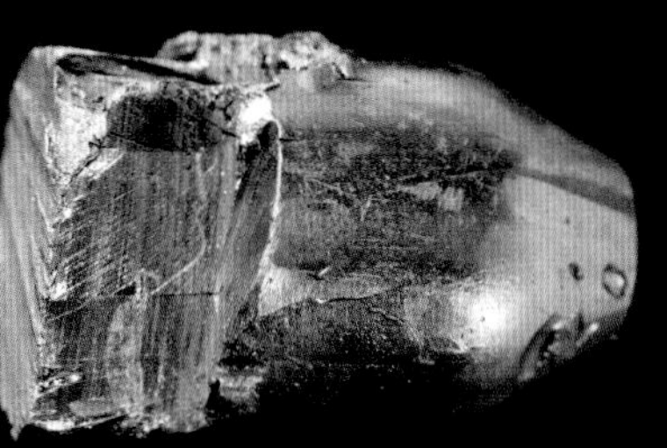

Solid erbium ingot torn apart to show the internal crystal structure.

Thulium

Tm 69

Thulium

JOHN EMSLEY, author of *Nature's Building Blocks* and a preeminent authority on the elements, called thulium "the least significant element there is" when we appeared together on a radio program. Strong words there. Will *someone* stand up for thulium? Not me, that's for sure. It's just one more rare earth, chemically interchangeable with the others and far less abundant. Thulium would probably make as fine a lighter flint as do lanthanum (57) and cerium (58), but it's more expensive and hard to purify—so why bother?

But no matter how obscure an element is, no matter how tempting it is to call it completely useless, there is always someone, somewhere, who will stand up for it, and I just had lunch with the man who will stand up for thulium.

When you're designing a high-intensity arc light, like my friend Tim, you build up the spectrum—the color—of the light it will emit by adding a mixture of elements into its arc tube. For example scandium (21) is popular because it provides a broad range of spectral lines for making nice white light.

Thulium's main purpose in life is to provide a broad range of green emission lines in an area of the spectrum not readily covered by any other element. Though the great majority of people have never even heard of thulium, lighting designers the world over would be lost without it. (You should have seen the look on Tim's face when I told him that I thought thulium was the most useless element.)

After its discovery in 1879, thulium remained so rare and so difficult to separate from the more abundant rare earths that not until almost 80 years later could you actually buy it commercially. And even then it was available only because an efficient new method of separating *all* the rare earths had been perfected. (The solvent extraction method described under praseodymium, element 59, is used to separate bulk quantities of fairly pure rare earths. Ion exchange methods can isolate very pure samples at higher cost.)

Thulium is now available at a quite reasonable price, and that price will remain reasonable right up until someone figures out a use that requires more than the small quantities needed for arc lighting, at which point thulium's rarity will drive the price through the roof.

Light of a different sort comes from the next element, ytterbium.

Dendritic crystals of pure thulium.

Thulium contributes a critical broad range of green emission lines to metal halide lamps.

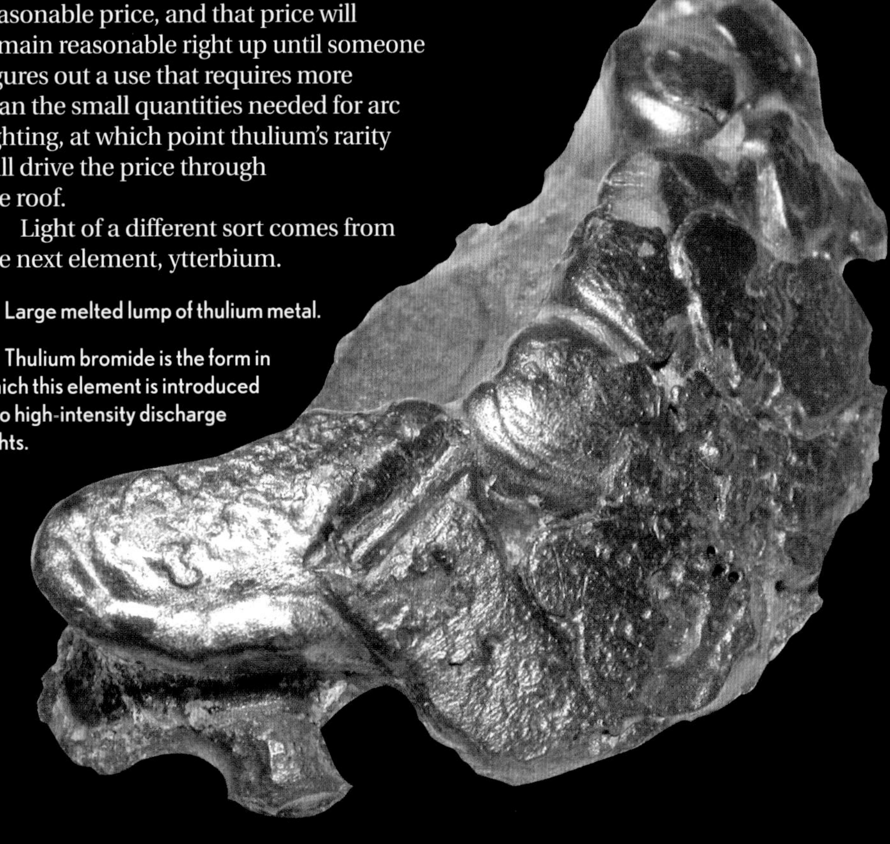

Large melted lump of thulium metal.

Thulium bromide is the form in which this element is introduced into high-intensity discharge lights.

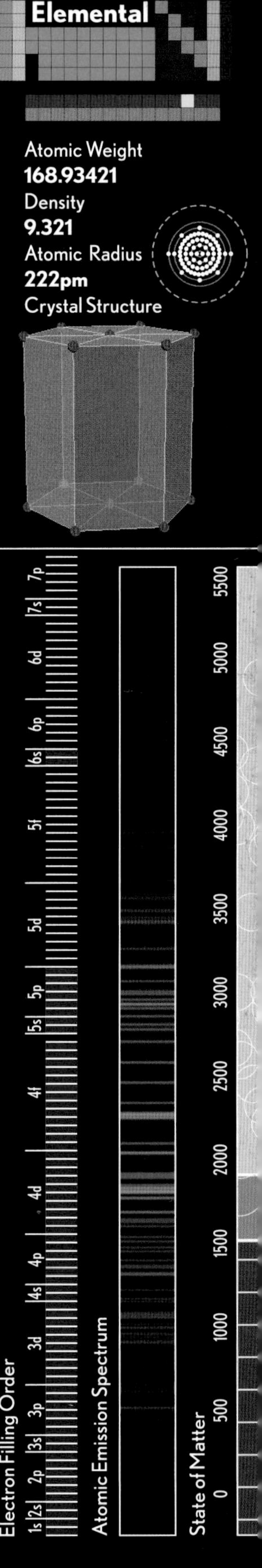

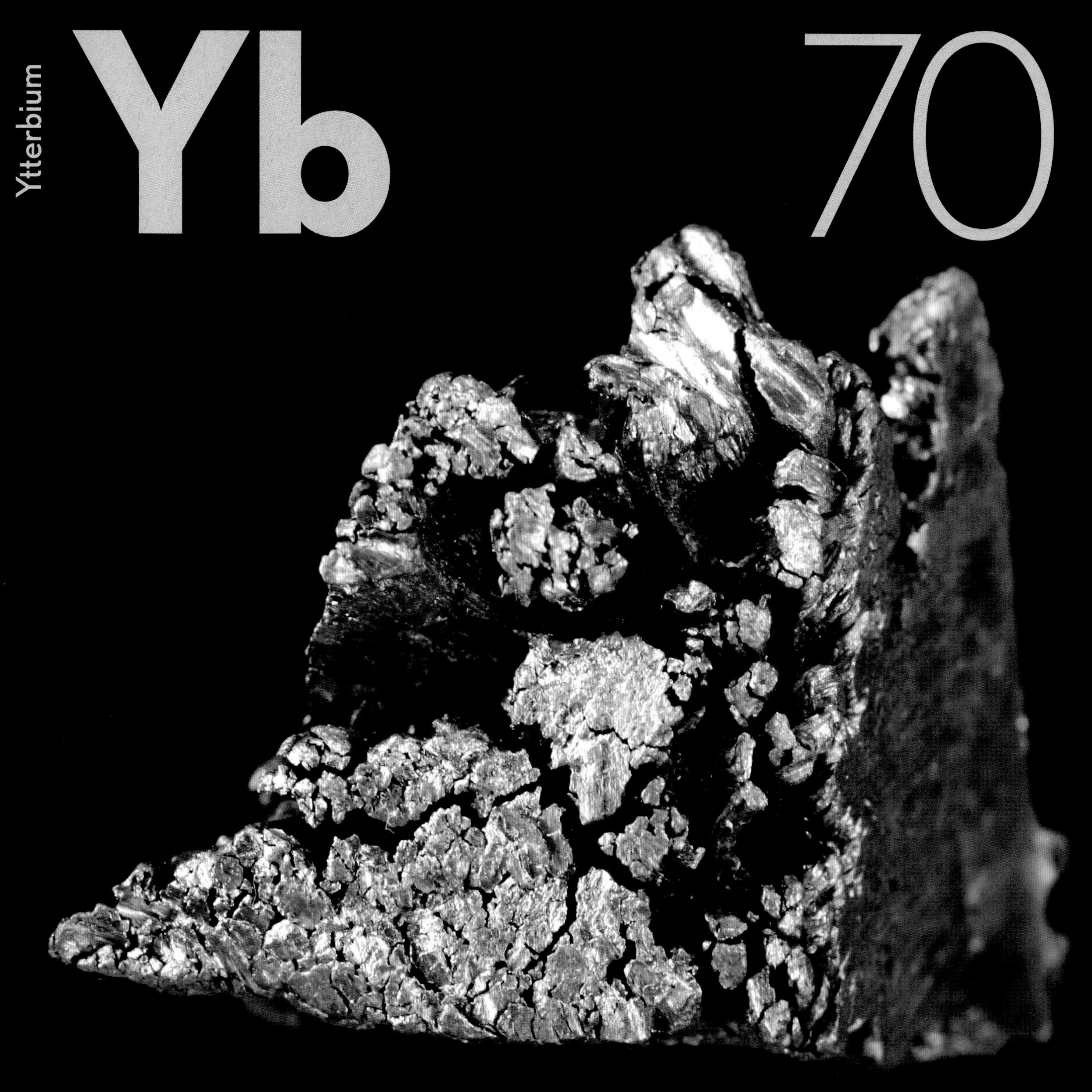
Ytterbium
Yb
70

Ytterbium

Atomic Weight
173.04
Density
6.570
Atomic Radius
222pm
Crystal Structure

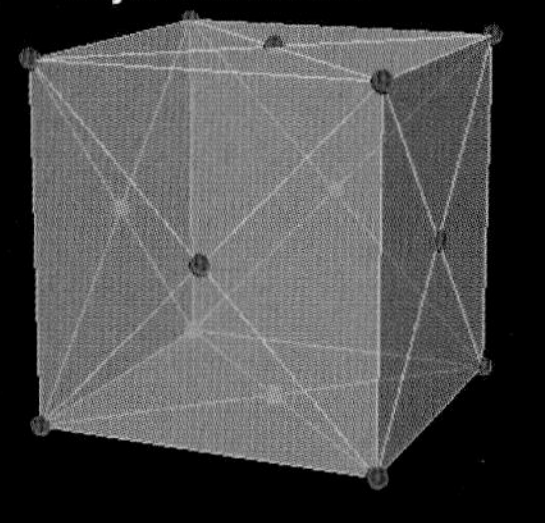

THE CITIES OF BERKELEY in California, Dubna in Russia, and Darmstadt in Germany had to work very hard to get elements named after them. In fact, each of them had to make their own elements from scratch in giant particle accelerators.

All three of these elements—berkelium (97), dubnium (105), and darmstadtium (110)—are pathetically short-lived laboratory curiosities. It must really grate, then, that Ytterby in Sweden has no fewer than *four* nice, stable elements named after it, all of which were just found lying around. Yttrium (39), terbium (65), erbium (68), and ytterbium were all found in the same mine outside the village of Ytterby!

Ytterbium's star application is as a doping agent in lasers, where it creates color centers that store energy in much the same way erbium does in fiber amplifiers, described under that element.

I'm just old enough to remember when lasers were new and exotic, and I still put them on the very short list of devices that if you don't stand slack-jawed in awe of their existence, you don't understand how they work.

Inside a laser's resonant cavity, vast numbers of atoms coordinate their actions with a perfection that can occur only at the quantum level: every photon of light the exact same wavelength, exactly in phase with every other photon, all traveling together as a single coherent beam. It's not just really well-focused light, it's a completely different kind of light, one that can be explained only by the utterly baffling laws of quantum mechanics.

I wish I could explain in a few paragraphs how a laser works, I really do. But it takes a couple of years of calculus and a semester or two of physics to even ask the question properly. Once you get there, the answer is absolutely worth it, deep and beautiful and so real you can taste it. This and the many answers like it are among the main reasons to study higher mathematics. Math is the language in which the secrets of the universe are written, and through its understanding comes enlightenment. So do your homework, OK? It's worth the journey.

Getting to lutetium, on the other hand, is worth the effort for an entirely different reason.

Ytterbium coin.

High-purity ytterbium bromide, used in the lighting industry.

The mineral xenotime, (YbY)PO4.

Torn dendritic crystals of pure ytterbium.

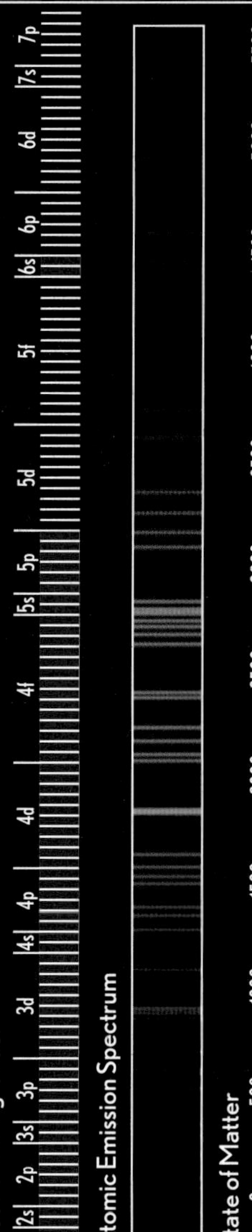

Lutetium

Lu

71

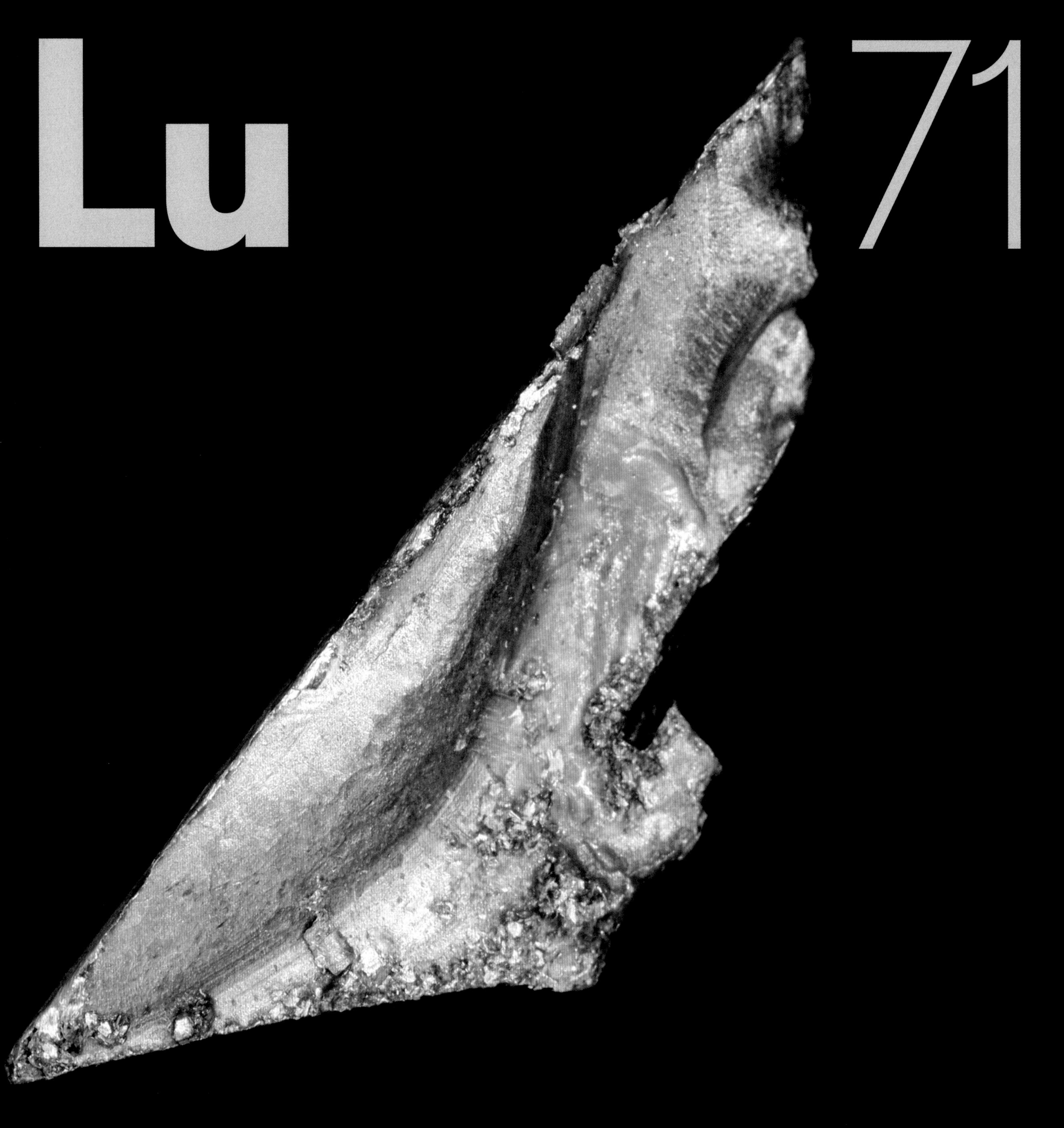

Lutetium

THE MOST WONDERFUL thing about lutetium, the last of the lanthanide series of rare earths, is that it's the last of the lanthanides. After lutetium we can get back to the dynamic and varied world of sixth-period transition metals, home of extremes in density (76, 77), temperature (74), and romance (79). But for now we're still stuck in the rare earths, of which lutetium is no standout.

So why, you might ask, are the rare earths all so similar, so nearly interchangeable that for years mixtures of several of them were thought to be pure samples of a single element?

The electrons in an atom of a particular element are arranged in concentric "shells." The weirdness of quantum mechanics means you're not allowed to think of an electron as having an actual location—they are more like clouds of possibility, known technically as probability distributions. But as a shorthand for understanding chemistry, you may imagine that some electrons spend their time closer in to the nucleus, while others live in shells further towards the outside.

Chemistry is primarily the story of the outermost shell of electrons. Elements with the same number of electrons in that outermost shell tend to have similar chemical properties. This is, in fact, the fundamental principle that gives shape to the periodic table—elements within a given column have the same number of these "valence electrons."

In most parts of the periodic table, when you move from one element to the next you are adding a new valence electron each time, giving each new element unique properties. But in the rare earth range, electrons are being added to an inner shell instead: Every one of the rare earths from 57 to 71 has a filled "6s" outer electron shell, and differing numbers of the deep "4f" shell, which contributes only minimally to chemical properties.

(Nothing with chemistry is quite that simple, so for example gadolinium puts one electron into a 5d orbital instead of a 4f, giving it slightly anomalous chemical and magnetic properties compared to its neighbors. Flip through the electron configuration diagrams on the right side of each page and you'll notice a few other out-of-order cases like this.)

Because the outer electron shell has the same configuration for each of the rare earths, their chemical properties are all similar. But magnetic properties follow an entirely different set of rules that involve all the electrons, not just the outer shell. So what they lack in chemical diversity, the rare earths make up for in imaginative magnetic properties.

You will sometimes find lutetium described as the most expensive element, or the most expensive rare earth, but that information is out of date. Lutetium is still not dirt cheap, but modern separation methods have made reasonable quantities of it available, and it's not as if anyone is beating down the door to buy the stuff. I would not be surprised to learn that element collectors are one of the larger markets for the pure metal.

There's not much more to say about lutetium, so let's just move on to hafnium.

A cut shape of pure lutetium.

The mineral euxenite, $(Y,Ca,Ce,Lu,U,Th)(Nb,Ta,Ti)_2O_6$.

If no one else will give an element a home, the lighting industry steps up. This lutetium bromide is lovingly created to the highest standards of purity for use in high-intensity discharge lights.

A coin made of pure lutetium, an unthinkable luxury a few decades ago but now quite practical. Well, not practical, since it serves no purpose, but at least not expensive.

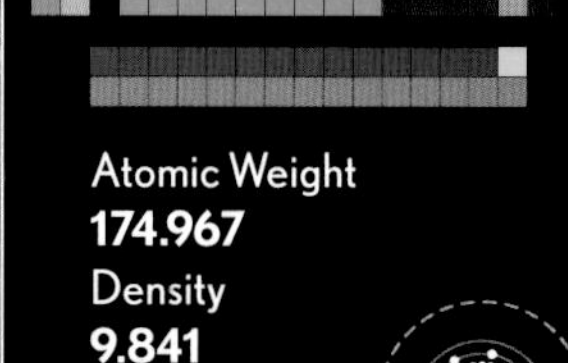

Atomic Weight
174.967
Density
9.841
Atomic Radius
217pm
Crystal Structure

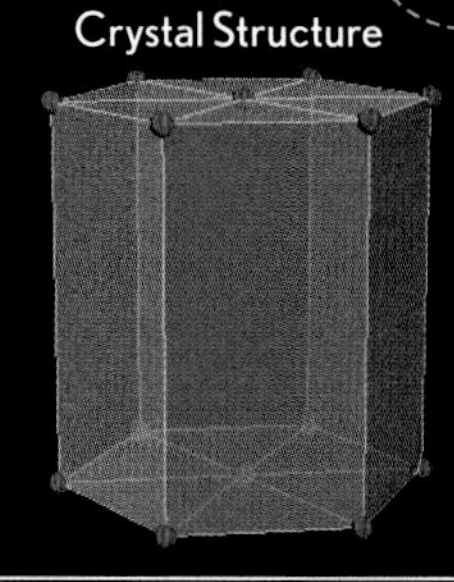

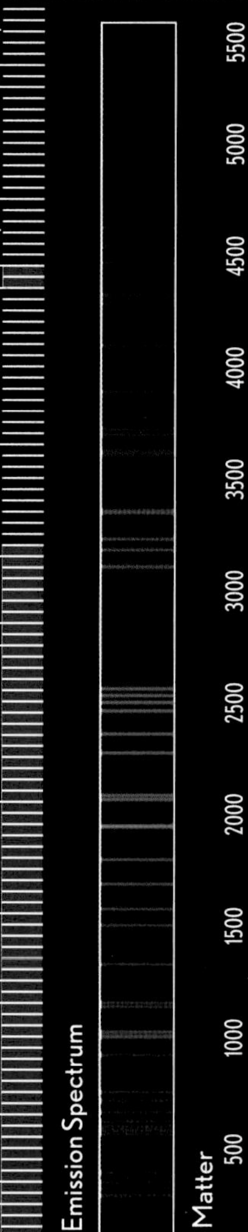

Hafnium

Hf 72

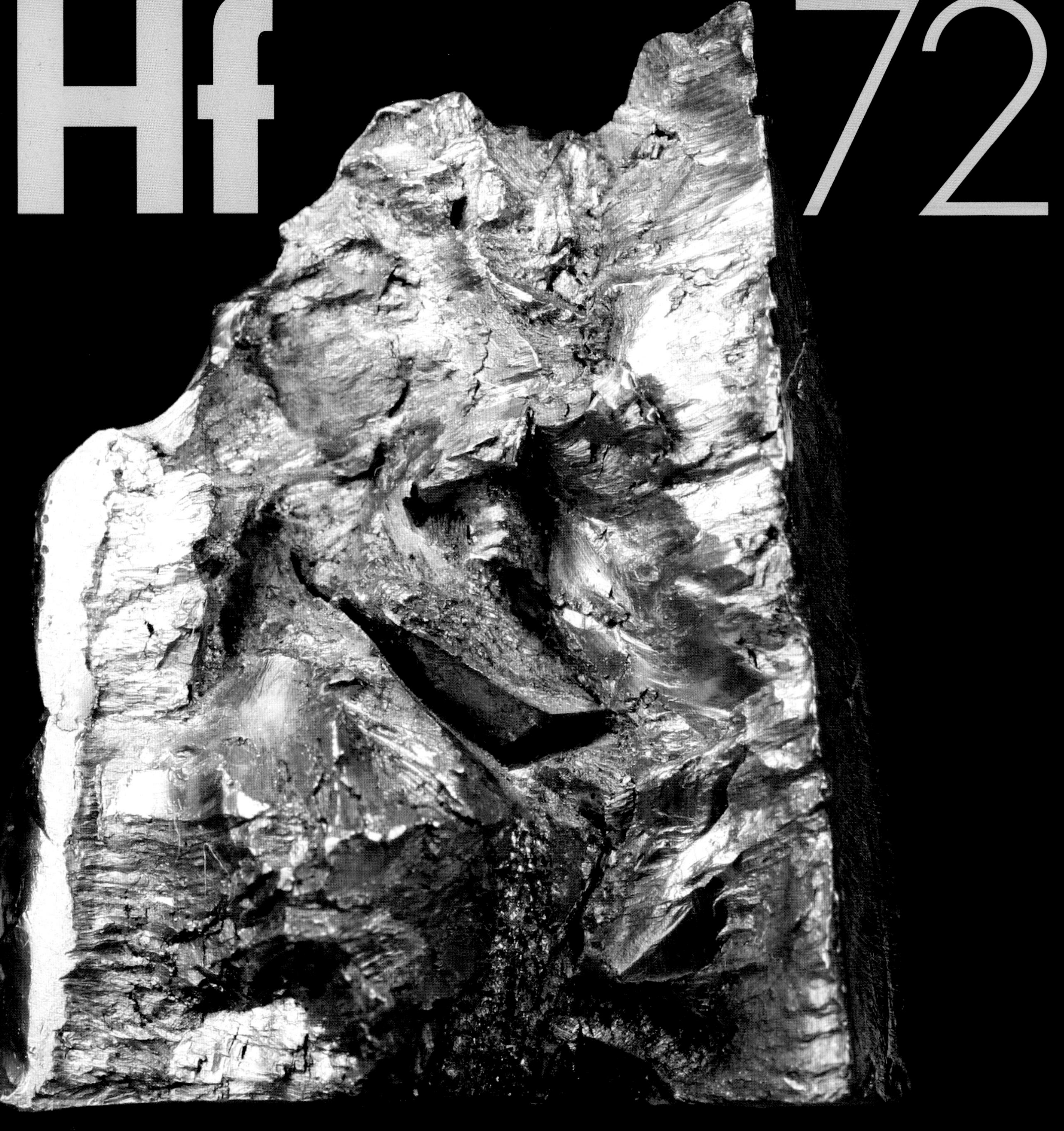

Hafnium

HAFNIUM IS A SPECIALIST: It does one thing and does it well.

In the past, cutting through steel required an oxyacetylene torch connected to two heavy and potentially dangerous cylinders of compressed gas. These days you can cut steel with a completely self-contained plasma torch that needs nothing more than electricity from a standard 120-volt wall outlet and the air around us.

A plasma torch contains an air compressor, some fairly complex control electronics, and a copper electrode with a tiny button of pure hafnium embedded in it. When you pull the trigger, the electronics initiate an arc from the hafnium button. A stream of compressed air blows the arc plasma out the tip of the torch and into the metal you want to cut.

Once steel has been heated to a high enough temperature, it will burn in air. The bulk of the cutting work is done by the stream of compressed air from the plasma torch, which contains oxygen that reacts with and burns away the steel. The arc doesn't actually cut the steel on it own, it simply supplies enough extra heat to keep the steel burning.

Why hafnium in the tip? Hafnium has a high melting point and is extremely resistant to corrosion even at very high temperatures, so it is able to withstand the conditions of the arc for extended periods. But other metals also have those properties. Hafnium's unique advantage is the ease with which it releases electrons into the air. When an electric spark leaves a metal surface to begin traveling through the air, it takes a certain amount of energy for the electrons to make the jump. With hafnium, the amount is minimal, allowing the electrode button to run cooler and the arc hotter.

The electronic circuitry that regulates the current in the arc of a plasma torch no doubt uses capacitors made of tantalum.

The plasma emitted from a hafnium button burns steel into a shower of sparks.

This remarkable image shows the inside surface of a huge high-purity hafnium crystal bar from Russia. It was created by the van Arkel process, in which hafnium tetraiodide vapor is broken down on a hot wire.

The hafnium button inside a copper plasma cutter tip.

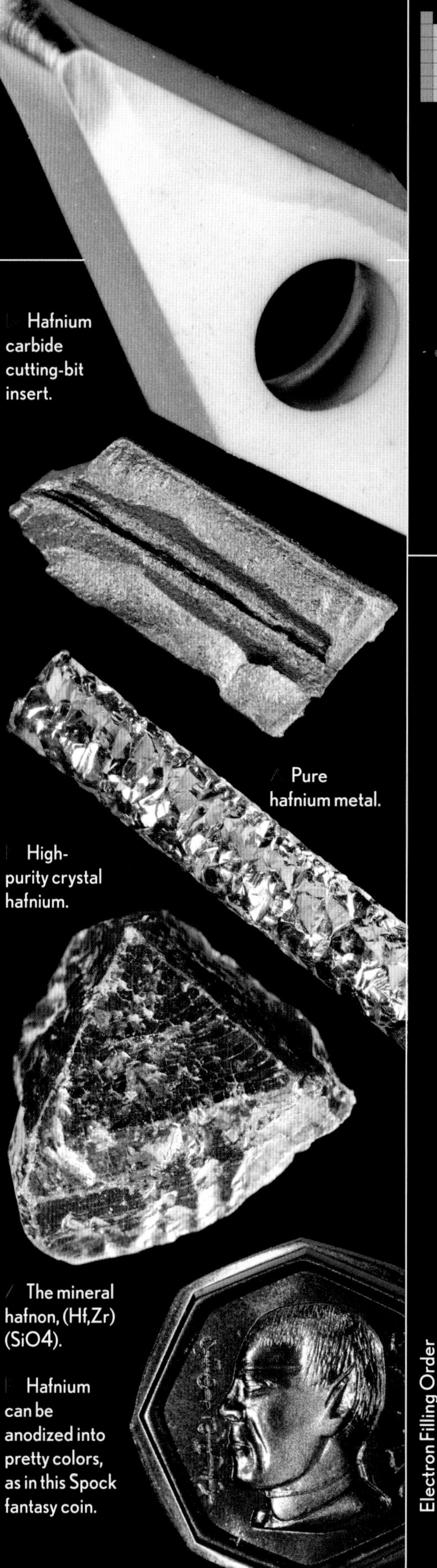

Hafnium carbide cutting-bit insert.

Pure hafnium metal.

High-purity crystal hafnium.

The mineral hafnon, (Hf,Zr)(SiO_4).

Hafnium can be anodized into pretty colors, as in this Spock fantasy coin.

Elemental

Atomic Weight
178.49
Density
13.310
Atomic Radius
208pm
Crystal Structure

Electron Filling Order: 1s 2s 2p 3s 3p 4s 3d 4p 5s 4d 5p 6s 4f 5d 6p 7s 5f 6d 7p

Atomic Emission Spectrum

State of Matter: 0 500 1000 1500 2000 2500 3000 3500 4000 4500 5000 5500

Tantalum

Ta 73

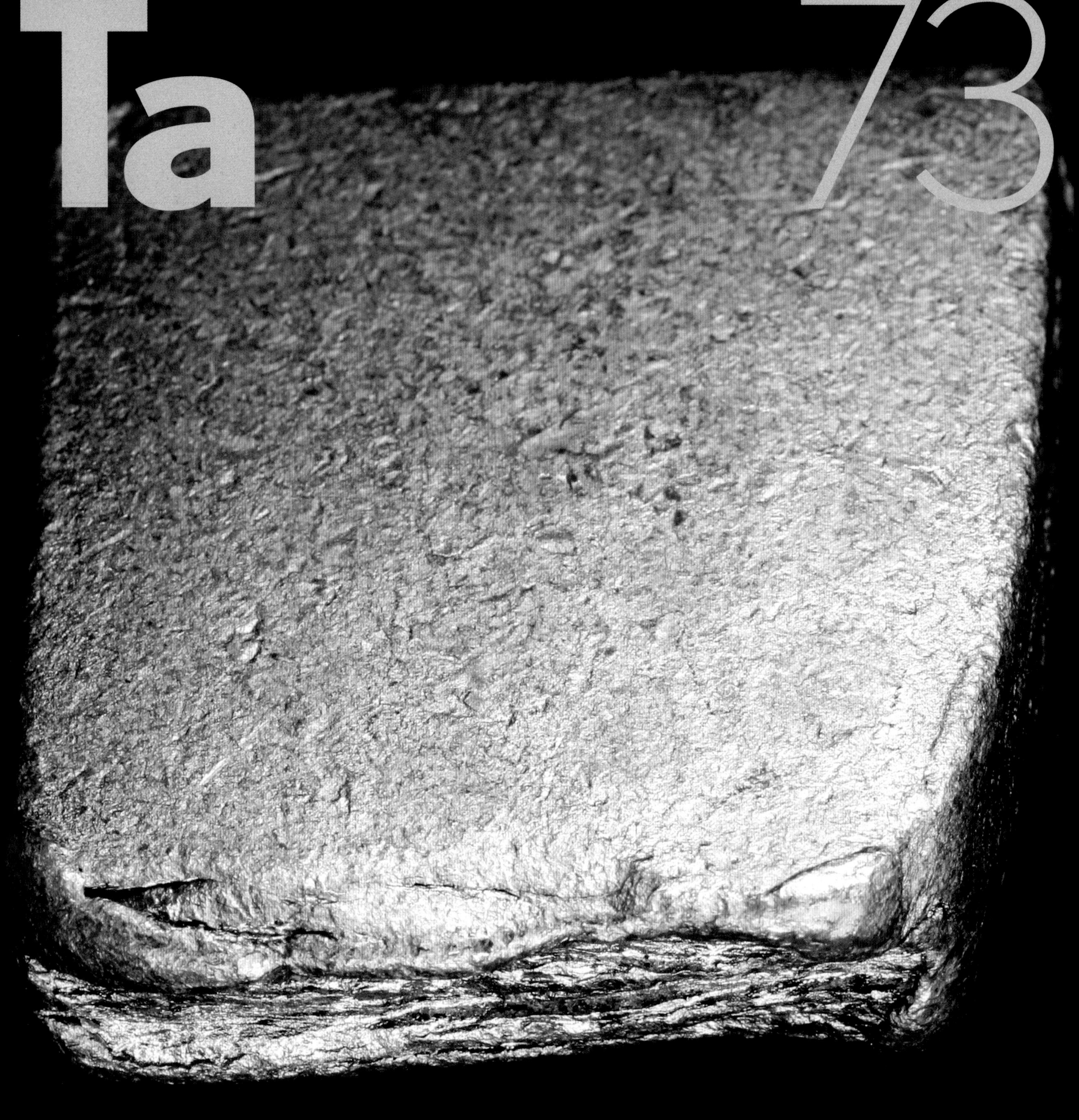

Tantalum

Antique tantalum-filament lightbulb.

Elemental

Atomic Weight
180.9479
Density
16.650
Atomic Radius
200pm
Crystal Structure

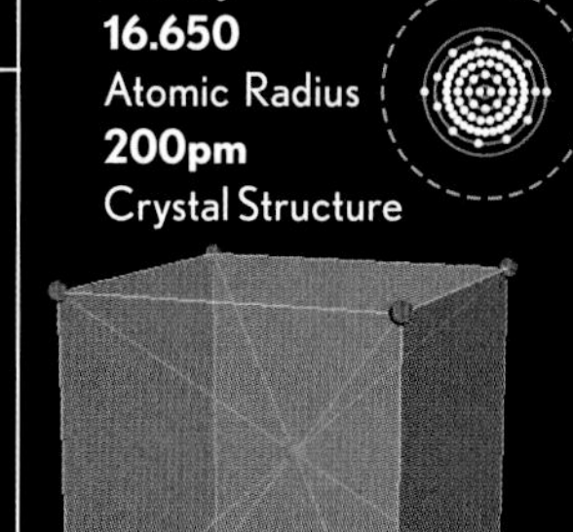

TANTALUM IS ONE OF TWO elements that have had boycotts organized against them. People are urged to boycott the other one, carbon (6), because sales of "conflict diamonds" support nasty local wars in the regions where the diamonds are mined. Tantalum is opposed for a similar reason, with the added motivation that it is mined in areas where endangered gorillas live: Gorillas are dying to fund guerilla wars, and all for tantalum.

How do you boycott an element as obscure as tantalum? Cell phones! Tantalum is not limited in applications, just in name recognition. Not just cell phones, but computers, talking dolls, medical equipment, radios, video games—virtually every device that contains digital electronics of any kind uses tantalum capacitors.

The advantages of tantalum capacitors over other types include their small size, their high capacity, and their high-frequency response. Digital circuits generate lots of high-frequency electrical noise, which can leak from one circuit to another through power and signal connections. Tantalum capacitors are especially effective at absorbing and dampening these noise spikes before they can cause trouble.

So in order to boycott tantalum, all you have to do is boycott . . . well, pretty much everything invented since 1982.

If it weren't for tungsten (74), you might also have to boycott lightbulbs. In the early history of incandescent light, tantalum-filament bulbs were commercially available. Indeed, among the many sophisticated technological advances advertised for the luxury liner Titanic was that its electric lights had tantalum filaments, which were so much more reliable than the older carbon-filament bulbs that the Titanic's night lights could actually be left on at night.

But all the early diversity in lightbulb filament material—including carbon, tantalum, osmium (76), even platinum (78)—fell by the wayside when it become possible to manufacture tungsten wire, the best (and, with luck, the last) incandescent bulb filament material.

Pressed tantalum powder capacitor cores.

Tantalum evaporation boat.

Tantalum skull plate.

Heavy solid tantalum slab, enough for thousands of capacitors.

Common tantalum capacitors.

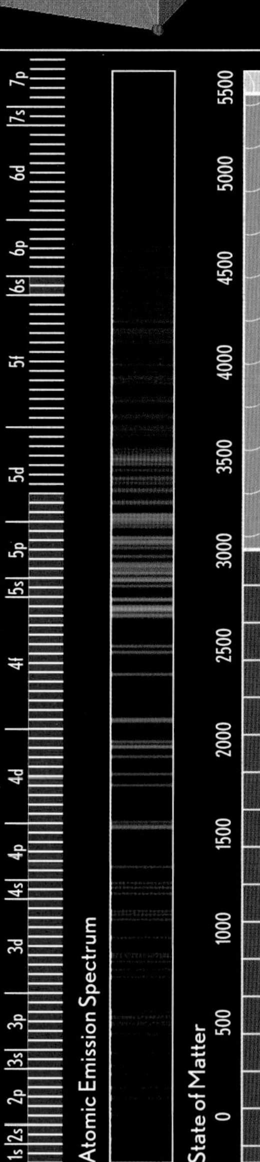

Tungsten

W

74

Tungsten

Tungsten shot is better than lead shot in many ways, and less harmful to the environment.

Elemental

Atomic Weight
183.84
Density
19.250
Atomic Radius
193pm
Crystal Structure

Electron Filling Order

Atomic Emission Spectrum

State of Matter

TUNGSTEN IS OVERWHELMINGLY associated with one application, the tragically inefficient incandescent lightbulb. These lamentable creations make light by electrically heating a very thin wire until it glows yellow-hot. Tungsten is the strongest metal at very high temperatures, and it's quite inexpensive, which makes it the best material for this application.

But tungsten's best is nowhere near good enough. A typical incandescent bulb converts only 10 percent of the electricity it uses into visible light. The other 90 percent is simply wasted, turned into heat and infrared radiation. You might as well call them electric heaters that happen to produce a minor by-product of light. Unless you're using one to heat a chicken coop, this is not a good thing.

If light is what you actually want, a far superior alternative is now available—compact fluorescent bulbs, which are several times as efficient as incandescent bulbs and last ten or twenty times longer. If you have any tungsten bulbs in your house, for the sake of the planet get rid of them! Every $2 compact fluorescent you install will cut more than a thousand pounds of carbon dioxide emissions just from the electricity it saves. And the light is much more pleasant, not depressing and yellow like tungsten light.

While the continued use of tungsten in lightbulbs is an abomination, tungsten carbide is widely and wonderfully useful in cutting tools and other things that need to stay sharp. It's tougher (more resistant to fracturing) than diamond and *much* harder than steel, making it superior for machining many materials.

The metals from tungsten through gold (79) are all very dense; indeed osmium (76) and iridium (77) are the densest of all elements. But tungsten is the cheapest by a factor of about a hundred, meaning it is used in many applications where it's necessary to put a lot of weight in a small space, including counterweights, fishing sinkers, darts, ear weights for dogs (seriously), shot puts, and so on.

With rhenium we enter the range of the expensive metals and begin our final approach to the pinnacle of the metals, gold.

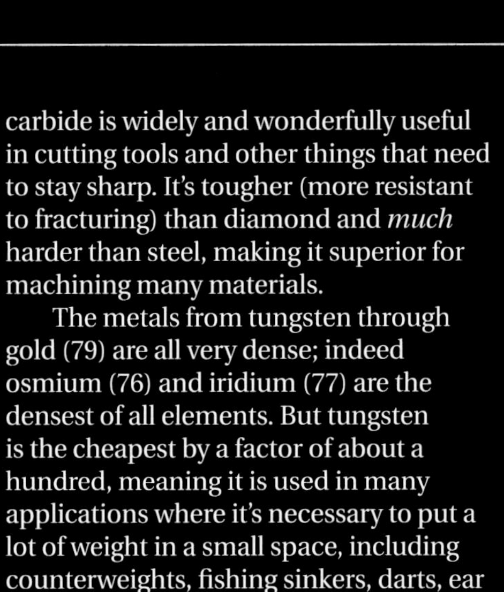

Tungsten carbide is the most common material for cutting-bit inserts.

Tungsten's density makes it useful as a compact, aerodynamic weight in this dart.

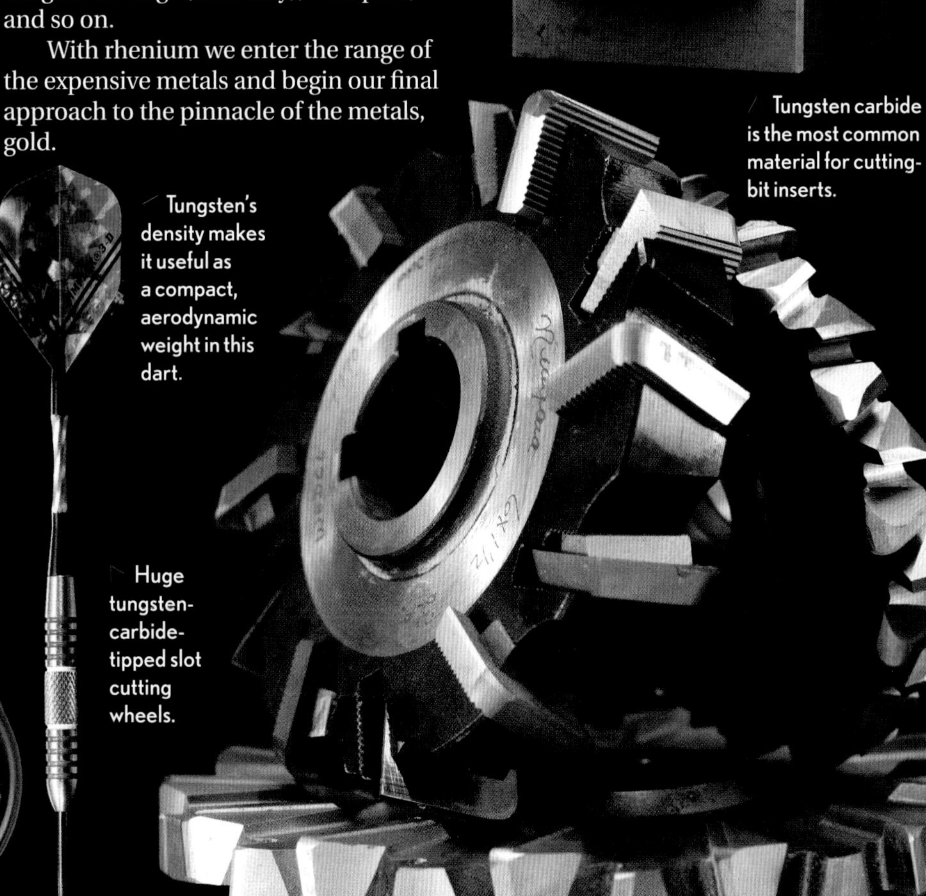
Huge tungsten-carbide-tipped slot cutting wheels.

Antique tungsten-filament bulb.

Filament from a tungsten incandescent lamp, hopefully soon to be an antique relic.

Faceted tungsten carbide has become popular for rings. Because it cannot be cut with conventional tools, doctors have developed a new method for removing stuck ones: Cracking them with a pair of vice grip pliers.

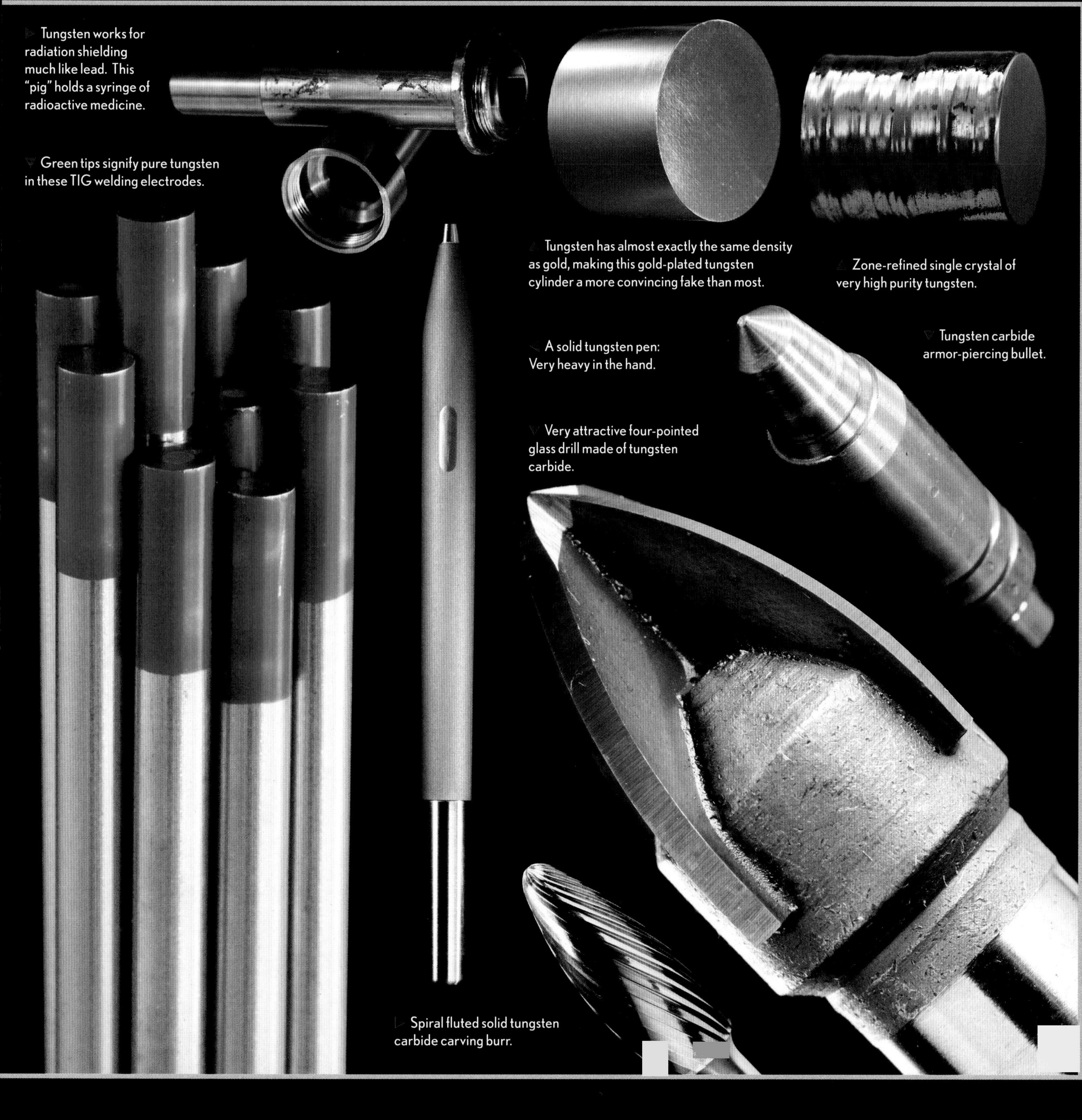

Tungsten works for radiation shielding much like lead. This "pig" holds a syringe of radioactive medicine.

Green tips signify pure tungsten in these TIG welding electrodes.

Tungsten has almost exactly the same density as gold, making this gold-plated tungsten cylinder a more convincing fake than most.

Zone-refined single crystal of very high purity tungsten.

A solid tungsten pen: Very heavy in the hand.

Tungsten carbide armor-piercing bullet.

Very attractive four-pointed glass drill made of tungsten carbide.

Spiral fluted solid tungsten carbide carving burr.

Re

Elemental

Rhenium

RHENIUM WAS THE LAST stable element to be discovered, in 1925 in Germany. It was probably discovered earlier, in 1908 in Japan by Masataka Ogawa, and it might be known as "nipponium" if only Ogawa had not thought, and claimed, that he had discovered the element directly above it in the periodic table—element 43, which we now know as technetium.

Elements in the same column of the periodic table share many chemical characteristics, so when Ogawa found an element that seemed very similar to manganese (25) but was heavier, it was quite reasonable of him to assume it must be element 43, a well-known hole in the periodic table in just the right place. Sadly, he was wrong. Technetium, the real element 43, is radioactive and does not occur in nature—something no one could have guessed in 1908.

Many years passed after rhenium's discovery before commercial quantities became available, and the stuff is still very expensive (we're talking several hundred dollars per troy ounce), due to its scarcity.

Most rhenium is used in nickel-iron supera[illegible] the turbine blades in fighte[illegible]ines. The latest single-crystal superalloys used in these state-of-the-art turbine blades contain about 6 percent rhenium. Even though there aren't that many fighter jets being made, they consume three quarters of the world's annual rhenium production.

Single-use flashbulbs for photography were typically filled with zirconium (40) wool, but old advertisements often boasted that they had "rhenium igniters" without even mentioning the zirconium. The ads were probably trying to tell us that they had electronic (tungsten-rhenium wire) igniters rather than percussion-explosive igniters like other types of flashbulbs, notably the GE MagiCubes that some of the older generation may fondly remember from their old Kodak Instamatics. (MagiCubes required no battery: A sharp knock from a rod connected to the shutter release set them off mechanically, while those with rhenium igniters required an electrical signal.)

Another old-fashioned instrument, the fountain pen, uses both of the next two elements, osmium and iridium.

Pressed button of rhenium powder, ready to be melted into a bead in an argon-arc furnace.

The rare mineral rheniite (rhenium sulfide).

Rhenium foil strips are used as evaporation filaments in mass spectrometers.

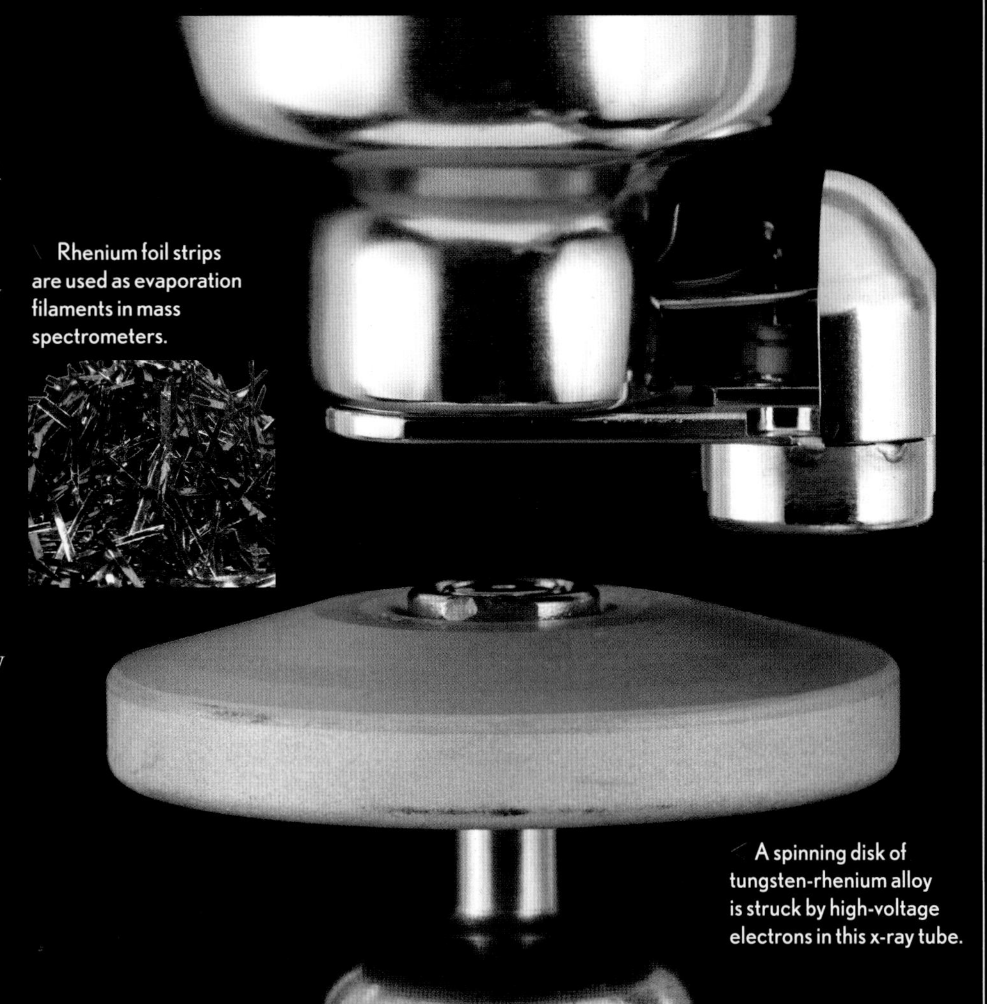

A spinning disk of tungsten-rhenium alloy is struck by high-voltage electrons in this x-ray tube.

A pound of pure rhenium—quite a valuable object, depending on the current market price.

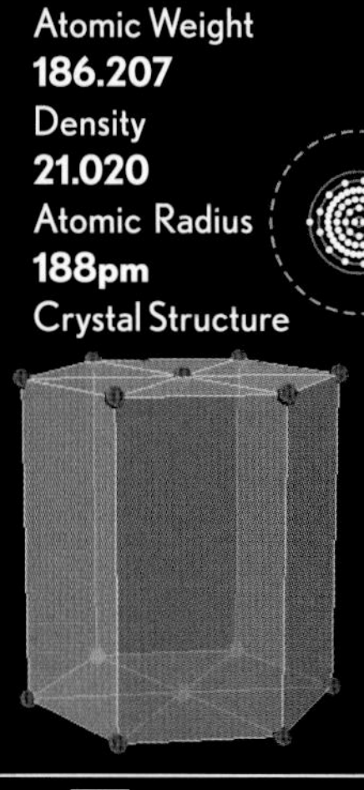

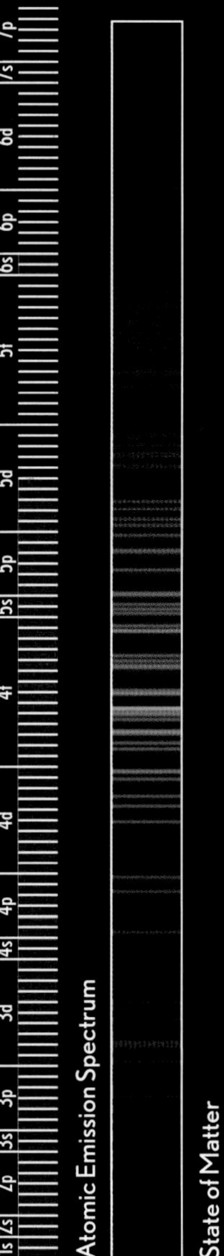

Osmium

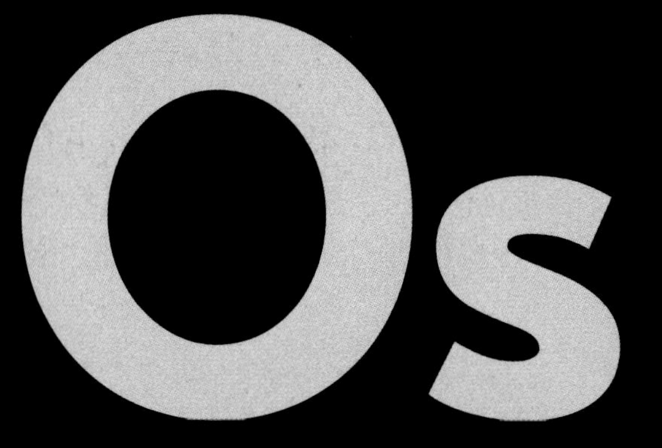

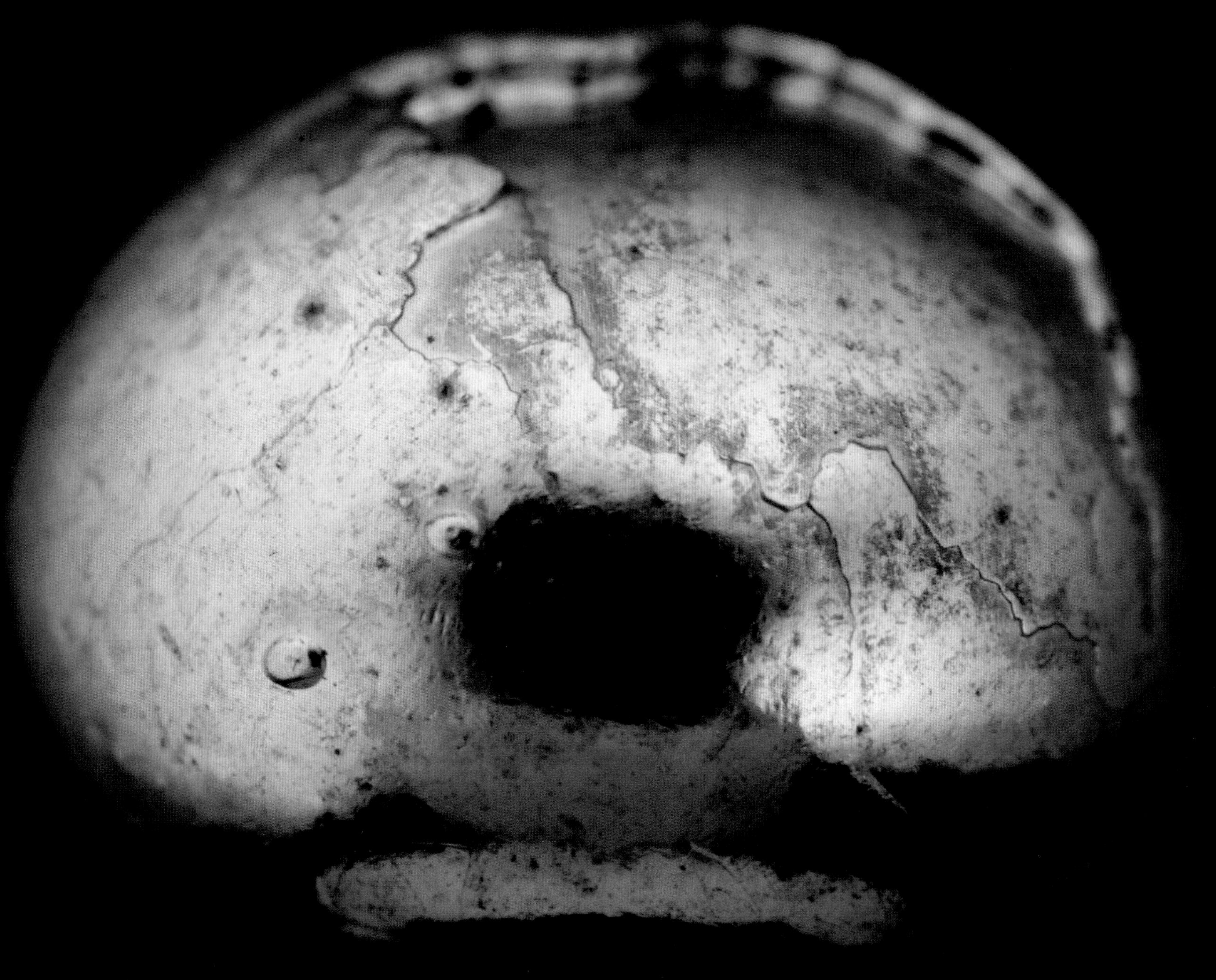

Osmium

OSMIUM ALMOST JOINS copper (29) and gold (79) on the extremely short list of metals that are not gray or silver in color, but its slight bluish tint is so faint you have to work to convince yourself you're really seeing it. Basically it's just another silvery metal.

Well, not *just* another silvery metal. Osmium is at least as expensive as rhenium (75), and on the Brinell scale of hardness (a measure of how far a ball penetrates a material when pushed in with a given force), osmium is the hardest metal element. (Not the hardest material, not the hardest element, but the hardest pure metal.)

Osmium is often found together with iridium (77) in a very rare but naturally occurring alloy known as osmiridium (or iridosmine or iridosmium, depending on who you ask). This extraordinarily hard and wear-resistant metal was (and still is) used in a number of applications found in every household several generations ago, including fountain pen nibs and phonograph needles, where a tiny button of osmiridium at the very tip of such an item was all that was needed of this expensive mixture to inhibit wear from long-term use.

Somewhat unusually for the generally oxidation-resistant metals in this region of the periodic table, finely powdered osmium metal will slowly oxidize in air, forming osmium tetroxide. Even more unusually for a heavy metal oxide, osmium tetroxide is volatile, sublimating at room temperature into a highly toxic vapor. It is said to smell a bit like ozone, but can kill or blind at concentrations well below the point at which it can be smelled, so information on this is understandably sketchy.

Despite its volatility, extreme toxicity, and high cost, osmium tetroxide is used more than you might think—as a stain for electron microscopy of tissue specimens, and as a reagent in chemical syntheses.

Osmium is special for one more reason. It is the most dense of all the elements. I leave this fact for last because if you look it up in virtually any reference source, online or printed, you will find a different answer. But they are all wrong—the densest element is *not* iridium.

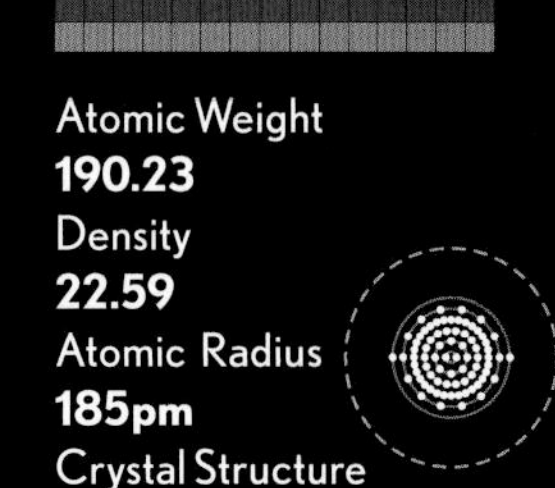

Atomic Weight
190.23
Density
22.59
Atomic Radius
185pm
Crystal Structure

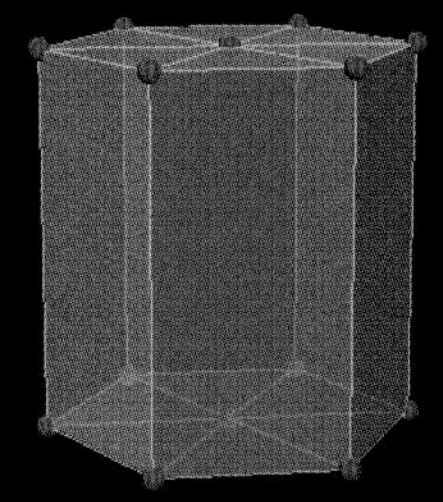

Electron Filling Order
Atomic Emission Spectrum
State of Matter

Osmium-tipped phonograph needle.

The package for this osmium-tipped phonograph needle proudly proclaims how hard this metal is.

A single bead of osmium with the most delicate hue of blue.

Beads of pure osmium take on distinct bluish hue in the right light.

Osmium tetroxide crystals are dangerously toxic and must be kept in a sealed glass ampule.

Iridium

Ir

77

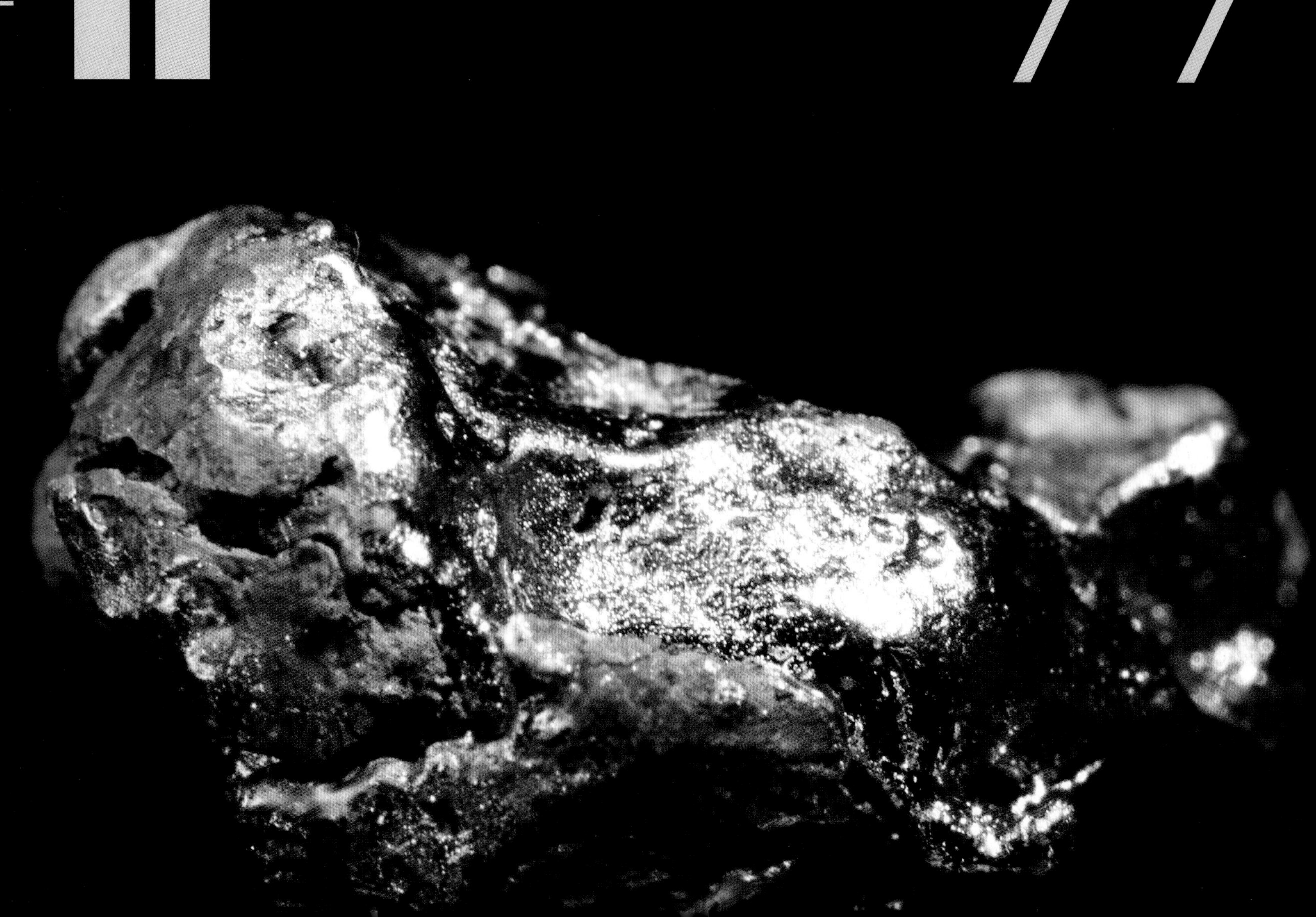

Iridium

THE MOST WIDELY quoted value for the density of iridium is 22.65 g/cm^3, while that for osmium (76) is 22.61 g/cm^3, making iridium the densest of all elements. But those values are simply wrong. The correct values are 22.59 g/cm^3 for osmium and 22.56 g/cm^3 for iridium, which gives the title of densest element to osmium, not iridium, though by less than one tenth of one percent.

You might think density would be an easy question to settle with a careful measurement. But when people talk about the density of an element, they mean the density of a perfect single crystal of a perfectly pure sample of the element.

Preparing such an idealized sample is of course impossible, and even getting close can be very difficult in some cases. A more accurate method is to use x-ray crystallography to measure the spacing between atoms in a sample containing tiny perfect crystals. If you know the spacing and the weight of each atom, you can calculate how much a perfect crystal of any size would weigh, and from this you can calculate the idealized density.

The problem is that when this was first done, the then-accepted values for the atomic weights of osmium and iridium were wrong. Those atomic weights have long since been corrected, but *no one ever went back to recalculate the densities.* All the reference works just copied one another back and forth for 70 years.

This situation could remain uncorrected so long because, other than students writing school reports, the values are seldom used. No actual sample of osmium or iridium you will ever hold will have a density equal to the theoretical value. Getting within even a few percent of it is pretty hard. Imperfect melting, voids created during cooling, and impurities all increase volume and thus decrease density. So in practice, the theoretical density of any element is of purely theoretical interest.

Due to its high cost, iridium is mostly used in places where you need only a very small amount. For example, some high-grade automobile spark plugs are equipped with tiny iridium tips that last up to 100,000 miles, far longer than those of conventional plugs.

But the largest use of iridium is in alloys with its far more popular neighbor, platinum.

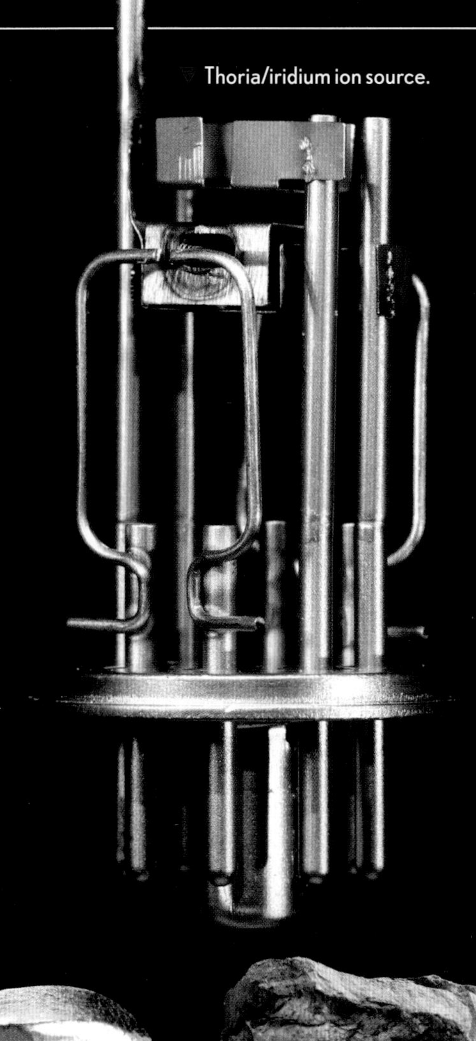

Thoria/iridium ion source.

Beads of pure iridium are incredibly shiny.

A tiny iridium-alloy wire lets this spark plug last up to 100,000 miles.

A thin layer of iridium-rich clay present all around the world marks the boundary between the Cretaceous and Tertiary periods. The iridium originated in a huge asteroid that wiped out the dinosaurs 65 million years ago.

Iridium is extremely hard to melt: This lump only made it about half way to being melted, hence its odd shape.

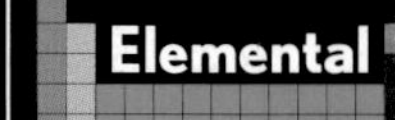

Atomic Weight
192.217
Density
22.56
Atomic Radius
180pm
Crystal Structure

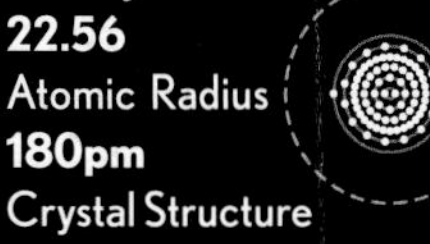

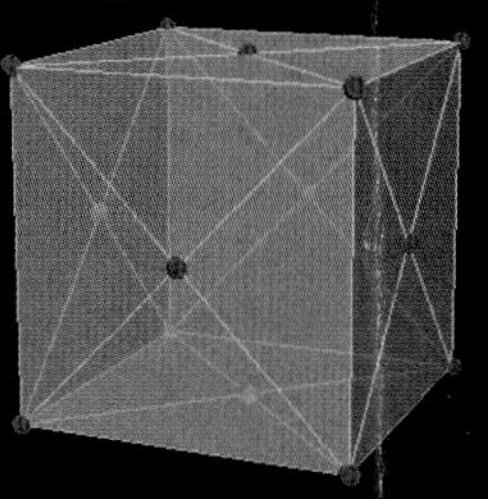

Electron Filling Order: 1s 2s 2p 3s 3p 3d 4s 4p 4d 4f 5s 5p 5d 5f 6s 6p 6d 7s 7p

Atomic Emission Spectrum

State of Matter: 0 500 1000 1500 2000 2500 3000 3500 4000 4500 5000 5500

Platinum

Pt 78

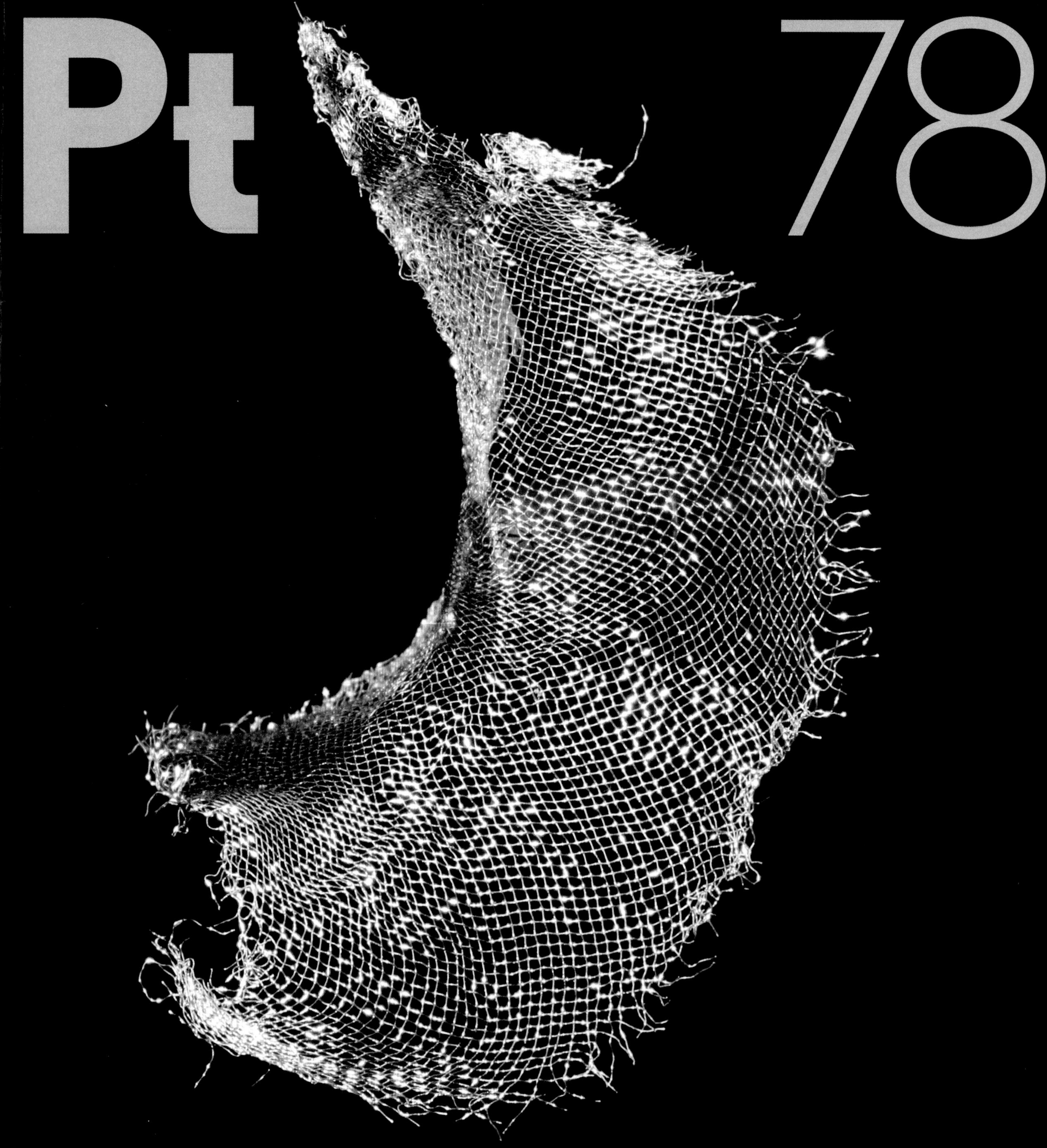

Platinum

PLATINUM IS *THE* MOST prestigious element, period. Sure, gold is great, but platinum is always better. Gold credit card? That's nothing compared to a *platinum* card. Platinum is more abundant in the earth's crust than the other platinum group metals such as rhodium (45), osmium (76), iridium (77), and even gold (79), but it is significantly more expensive because demand is so high.

Platinum is so important in laboratory and industrial applications that despite their absolutely outrageous cost you can buy things like solid platinum bowls, crucibles, filter holders, and electrodes. More than any other metal, platinum is able to withstand powerful acids and high temperatures—almost anything you can throw at it—without so much as staining.

Equally important as its resistance to corrosion is platinum's ability to catalyze reactions, such as those crucial to the refining of crude oil into gasoline. (Anything used in oil refineries is automatically a huge business.) At the very end of its life cycle, petroleum products often meet platinum again—in the catalytic converters of the world's gasoline and diesel vehicles. Assisted by platinum, unburned hydrocarbon fragments in exhaust gases are oxidized to carbon dioxide and water.

All the fundamental units of measure, including time (see cesium, element 55) and distance (see krypton, element 36), are defined using basic properties of matter that anyone can measure—with one exception. Mass is defined by the International Prototype Kilogram, a particular cylinder of platinum (plus 10 percent iridium) made in 1879 and housed in a special room in Paris. This cylinder has a mass of one kilogram by *definition*.

It's not a very good definition. The cylinder changes weight regularly when it's cleaned or handled, and is known to have been drifting by tens of micrograms—eventually a better definition must be developed. Most likely the kilogram will end up defined as a certain number of atoms of one element or another, or in terms of the magnetic force generated by a precisely controlled electric current.

The problem with platinum jewelry is that it looks pretty much like silver (47), palladium (46), or even lowly chromium (24)—in short, it's shiny silver-white, just like nearly every other metal. If I were going to pay big money for some scraps of metal around my finger, I'd at least want a bit of color to it, which leads us inexorably to gold.

Platinum spark plugs are now quite common and last nearly the life of a car.

Elemental

Atomic Weight
195.078
Density
21.090
Atomic Radius
177pm
Crystal Structure

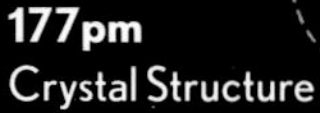

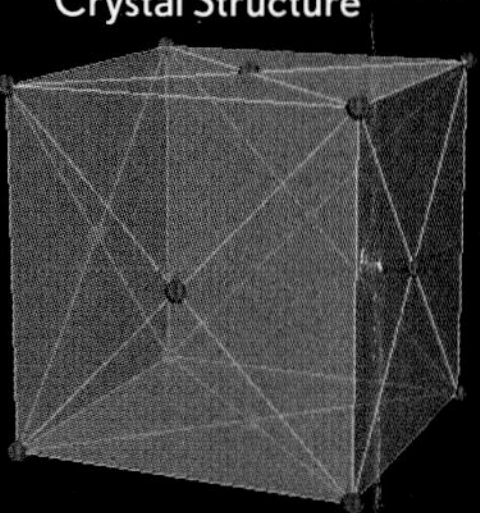

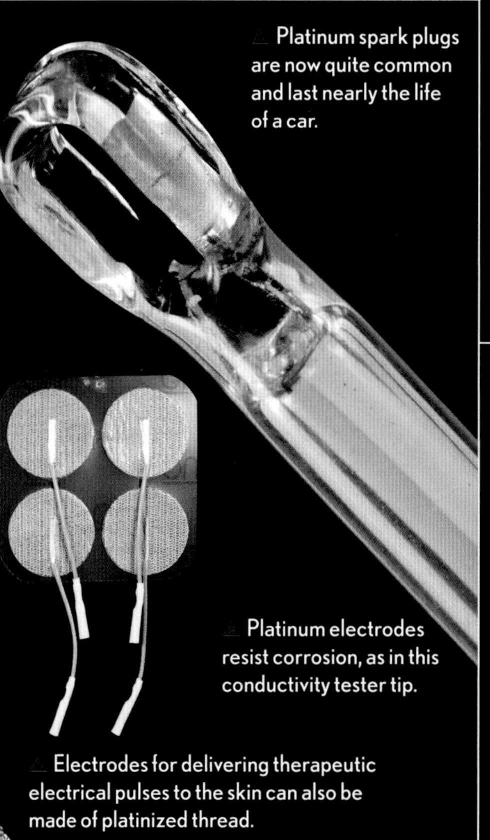

Platinum electrodes resist corrosion, as in this conductivity tester tip.

Electrodes for delivering therapeutic electrical pulses to the skin can also be made of platinized thread.

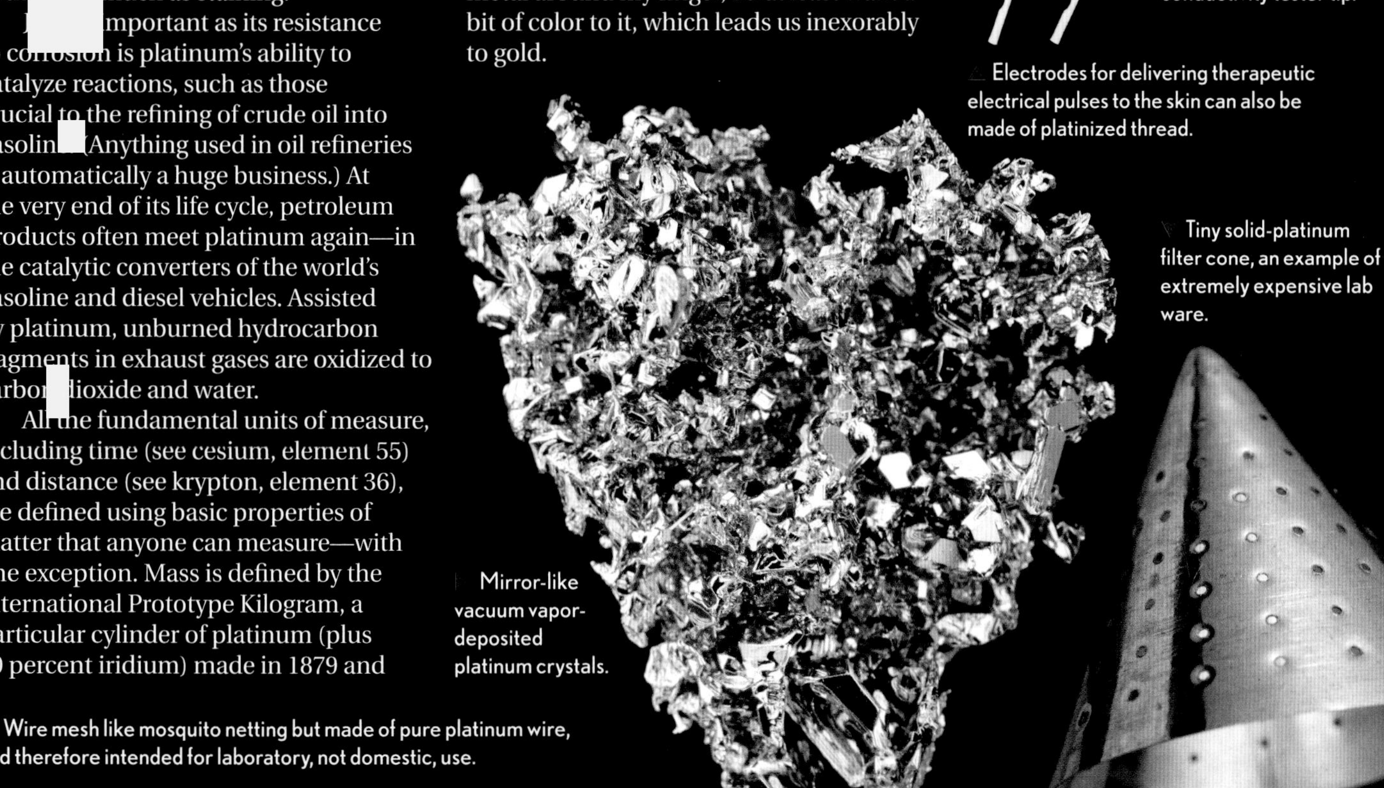

Tiny solid-platinum filter cone, an example of extremely expensive lab ware.

Mirror-like vacuum vapor-deposited platinum crystals.

Wire mesh like mosquito netting but made of pure platinum wire, and therefore intended for laboratory, not domestic, use.

Electron Filling Order

Atomic Emission Spectrum

State of Matter

Gold
Au
79

Gold

Atomic Weight
196.96655
Density
19.3
Atomic Radius
174pm
Crystal Structure

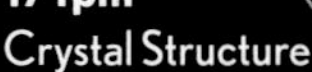

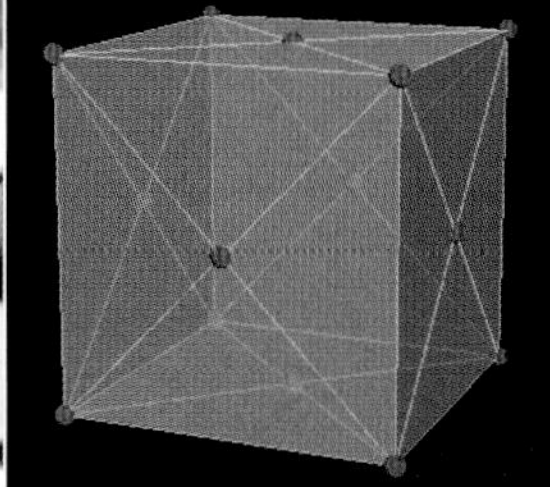

GOLD IS THE GOLD standard of metals. Rhodium (45) may be more valuable, but no one lusts after it the way they lust after gold. Only carbon (6), in the form of diamond, inspires the same feverish desire, but diamonds are temporary, easily destroyed by heat, and soon to be worthless when large synthetic diamonds become available.

Diamonds are a fraud, but gold is the real thing, richly deserving the adoration it inspires.

Gold is inherently valuable. There is very little of it around—all the gold ever mined in the history of the human race would fit into a cube about 60 feet on edge. (And if you meet one of the nuts suggesting that our money should go back on the gold standard, you might point out that, at current prices, this is worth only a few trillion dollars, significantly less than the money in circulation. There simply isn't enough gold to go around.)

Gold is undeniably beautiful. Of all the metals it is the only one that is both colored and whose color keeps its shine and beauty forever. You can find a piece of gold lying on the ground where it has been for a million years, pick it up, dust it off, and it will shine for you as if it's been waiting the whole time for this moment. Billions of years from now, when aliens come to rescue the last artifacts from earth before our sun explodes, King Tut's solid-gold mask will be just as shiny as it is today—which is just as shiny as it was 3,300 years ago when it was new. Not skin-deep, not temporary, the beauty of gold is built into its very atomic structure.

Gold is terrifically useful. It is a good conductor of electricity that absolutely does not tarnish, making it the best material for electrical contacts. Where conductors join two circuits merely by touch, any corrosion on either surface could interfere with the connection. So much gold is used in electronic devices that recycling them to recover the gold is a big business.

Gold has fascinated and inspired us since before those words existed. Wonder and fascination of a very different kind have been inspired almost as long by an element known to the ancients as the living, or "quick," silver: mercury.

High-purity vacuum-vapor-deposited crystals of gold—the purest, shiniest gold there is, bar none.

Gold on quartz

Cheap mall jewelry can be plated with a thin layer of real gold, which makes it just as pretty as the solid alternative.

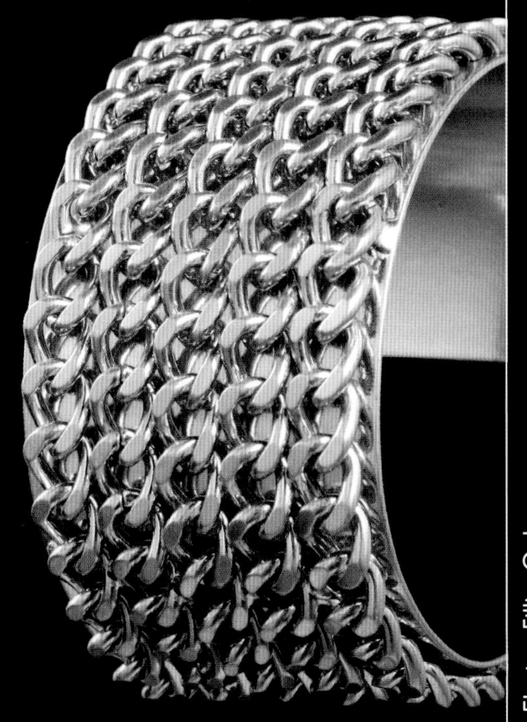

"Healey gold" is made with a plating process that uses uranium, but no radioactivity is left in the final article.

Gold paint may or may not contain real gold leaf, depending on how old or how expensive it is.

This 1-ounce nugget of pure gold was found in Alaska in 1890 by Hormidas O. Marion while on a trip to sell shoes to Eskimos. Seriously.

Electron Filling Order
Atomic Emission Spectrum
State of Matter

Gold 79

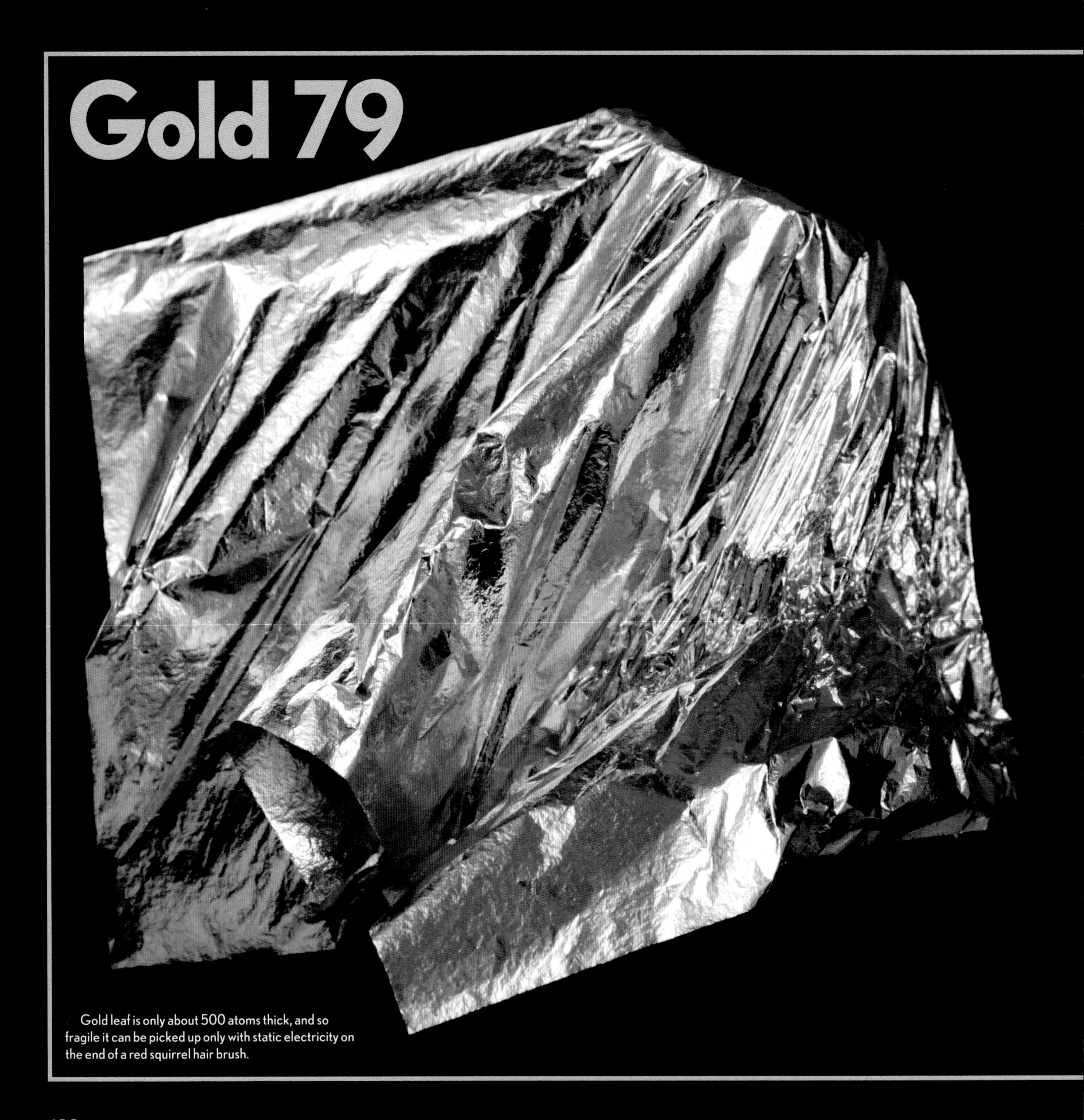

Gold leaf is only about 500 atoms thick, and so fragile it can be picked up only with static electricity on the end of a red squirrel hair brush.

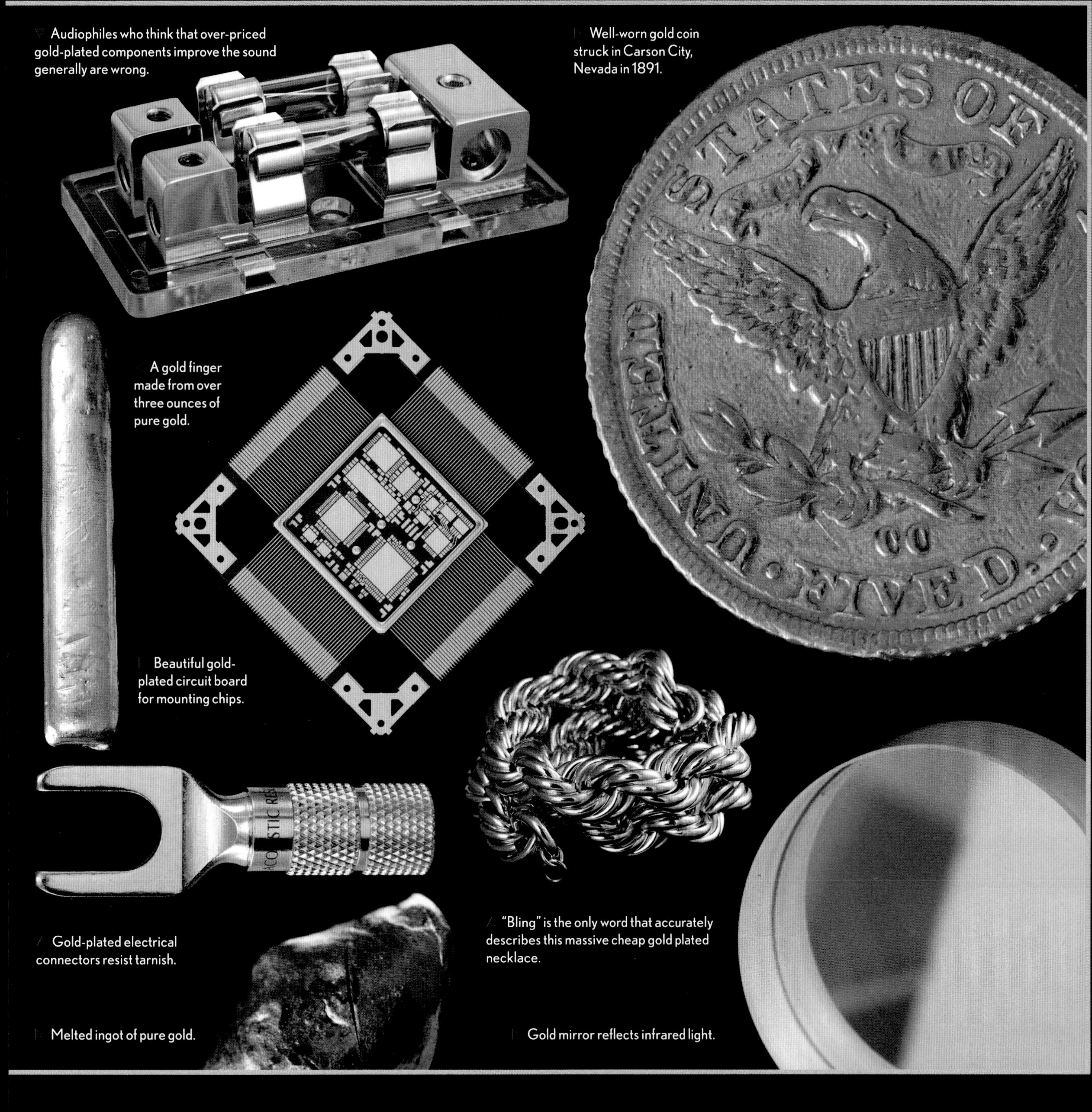

Audiophiles who think that over-priced gold-plated components improve the sound generally are wrong.

Well-worn gold coin struck in Carson City, Nevada in 1891.

A gold finger made from over three ounces of pure gold.

Beautiful gold-plated circuit board for mounting chips.

Gold-plated electrical connectors resist tarnish.

"Bling" is the only word that accurately describes this massive cheap gold plated necklace.

Melted ingot of pure gold.

Gold mirror reflects infrared light.

Mercury

Hg

80

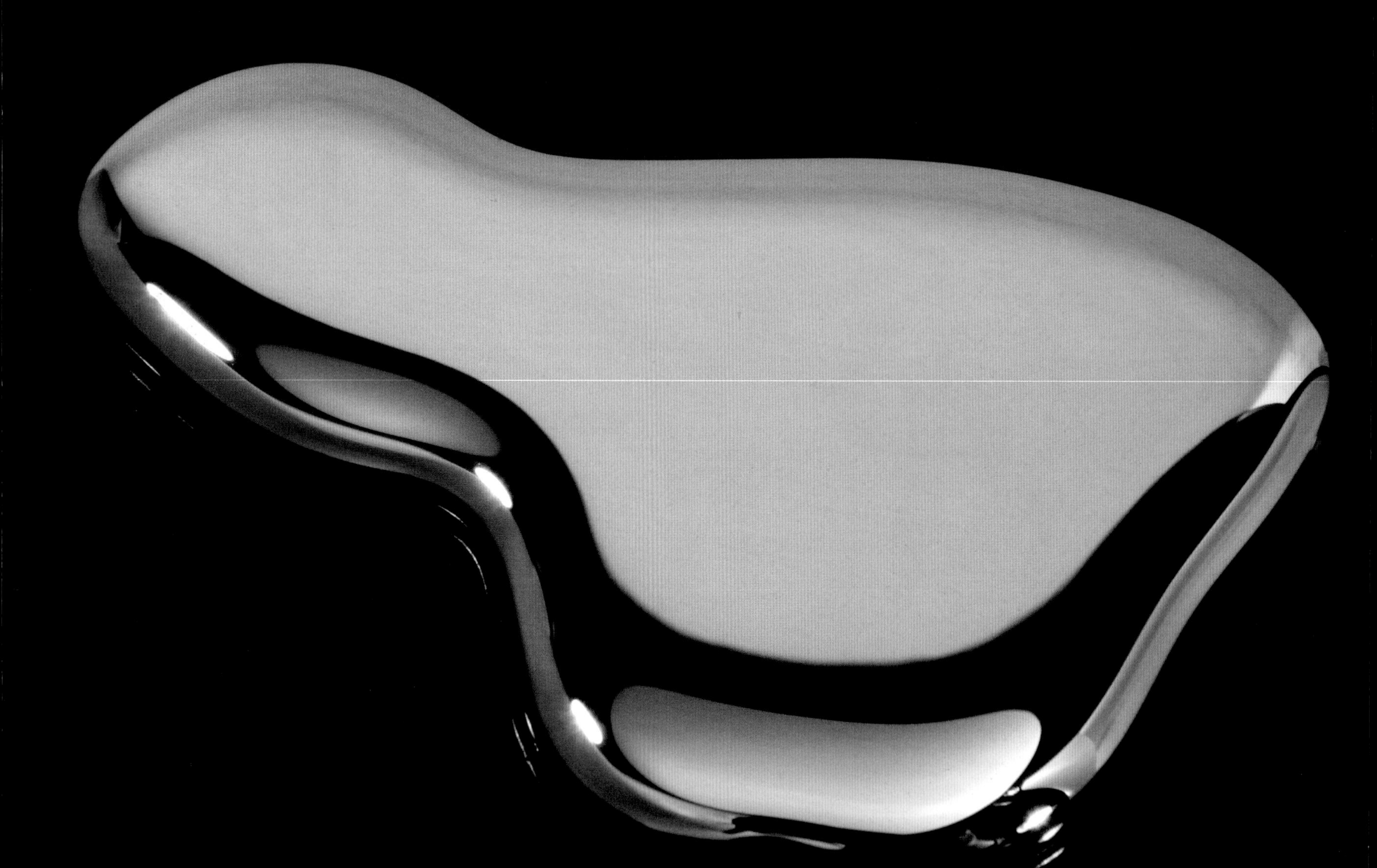

Mercury

Mercury thermostat switch. When the mercury rises to the level of the second contact wire, the circuit is closed.

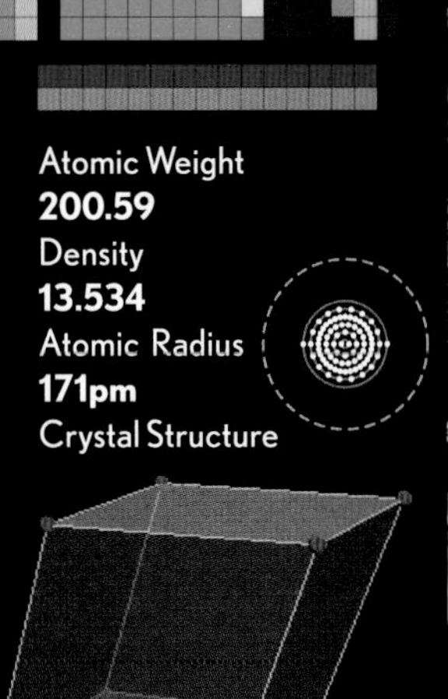

Atomic Weight **200.59**
Density **13.534**
Atomic Radius **171pm**
Crystal Structure

LIQUID MERCURY literally drips from the walls of caves in the ancient mines at Almadén, Spain. How magical a liquid metal must have seemed, when no framework existed to understand it or place it in context!

Oh piffle, mercury is every bit as magical today, no matter how much you know about it. And the more you have, the more magical it gets. I've got enough to fill a salad bowl, which lets me float a small cannonball or (wearing rubber gloves) feel the incredible pressure on my fingers plunged a few inches down into it. Even lead (82) will float on mercury—it's incredibly dense, the first thing you notice when you pick up a bottle of it. People who have a lot more, such as the miners at Almadén, can float themselves on it: A person trying to take a bath in a pool of mercury sinks in just a few inches, practically floating on the surface.

But is a liquid metal really such a surprising thing? After all, if you get any metal hot enough, it turns liquid. That's why you can cast lead or iron (26) in molds. Mercury is actually a perfectly ordinary metal, just one that happens to be shifted into a different temperature range. Sure enough, if you cool mercury in liquid nitrogen, it turns into a tough, malleable metal quite similar to tin.

The tragic thing about mercury is how toxic it turned out to be. For thousands of years it was treated as a marvelous thing to play with, to experiment with, to use for whatever it seemed useful for. But all that time it was insidiously, slowly, and invisibly poisoning everyone who came in contact with it, damaging the central nervous system and eventually leading to madness. Mercury is the worst kind of poison—the kind you don't notice until years after the damage has been done. No wonder it took literally centuries to put the pieces together.

We now know that mercury, particularly in the form of organic compounds such as methyl mercury, gets into the food chain and stays there, collecting and concentrating in larger and larger animals until it reaches tuna fish.

The delay between exposure and symptoms kept us from noticing mercury's toxicity for hundreds of years. Thallium toxicity went unnoticed for quite a while, also, despite its much faster action.

A pool of mercury carefully lit and lovingly photographed by the author.

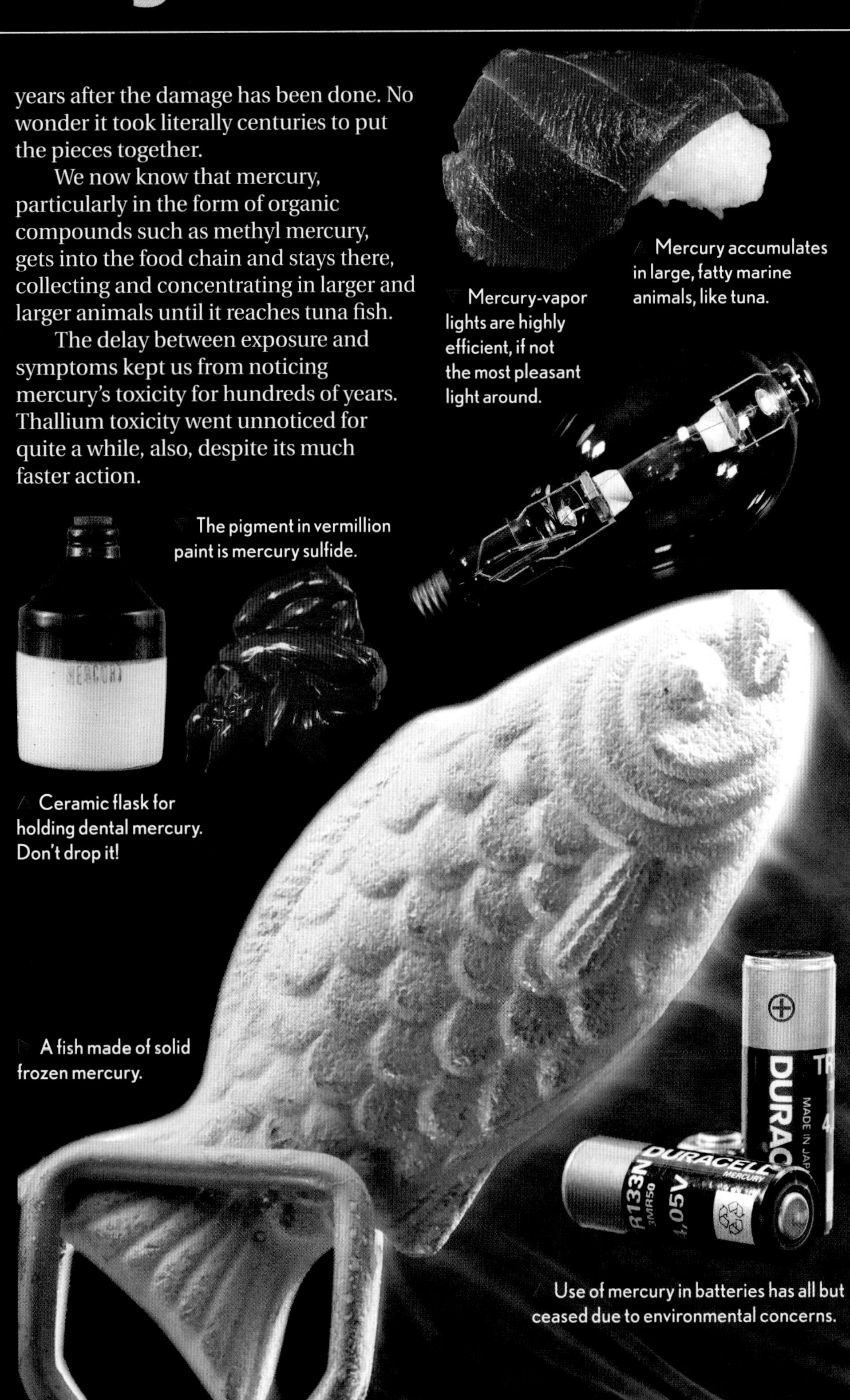

Mercury accumulates in large, fatty marine animals, like tuna.

Mercury-vapor lights are highly efficient, if not the most pleasant light around.

The pigment in vermillion paint is mercury sulfide.

Ceramic flask for holding dental mercury. Don't drop it!

A fish made of solid frozen mercury.

Use of mercury in batteries has all but ceased due to environmental concerns.

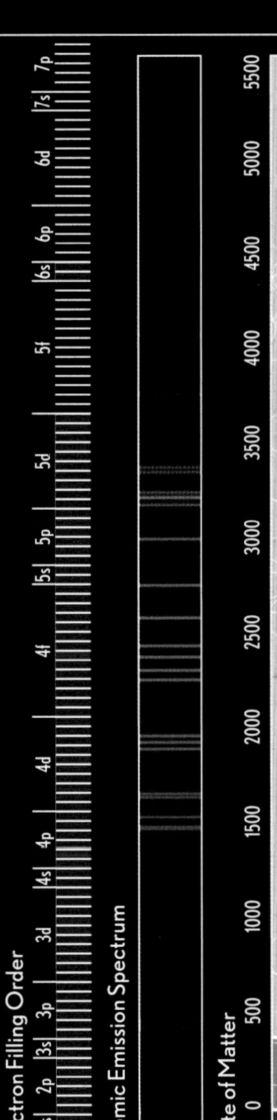

Thallium

Tl

Thallium

THALLIUM IS the first acutely toxic element since arsenic (33). Sure, selenium (34), cadmium (48), mercury (80), and a few others are not good for you, but they won't flat out knock you dead. In other words, none of them is a good murder weapon, not like thallium.

The trick to poisoning someone and getting away with it is to find a new kind of poison whose symptoms no one recognizes, and that no one knows how to detect. If you're lucky, people may not even realize that a murder has been committed. (Admittedly this worked a lot better a hundred years ago, when dying for no explainable reason was pretty common.)

Arsenic became a victim of its own success as a murder weapon. It was so wide[illegible] as "succession powder" that the symptoms became commonly known. A sensitive chemical test developed in 1836 was the beginning of the end of arsenic as a stealth poison.

Thallium, on the other hand, stayed obscure much longer. The most famous thallium murders occurred in the 1950s, but thallium-poisoning cases, both intentional and accidental, occasionally confuse police even today. Tests are of course available to prove the presence of thallium in a victim, but police would have to suspect it before they'd think to test for it, and in many cases it has taken months or years before investigators were able to put all of the pieces together.

If you'd like to check whether you are the victim of a thallium poisoning, the symptoms include vomiting, hair loss, delirium, blindness, and abdominal pain—each of which, you will notice, is also a symptom of a hundred other conditions.

The signs of murder by lead are generally much more easily recognizable.

A large piece of thallium metal, kept in a safe due to its potential for poisoning hundreds.

The mineral weissbergite, TlSbS2.

Thallium brand perfume that one can only hope does not contain actual thallium.

The health claims for Himalayan sea salt are undermined to some degree by the marketing claims that it contains thallium. While it is more likely that there is measurable thallium in an unrefined product like sea salt than in manufactured Thallium perfume, the amount is probably too small to be significant. Which does make you wonder why a company would list an acute toxin in their ingredients when doing so is not required.

Atomic Weight
204.3833
Density
11.850
Atomic Radius
156pm
Crystal Structure

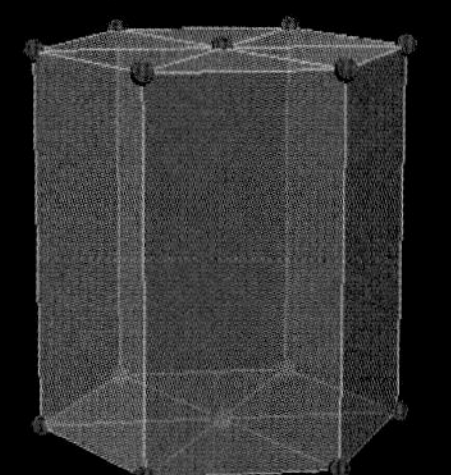

Electron Filling Order
1s 2s 2p 3s 3p 4s 3d 4p 5s 4d 5p 6s 4f 5d 6p 7s 5f 6d 7p

Atomic Emission Spectrum

State of Matter
0 500 1000 1500 2000 2500 3000 3500 4000 4500 5000 5500

Lead
Pb
82

Lead

AS FEW AS TWO grams of lead is a lethal dose, when it's delivered from the barrel of a gun.

Lead is the preferred metal for bullets because it's quite dense, allowing a lot of mass to fit into a small space, reducing air resistance. It's also soft enough to fit tightly in a gun barrel without scratching the barrel or getting stuck. People often think of lead as extremely dense, but it's actually only half as dense as osmium (76) or iridium (77). Those metals are too expensive even for the United States military to use in bullets, but tungsten (74) and depleted uranium (92) are 75 percent more dense than lead, and cheap enough that they are used in special armor-piercing rounds (discussed under uranium).

Another time-honored method of lead-based murder was popularized by the board game Clue: a lead pipe. This may sound exotic to modern ears—we now use cast iron (26), copper (29), and plastic for household pipes—but for well over 2,000 years, lead pipes were the norm.

Drainpipes made of lead have been in service in the city of Rome since Roman times. I don't mean they have been using that *type* of pipe for 2,000 years, I mean the same actual pipes have been there for 2,000 years—they last virtually forever. Lead is ideal as pipe material because it is so soft it can be beaten into sheets and welded into tubes by hammering it together. Leaks can be patched with a bit of hammering, or by means of some molten lead dripped onto the offending section. Its melting point is low enough that lead can be liquefied easily over a wood fire, making it popular also for pouring onto enemies from castle fortifications.

Considering how much we've discussed poison over the last few elements, it should come as no surprise that lead, too, is toxic. Indeed, it is the prototypical heavy metal poison, and like mercury (80) it has been responsible for some of the worst environment contamination of the modern age. Thank goodness it is no longer added as a performance booster to gasoline!

What *is* surprising is that after the three worst heavy-metal poisons in a row, we arrive at bismuth, which people drink in large amounts to soothe their upset stomachs.

This exotic six-way union was hammered out of lead sheet by an apprentice pipefitter, and it duly impressed the master.

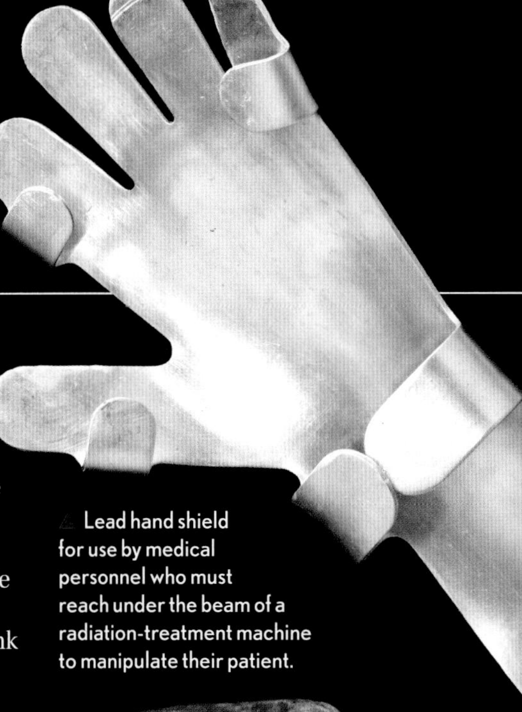

Lead hand shield for use by medical personnel who must reach under the beam of a radiation-treatment machine to manipulate their patient.

Antique lead smoking pipe.

Lead bullets have been used since before the invention of guns—above are Civil War–era rifle and musket bullets, and at left is a Roman sling bullet.

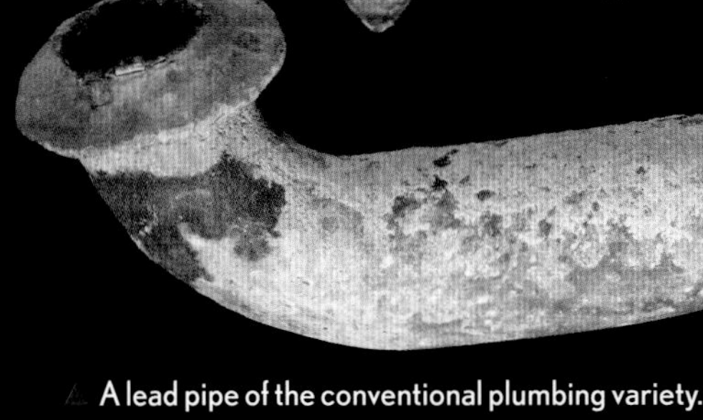

A lead pipe of the conventional plumbing variety.

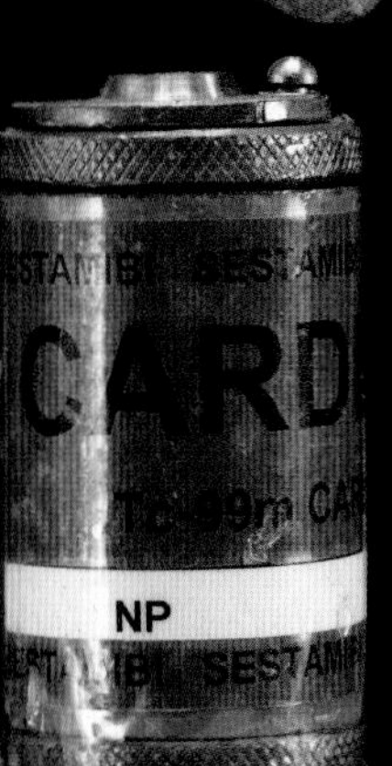

A lead "pig" provides shielding for radioactive drugs.

Leaded glass typically contains between 20 and 30 percent lead, yet it is completely transparent.

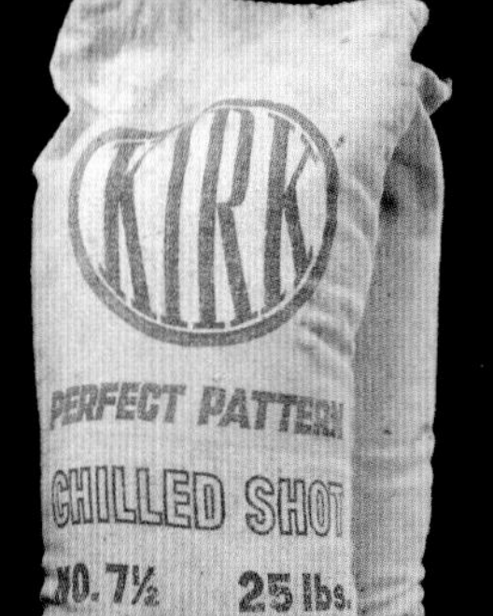

Lead shot is falling out of favor for environmental reasons.

Elemental

Atomic Weight
207.2
Density
11.340
Atomic Radius
154pm
Crystal Structure

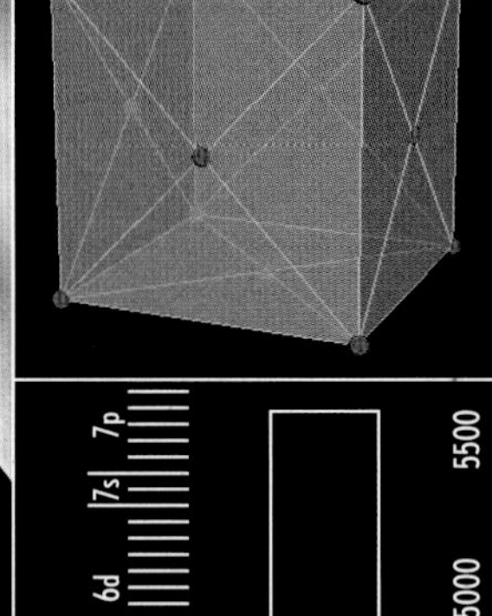

Electron Filling Order

Atomic Emission Spectrum

State of Matter

Lead 82

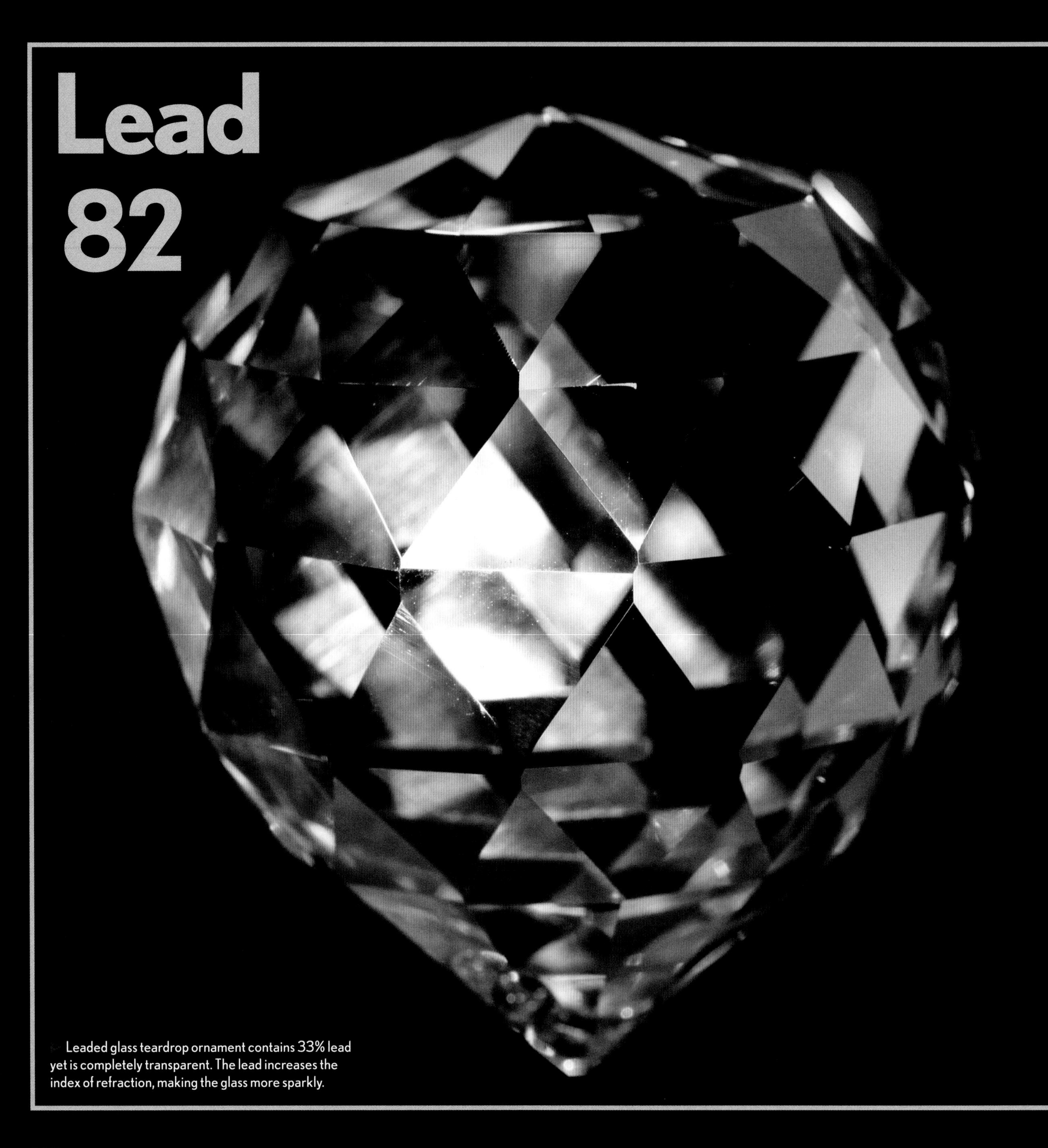

Leaded glass teardrop ornament contains 33% lead yet is completely transparent. The lead increases the index of refraction, making the glass more sparkly.

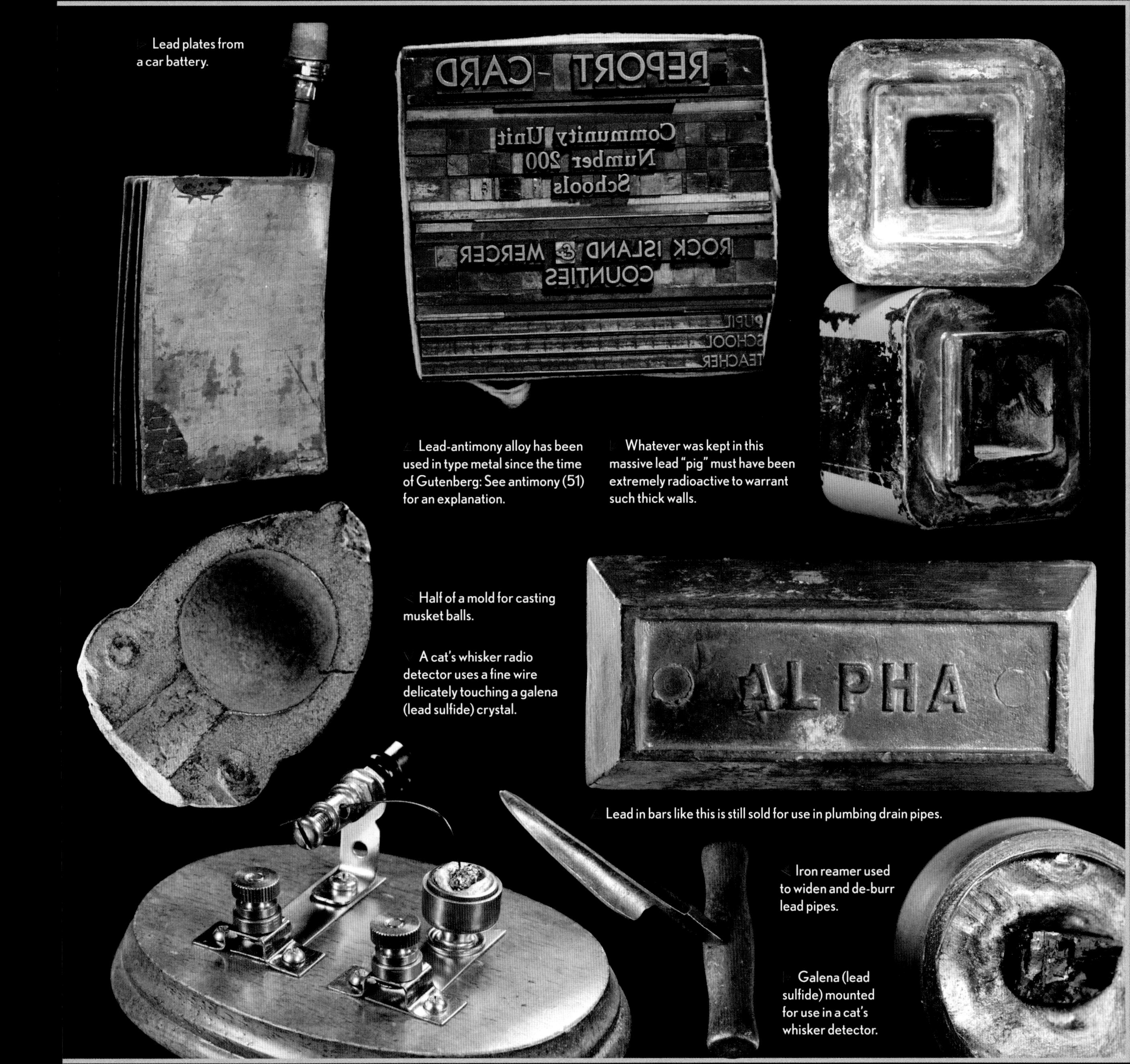

Lead plates from a car battery.

Lead-antimony alloy has been used in type metal since the time of Gutenberg: See antimony (51) for an explanation.

Whatever was kept in this massive lead "pig" must have been extremely radioactive to warrant such thick walls.

Half of a mold for casting musket balls.

A cat's whisker radio detector uses a fine wire delicately touching a galena (lead sulfide) crystal.

Lead in bars like this is still sold for use in plumbing drain pipes.

Iron reamer used to widen and de-burr lead pipes.

Galena (lead sulfide) mounted for use in a cat's whisker detector.

Bismuth

Bi 83

Bismuth

THE ACTIVE INGREDIENT in Pepto-Bismol brand upset-stomach medicine is 57 percent bismuth by weight. This is really [illegible]uite odd when you consider that the [illegible]t to the left of bismuth is lead (82[illegible]c that entire toy industries have been turned on their heads trying to eliminate it, and the element to the right is polonium (84), a deadly radioactive poison used in recent times by Russian bad guys to eliminate inconvenient people.

Despite the fact that bismuth sits smack in the middle of the toxic heavy metals, so far as we know the metal form is completely nontoxic. (If you consume enough soluble bismuth salts, there are some side effects, such as your gums turning black, but this is very rare.)

Bismuth is known as the very last stable element: No element above 83 has even a single stable isotope. But bismuth is only *culturally* stable. By which I mean that everyone thinks of it as stable, and for all practical purposes it might as well be stable, but strictly speaking there are no stable isotopes of bismuth either. On the basis of theoretical calculations, people thought for years that the "stable" isotope, ^{209}Bi, should be unstable, but it wasn't until 2003 that its half-life was finally measured and found to be 1.9×10^{19} years. (To put this in perspective, 19,000,000,000,000,000,000 years is about a billion times longer than the age of the universe. The stuff isn't going anywhere anytime soon.)

It is with some regret that we leave the realm of the stable elements. From here on out, the elements are touchy to have around and highly regulated, for health and national security reasons. But that doesn't mean you can't buy many of them. You can find at least one in the grocery store.

Our brave new radioactive path begins with polonium, a real doozy of a radioactive element.

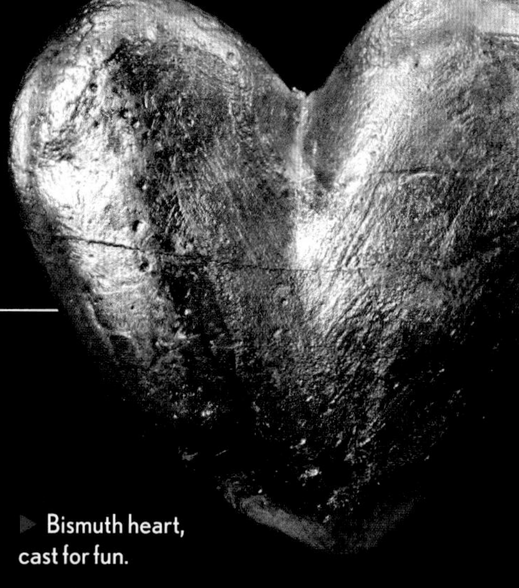

Bismuth heart, cast for fun.

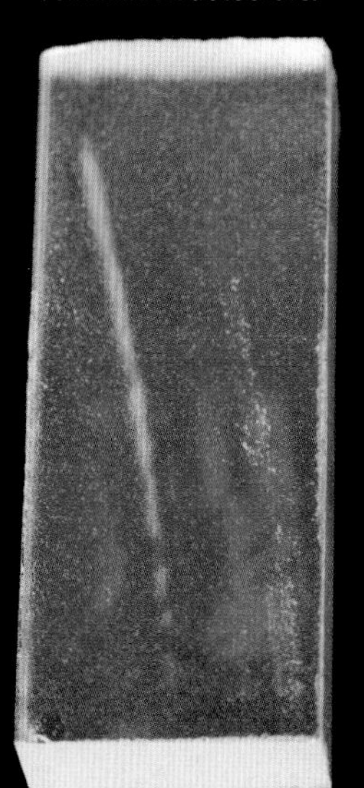

Bismuth germanate, Bi4Ge3O12, used in scintillation detectors.

Pepto-Bismol did not get its name by accident: The active ingredient is bismuth subsalicylate.

The author's multimetallic chain contains one link cast from 99.99 percent pure bismuth.

Thi[illegible]pound ingots of pur[illegible]uth are how the metal is o[illegible]d commercially. Broken in half[illegible]y show beautiful internal crystals.

[illegible]uth spontaneously forms large "hopper" crystals on cooling. When very pure bismuth is cooled very slowly, these can grow to huge sizes. This one is more than four inches tall.

Elemental

Atomic Weight
208.98038
Density
9.780
Atomic Radius
143pm
Crystal Structure

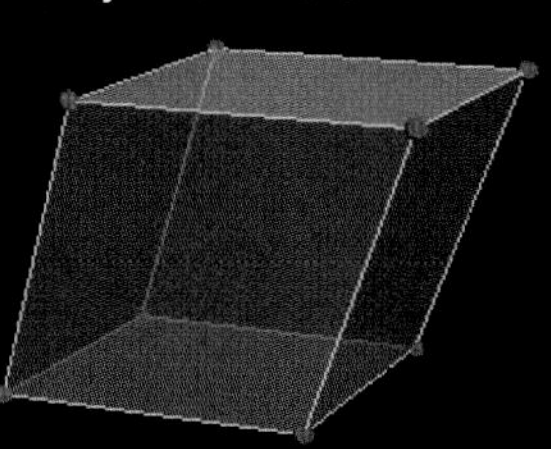

Electron Filling Order: 1s 2s 2p 3s 3p 4s 3d 4p 5s 4d 5p 6s 4f 5d 6p 7s 5f 6d 7p

Atomic Emission Spectrum

State of Matter: 0 500 1000 1500 2000 2500 3000 3500 4000 4500 5000 5500

Polonium Po 84

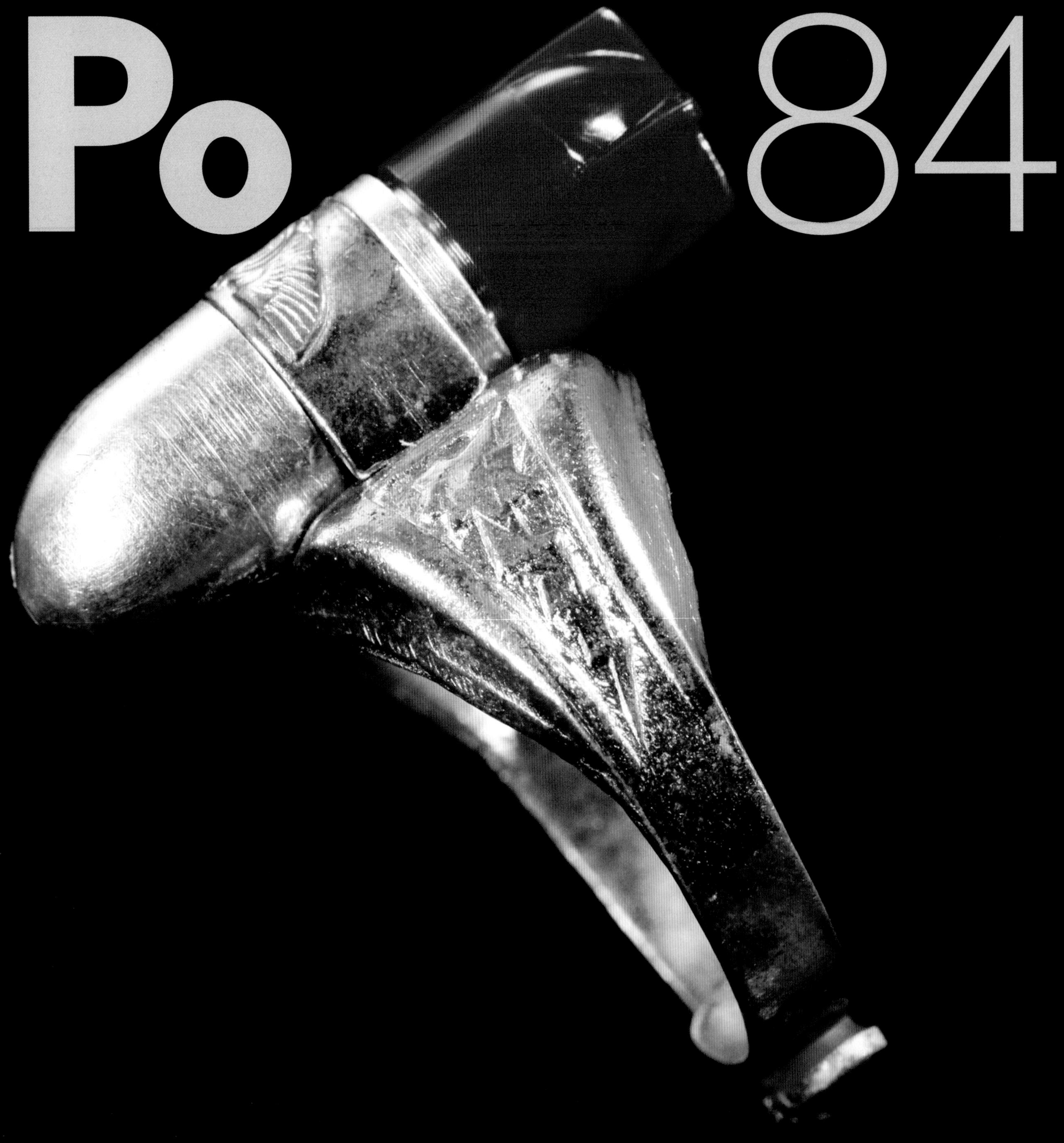

Polonium

▷ A commemorative coin celebrates Madam Curie's discovery of polonium and radium. If made of either of these elements instead of silver, it would kill everyone in the room.

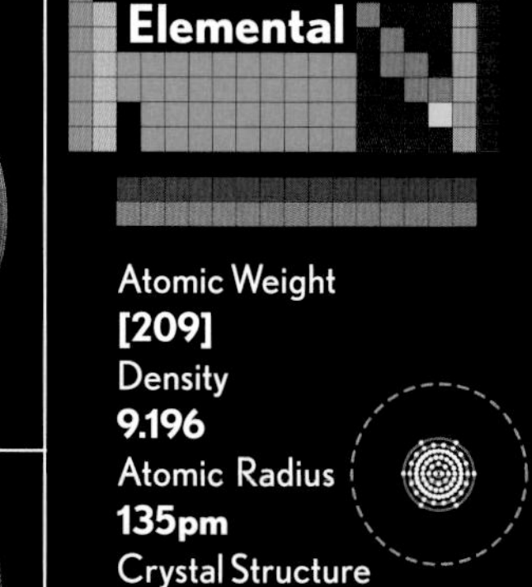

Atomic Weight
[209]
Density
9.196
Atomic Radius
135pm
Crystal Structure

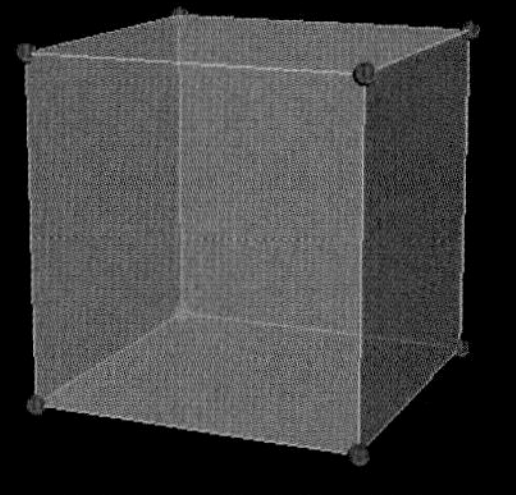

POLONIUM, DISCOVERED by Marie and Pierre Curie and named for their nat[illegible]land, occurs naturally in ura[illegible](92) ores, but these days it is created fresh for its most common application[illegible]antistatic brushes.

These brushes are used on phonograph records and film negatives to dissipate static charges that attract dust. Mounted just behind the bristles is a gold-colored strip of foil containing the polonium[illegible]ates a field of ionized air that conducts the charge away. The strip is mad[illegible] silver (47) with a thin plating of gold (79) over it. Between the silver and gold is a very thin layer of polonium.

Interestingly, these strips are not created by encasing polonium between silver and gold. Instead the polonium is *created in place* after the foil has been fully assembled. Silver foil is plated first with bismut[illegible]83), then with gold. The foil [illegible]hen run under an intense neutron bea[illegible]hat transmutes some of the bismuth into polonium. This is extremely clever: Polonium never exists in the open, which is a good thing since as little as ten nanograms (ten one-billionths of a gram) can be fatal.

It's also one reason people were immediately suspicious when the Russian ex-KGB agent Alexander Litvinenko died of polonium poisoning in London in 2006. H[illegible]ad been given so much poloniu[illegible]out ten micrograms (ten one-millio[illegible] a gram), that the only plausible source was a government with a nuclear weapons industry.

These things always come out in the end. No doubt fifty years from now all the details will be known, and far be it from this humble book of the elements to claim to know who really killed Litvinenko. But the fact that the Russian government effectively controls the world's supply of polonium and wanted Litvinenko dead does not look good.

The most common isotope, ^{210}Po, is so radioactive that a solid chunk of it glows from the excitation of the surrounding air: A single gram puts out about 140 watts of power continuously. But that's nothing compared to astatine.

▷ Polonium spark plugs were a gimmick and have completely lost their radioactivity by now.

◁ Spinthariscopes from the 1940s through 1960s often contained polonium sources.

▷ Polonium is still widely used today in antistatic brushes, but the half-life is only 138 days, so old ones like this are useless.

◁ This Lone Ranger Atomic Bomb Spinthariscope Ring was of[illegible] as a Kix cereal premium in 1947. Fifteen cents back th[illegible]day these sell for more than $100 as proof of how different attitudes toward radiation and the bomb once were.

△ The foil inside an antistatic brush contains polonium sealed under a thin gold layer over a silver base.

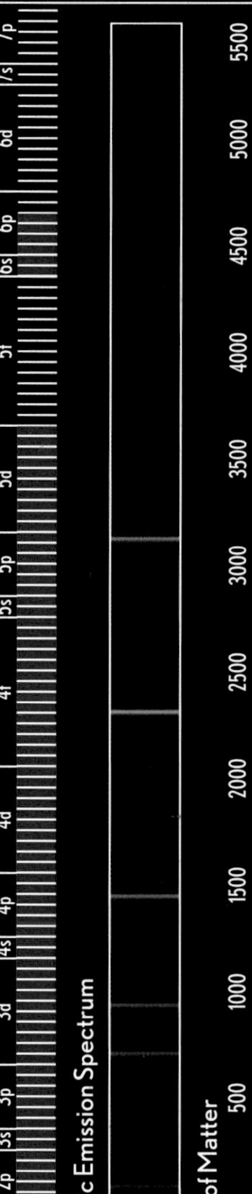

Astatine

At

Astatine

ASTATINE IS THE FIRST of the four really frustrating elements for element collectors. The others are francium (87), actinium (89), and protactinium (91). Radon (86) is slightly annoying too, but not quite as much.

Astatine is considered naturally occurring, as are all the elements from hydrogen (1) to uranium (92), except technetium (43). But its half-life is only 8.3 hours, which means that whenever astatine occurs naturally it doesn't stick around very long. A rough estimate indicates that at any given time there's about an ounce of it present in the entire earth, and it's never the same ounce from day to day: The supply is constantly refreshed by the slow radioactive decay of uranium and thorium (90), of which there is vastly more.

The typical solution for element collectors is to display a radioactive mineral specimen containing uranium or thorium and wave their hands around talking about how there's probably an atom or two of astatine in it. Maybe, but it's much more likely that there isn't. In all of the North American crustal plate to a depth of ten miles, there are roughly a trillion naturally occurring astatine atoms at any one time. What do you think are the chances that you've got one of them in that little rock of yours?

Despite its short half-life, astatine is being studied for use in radiation therapy for cancer. This seems less surprising when you consider that $^{99}Tc_m$, discussed under technetium, is widely used in medicine and has just as short a half-life. The trick is to develop a compact device that hospitals can keep on-site to generate the substance on demand.

Although its half-life is only a few times longer than that of astatine, radon is far more abundant, to the point of being a household name in many parts of the world.

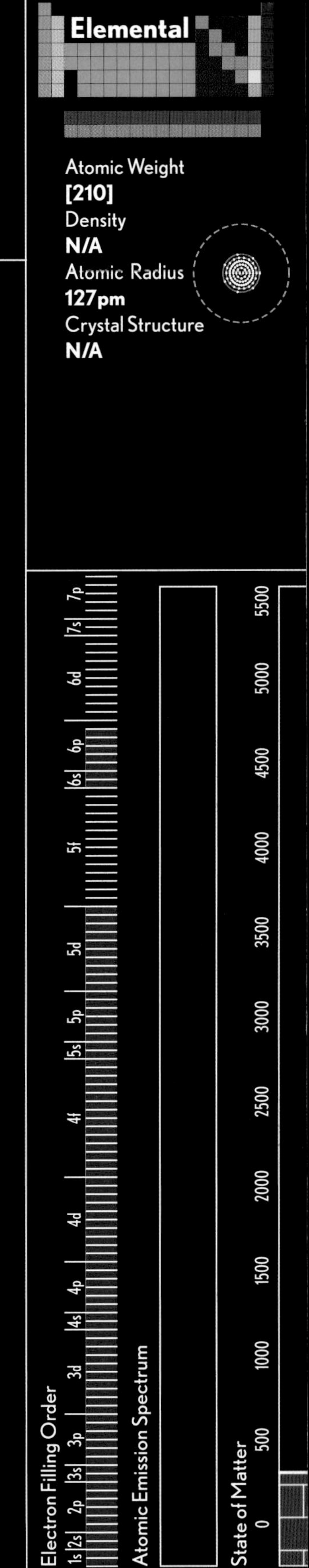

◁ This beautiful fluorescent uranium mineral, autunite, $Ca(UO_2)2(PO_4)_2 \cdot 10H_2O$, may or may not contain an atom of astatine at any one time.

Radon

Rn 86

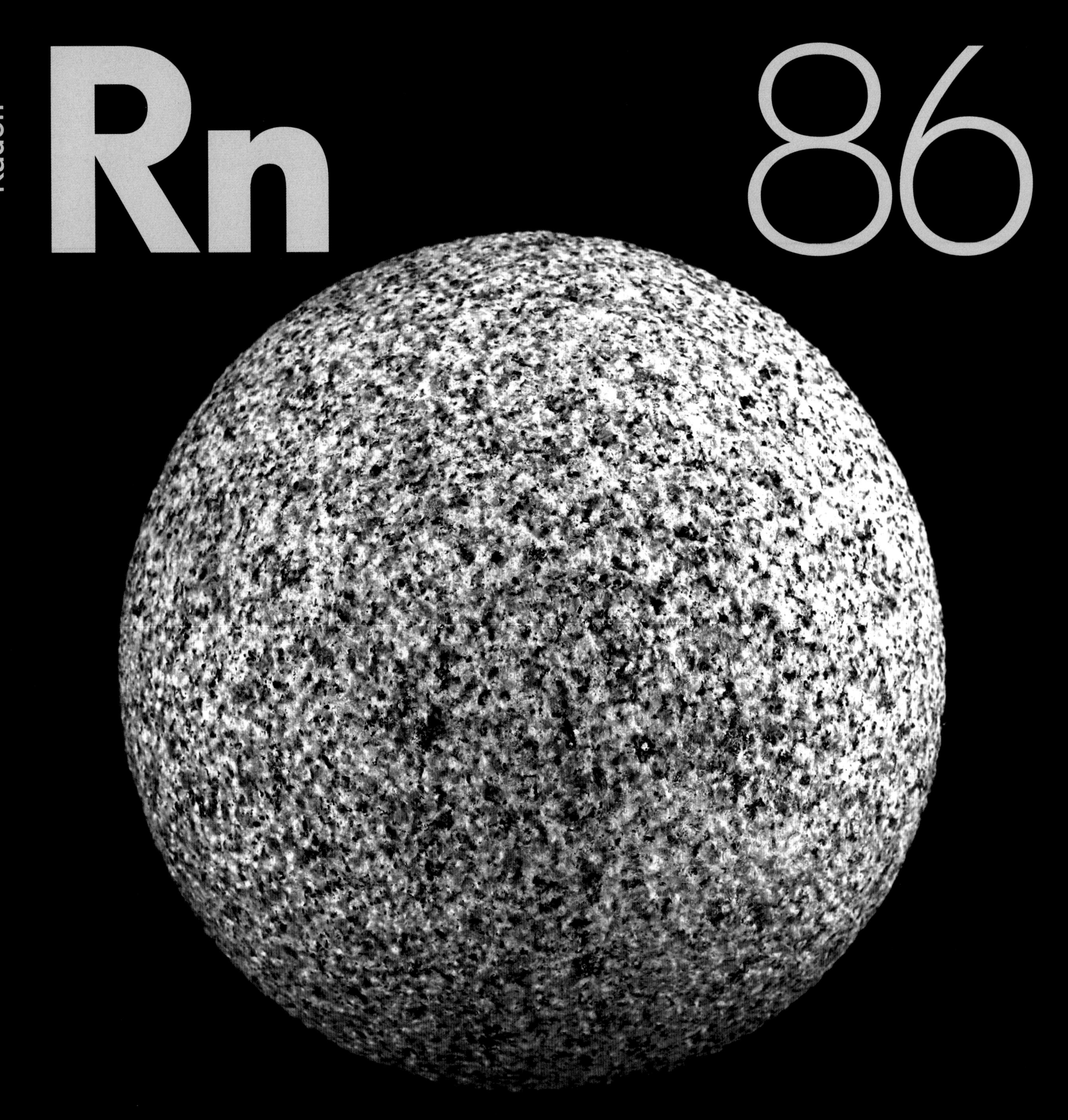

Radon

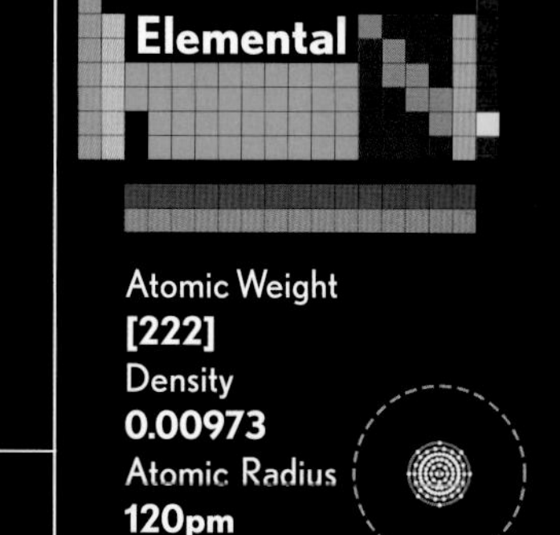

Atomic Weight
[222]
Density
0.00973
Atomic Radius
120pm
Crystal Structure
N/A

RADON IS A HEAVY radioactive gas with a half-life of only 3.2 days, but there is quite a lot of it around because it is a princi[illegible]ment in the decay chains of both [illegible]um (92) and thorium (90), both of which exist in large quantities, particularly in granite bedrock. (Granite buildings emit significant amounts of r[illegible]tion—Grand Central Station in N[illegible]York is famously radioactive for this reason.)

Radon seeping up from the ground and collecting in the basements of buildings is of concern to many people. A whole industry exists to detect and mitigate radon. (Your friendly neighborhood radon-abatement service will install expensive underground air channels and fans to draw the radon out from underneath your house before it gets inside.)

Ironically, while some people are paying big bucks to get rid of radon, oth[illegible] gather in cave spas near uranium de[illegible]its to breathe the rich, radon-laden air, believing it to be healthy. This belief, more popular a hundred years ago than today, originated with the discovery that many hot springs are quite radioactive (the water in them is hot because it flows near hot rocks, which are hot because of uranium and thorium decay deep in the earth).

A hundred years ago, when radioactivity was first being investigated, no one had any reason to suspect it was dangerous. Everyone knew hot springs were healthy, but no one could really say why. When it was discovered that many famous hot springs are radioactive, the answer seemed obvious: It must be this great new radiation thing!

The result was a decades-long health fad for all things radioactive, which ended only after the spectacular death of a famous proponent, which you can read about under thorium.

If people back then had known about francium, I'm sure someone would have been selling francium foot warmers.

For people who are afraid they may have high levels of radon in their basements, inexpensive mail-in radon test kits provide an answer in a few days.

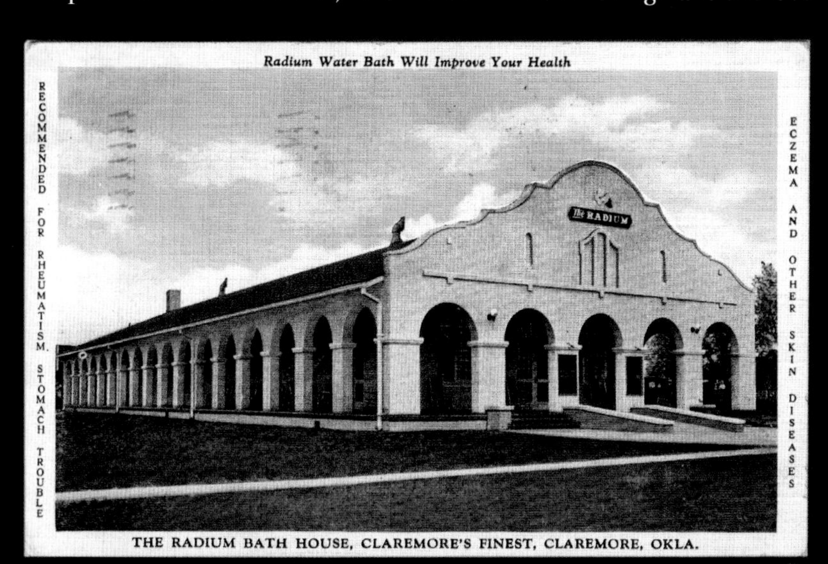

For those who felt they were not getting enough radon in their lives, this radium bathhouse offered soaks in radon-laced hot-spring water (actual radium is too expensive for this application, and water from many hot springs is radioactive due to the radon gas emitted by uranium and thorium decay deep in the earth).

This granite ball represents the major source of radon: uranium and thorium in bedrock.

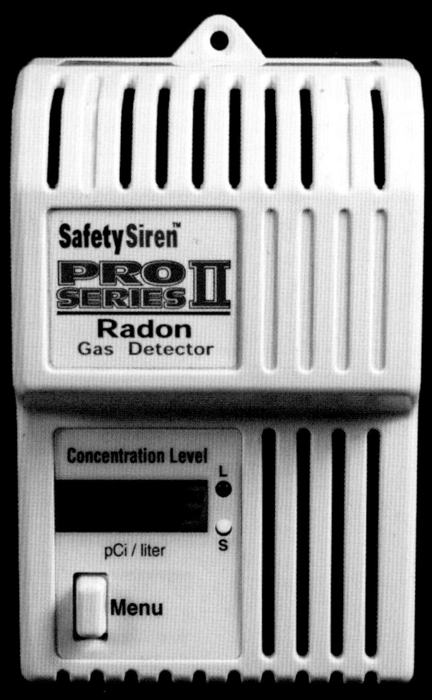

For those truly worried they are getting too much radon in their lives, continuous electronic monitors are available.

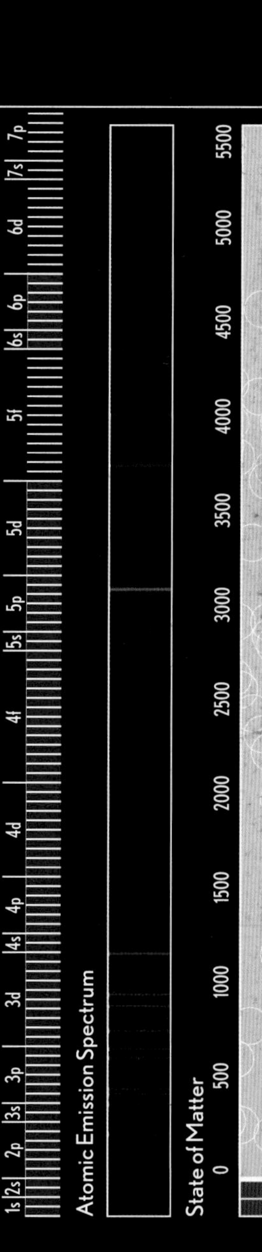

Francium

Fr 87

Francium

FRANCIUM IS THE least-stable naturally occurring element (with a half-life of 22 minutes) and the last element discovered in nature (1939 in, you guessed it, France).

You may recall I said something similar about rhenium (75), but that was the last *stable* element discovered, while this is the last *naturally occurring* element, including unstable ones. The last nonnaturally occurring element discovered at the time of this writing is element 115, which has not yet been given a name. (No doubt more will be discovered over time: There is no absolute upper limit to the number of elements.) And finally, just to complete this body of trivia, astatine (85) is the last naturally occurring element to be discovered. Wait, didn't I just say that about francium? Subtle difference: Francium was discovered in nature, while astatine is naturally occurring but was first discovered when it was created artificially—no trace of it was found in nature until three years later.

Its 22-minute half-life makes francium impractically radioactive. There are no commercial applications for it, even in medicine, which uses an amazing array of other wildly radioactive isotopes.

If you ever managed to put together a lump of it, the thing would evaporate itself violently through the tremendous heat generated by its radioactivity. But if you could somehow delay that for just a few seconds—boy, could you have some fun with it!

You see, francium is the last of the alkali metals, all of which—but particularly sodium (11)—are fun to throw in lakes because they react explosively with the water. According to the systematic trends of the periodic table, francium should be the most reactive of them all. If you could throw a hundred grams of it in a lake, the result would be a truly monumental explosion.

The other result would be a truly monumental radioactive mess, not unlike those once created by the radium industry.

◄ This piece of the mineral thorite, (Th,U)SiO, might contain an atom of francium, if you watch closely.

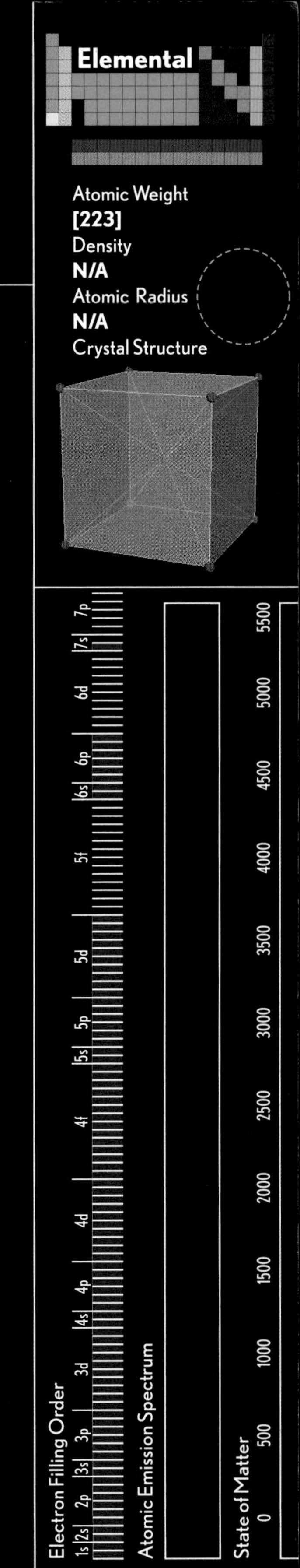

Ra 88

Radium

Radium

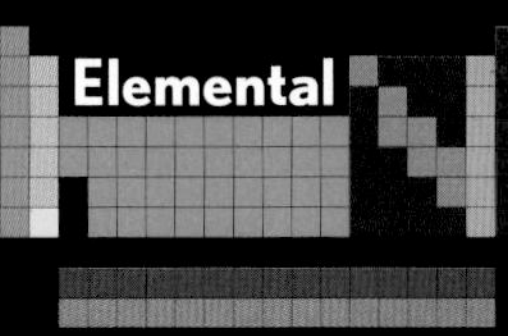

Atomic Weight
[226]
Density
5.0
Atomic Radius
215pm
Crystal Structure

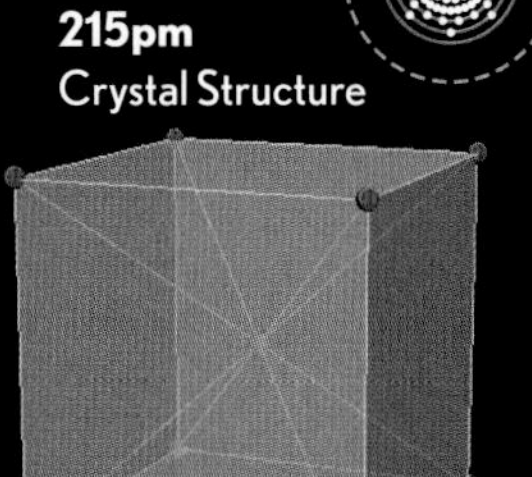

RADIUM WAS THE titanium (22) of the early 1900s. It was the brilliant, shiny, powerful element everyone wanted to associate their products with, whether or not they actually contained any radium. Just as many "titanium" products today contain no actual titanium, many "radium" products of a century ago, such as radium furniture polish and radium toothpaste, contained no radium.

Other products, such as radium suppositories and the supremely scary Radiendocrinator, *did* contain radium, in some cases substantial quantities of it. (The Radiendocrinator was designed to be worn by men, with the radiation source pointed at a certain private area that contains many rapidly dividing cells. Based on the mistaken belief that high doses of radiation to the reproductive organs would promote health and virility, this was in fact an *extremely bad idea*. Today special lead shields are made to protect that particular area from even the smallest doses of radiation during x-ray procedures.)

The best-known application of radium, and the reason you can still buy it easily on eBay, was for luminous watch hands. Paint containing a combination of zinc (30) sulfide and radium glows in the dark for many years. Sadly the zinc sulfide breaks down, and most historical radium watch hands no longer glow. (The radium itself is still just as radioactive as ever; its half-life of 1,602 years guarantees that those watches will be hot for a very long time.)

Radium watches and clocks were painted by hand, using tiny brushes that the women doing the painting would sharpen to a fine point by licking them. This was not a very good idea when you consider the radioactive paint on those brushes, and it was the progressive and ultimately undeniable radium-related sickness and death of these women that finally convinced many people something had to be done about radiation safety.

The case of the "radium girls" was a landmark in labor law, establishing the right of workers to sue for damage caused by unsafe and abusive working conditions. (Intentionally withholding information about the dangers of licking radioactive paintbrushes is at the high end of that scale.) But it took still more death and dismemberment before radioactive "health" products finally fell out of favor, a story you can read about under thorium (90), after you make it past another annoyingly short-lived element, actinium.

Radium paint carefully hand-painted on watch dials led to the establishment of modern labor laws.

Radium shoe polish contained no radium.

Radium starch contained no radium.

The Radiendocrinator contained large amounts of real radium, and is thus among the most dangerous of the radium era's products.

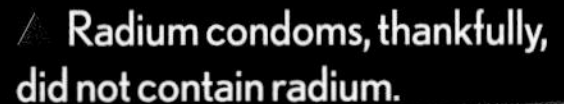

Radium condoms, thankfully, did not contain radium.

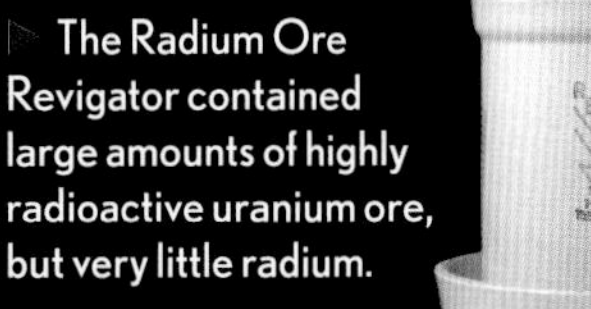

The Radium Ore Revigator contained large amounts of highly radioactive uranium ore, but very little radium.

Beautiful brass spinthariscopes, such as this one, contained radium and are thus still radioactive to this day.

Electron Filling Order
Atomic Emission Spectrum
State of Matter

Actinium

Ac 89

Actinium

ACTINIUM IS THE first of the actinide series of rare earths, the ones placed in the very bottom row of a standard periodic table arrangement. As with the lanthanide series—lanthanum (57) to lutetium (71)—all the elements in the actinide series—actinium (89) to lawrencium (103)—share chemical properties with each other, though they are more diverse than the often nearly indistinguishable lanthanides.

The biggest difference between the lanthanides and the actinides is, of course, that while all but one of the lanthanides are stable elements, every one of the actinides is radioactive—so radioactive that only three of them are tame enough for you to actually hold a substantial lump in your hand and live to tell the tale.

Actinium, with a half-life of 21.8 years, is not one of those three. It is so radioactive that it glows on its own without the benefit of a phosphorescent screen (which you need to see a glow from less radioactive elements, such as radium, element 88).

While actinium occurs naturally in uranium (92) ore, there is so little of it there that when people actually want some, they make it in a nuclear reactor by bombarding ^{226}Ra with neutrons, turning it into ^{227}Ra, which then decays with a half-life of 42 minutes into ^{227}Ac, the longest-lived isotope of actinium.

This kind of nuclear alchemy—literally, the transmutation of one element into another—is quite commonly done these days to synthesize useful elements and isotopes. The alchemists weren't wrong in trying to transmute base elements into gold, they just didn't have the right technology—a nuclear reactor—to pull it off.

While there are some experimental applications of actinium, very little of it is made or used. In contrast, the most abundant radioactive element of all is thorium.

◀ This sample of vicanite, (Ca,Ce,La,Th)15As(AsNa)FeSi6B4O40F7, from the Vica Complex in Tre Croci, Italy, probably doesn't have any actinium in it right now, but once in a while might have an atom or two.

Thorium

Th

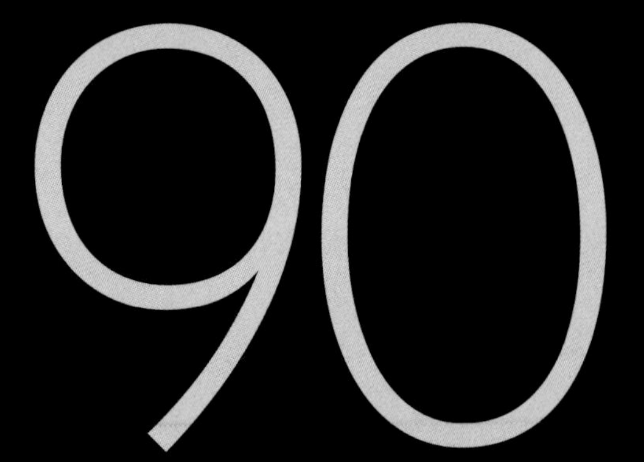

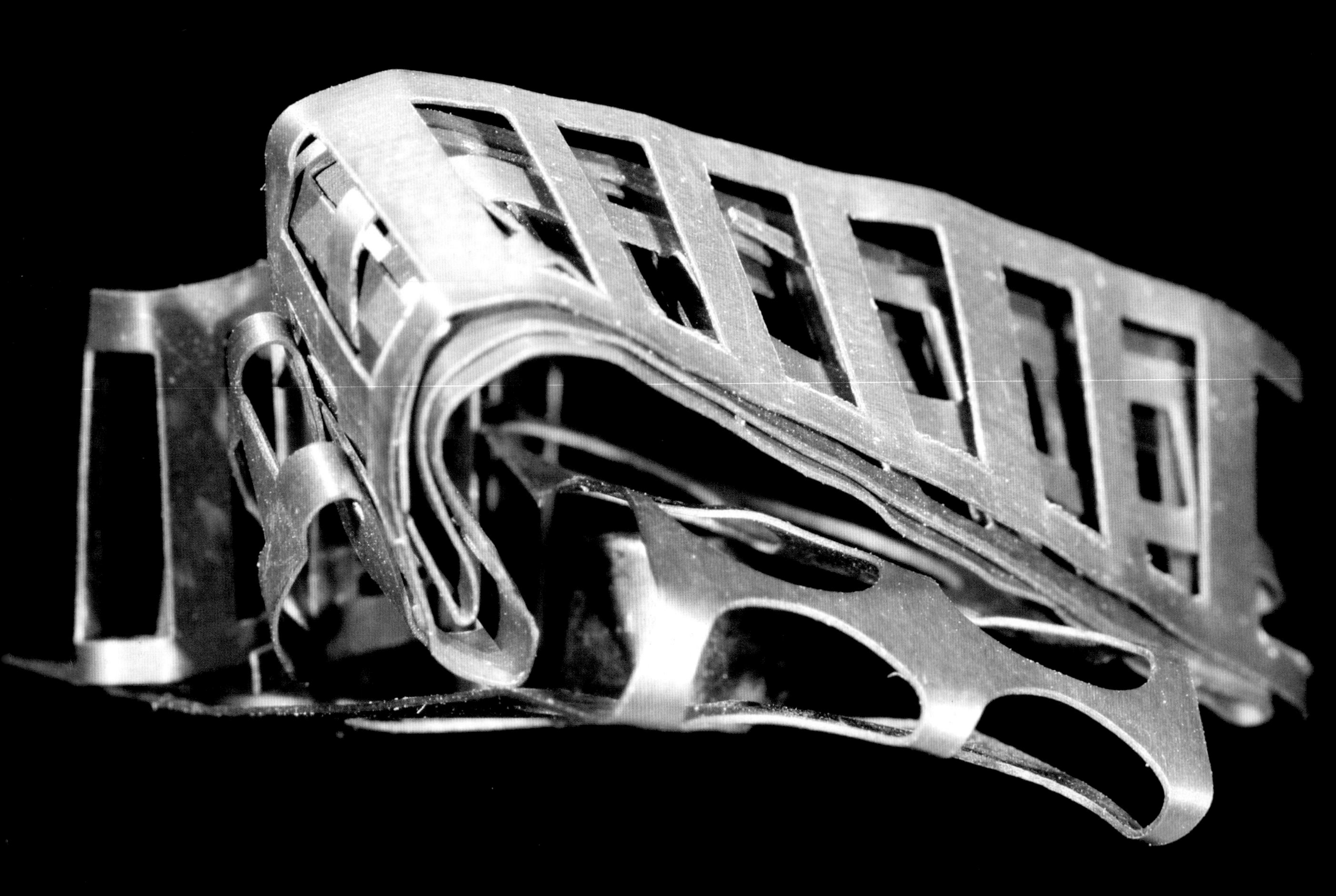

Thorium

Chips of pure thorium metal.

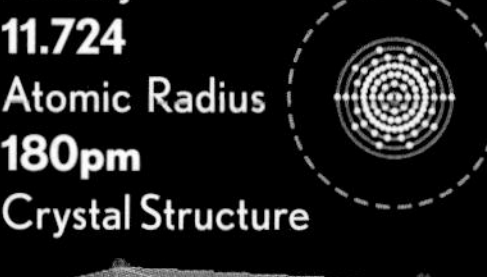

Atomic Weight
232.0381
Density
11.724
Atomic Radius
180pm
Crystal Structure

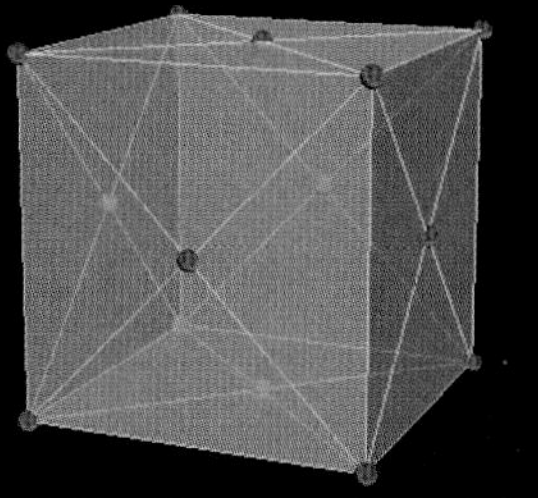

THERE IS MORE THORIUM in the earth's crust than there is tin (50)—almost three times as much. There's also more than three times as much thorium as there is uranium (92), one reason considerable work (i.e., billions of dollars) was put into research to develop a viable thorium-based nuclear power reactor. It was abandoned, but not before researchers created a huge stockpile of high-grade thorium metal lusted after by element collectors.

One result of thorium's abundance is that it was used for many years purely for its chemical properties with complete disregard for the fact that it is radioactive. Thorium oxide was used until recently in camping-lantern mantles, where it glows brightly when heated by a gas flame. Plenty of other oxides work just as well, but thorium oxide is cheap, and for a long time people didn't see a problem with thorium's relatively low level of radioactivity. Even today you can buy thoriated tungsten (74) welding rods, which contain as much as 2 percent thorium to help strike the arc.

A "health" drink called Radithor, which contained substantial amounts of radium (88) and thorium, is what finally put a stop, in 1932, to widespread use of radioactive quack medical products. Eben Byers was a wealthy playboy industrialist who drank three bottles of Radithor a day. On his death the headline in the *Wall Street Journal* read "The Radium Water Worked Fine until His Jaw Came Off." That incident contributed to strengthened FDA control of cosmetics and medical devices, but it's not the strangest story about thorium.

In the heat of the Second World War, Allied intelligence services were freaked out by the discovery that Auergesellschaft, a German military contractor, had confiscated a huge stockpile of thorium from a company in occupied Paris and taken it back to Germany. Nuclear scientists working on the Allied bomb realized that if the Germans had figured out they needed thorium, they must be pretty far along in their own nuclear bomb program. Actually the Germans had made pathetically little progress on the bomb. No, Auergesellschaft just had a secret plan to launch a brand of thoriated toothpaste after the war—they had hoped it would become as popular as radium toothpaste and wanted to be sure they had plenty of thorium on hand.

No such plans were ever made for protactinium.

Pure thorium foil from which arc-initiating buttons have been punched.

Antique lantern mantles contained thorium oxide, which luminesces beautifully when heated with gas.

A solid sheet of thorium metal is very hard to come by—while legal to own, just try to find anyone willing to sell you one.

Thoriated toothpaste is no longer made, thankfully.

The cork on this empty bottle of Radithor still reads over a thousand counts per minute on a Geiger counter.

Welding electrodes containing 2 percent thorium are in use today and widely available.

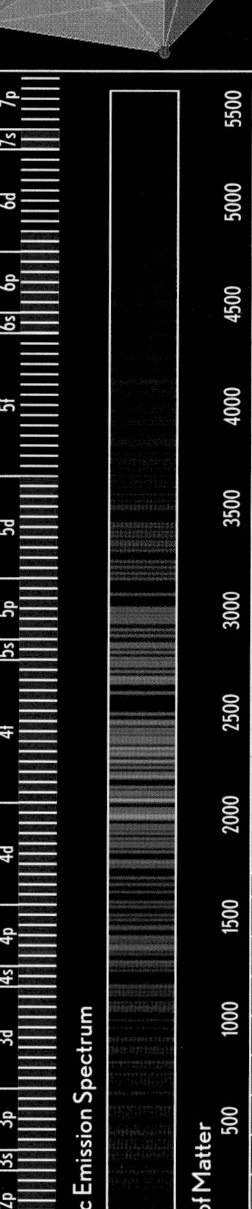

Protactinium
Pa
91

Protactinium

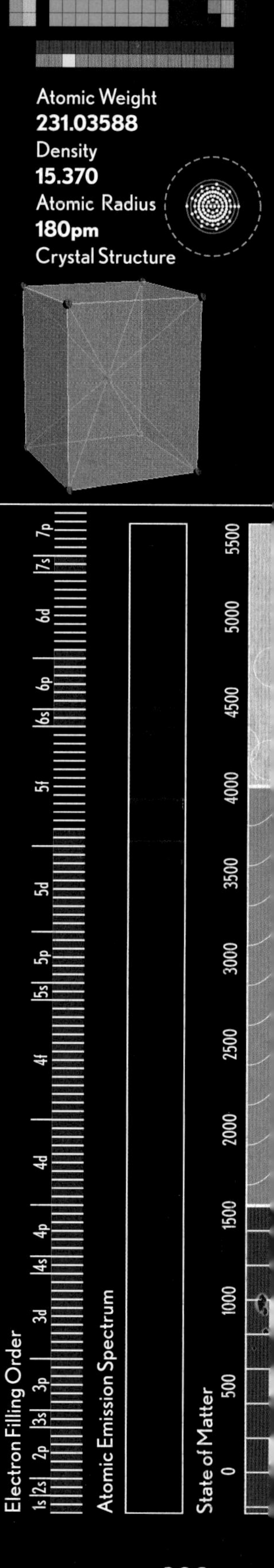

PROTACTINIUM IS THE LAST of the naturally occurring elements that seriously annoy element collectors. Unlike the others—astatine (85), francium (87), and actinium (89)—protactinium's half-life is long enough—32,788 years—that a lump large enough to see, while dangerous, would be entirely practical to show off in a nice lead-lined display case. This makes its unavailability all the more frustrating.

About 125 grams of it was put together in the 1960s and distributed to laboratories wishing to study it for potential applications. Apparently that didn't work out very well, because to this day there are none. I'm still waiting for some leftover protactinium to show up on eBay.

Protactinium in the form of the very short-lived isotope $^{234}Pa_m$ (half-life: 1.17 minutes) was discovered by Kasimir Fajans and O. H. Göhring in 1913. The much longer-lived isotope ^{231}Pa was independently discovered in 1918 by Frederick Soddy and John Cranston in Scotland and by Otto Hahn and Lise Meitner in Germany. We'll be hearing more about Hahn and Meitner under meitnerium (109), but the fact that we can talk about different isotopes at all was due to one member of the other team, Frederick Soddy.

Soddy discovered that it's possible for different atoms of the same element to have different masses, and lived to regret it.

An *element* is defined as a substance whose nuclei contain a specific number of protons (this is the atomic number you see in large type on every periodic table tile). But all nuclei (except ^{1}H) also contain a bunch of neutrons as well as protons. Each *isotope* of an element has the same number of protons, but a different numbers of neutrons. So, for example, the isotope ^{234}Pa contains 91 protons (because protactinium is element 91) and 143 neutrons, the result of 234 minus 91. On the other hand, ^{231}Pa also contains 91 protons, but only 140 neutrons.

The number of neutrons is of virtually no consequence to the chemical behavior of atoms but turns out to be critical to their nuclear stability. Nuclei without a suitable number of neutrons tend to be unstable, and eventually fly apart in what is known as radioactive decay.

When atoms disintegrate, called fission, a *huge* amount of energy is released—energy that is the basis for both nuclear power plants and nuclear bombs. Frederick Soddy realized just how much energy could be generated this way and started preaching that mankind could now look forward to a clean and beautiful future of unlimited energy. But after seeing how scientists contributed to the bloodbath of the First World War, he turned away from nuclear science and began warning people of the terrible consequences of continued research into the nucleus.

Though there could not have been any pleasure in it for him, he lived to see his worst nightmares realized when the atomic bomb known as "Little Boy" was dropped on the Japanese city of Hiroshima on August 6, 1945.

That bomb was made with uranium.

Torbernite, $Cu(UO_2)_2(PO_4)_2 \cdot 8\text{-}12H_2O$, is a lovely green uranium mineral I've chosen to represent protactinium out of sheer desperation. There is no practical way to get or photograph actual protactinium, but some atoms of it might be in this rock from time to time.

Uranium

U

92

Uranium

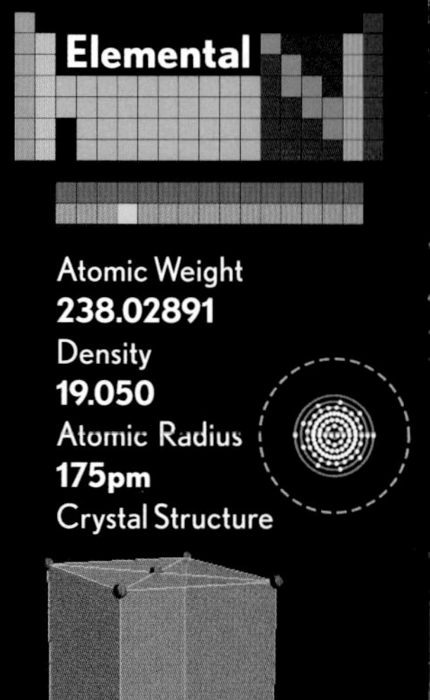

IT IS IMPOSSIBLE to discuss uranium without acknowledging that the first atomic weapon used in anger was a uranium fission bomb, built in secret deep in the deserts of New Mexico and detonated over the unsuspecting city of Hiroshima on the Honshū island of Japan. The Great Wall of China and the Apollo moon shot were large undertakings, but measured by its irreversible consequences for the whole of the planet, and by the sheer leap of faith required to attempt it, nothing yet wrought by the hand of man compares to the Manhattan Project.

The scientists who built the uranium bomb were so sure it would work, they didn't even bother testing it (also, they had only enough ^{235}U for one bomb, a fact that was one of the most closely guarded of all nuclear secrets). The Trinity test, carried out at Alamogordo twenty-one days earlier, proved the feasibility of the more complex plutonium-based design of the "Fat Man" bomb dropped on Nagasaki three days after Hiroshima.

Whether mankind will survive the invention of the atomic bomb remains an open question.

Though nuclear weapons have been used in war only twice, uranium itself has become commonplace in recent wars around the world. Naturally occurring uranium is 99.28 percent ^{238}U and 0.71 percent ^{235}U. Both isotopes are radioactive, but only the ^{235}U can be used to make fission bombs. When uranium is processed for bomb making, about two-thirds of the ^{235}U is taken out. What's left is called "depleted uranium," or DU.

DU is widely used not for its remaining radioactivity but simply because it is a very hard, very dense metal that makes excellent armor-piercing projectiles. Tungsten (74), similarly dense, can be used as well, but if you're the government of a nuclear state, you have a lot of leftover DU on hand from making bombs. And DU has the added advantage of catching fire on impact.

When not dealing death, uranium can be found on eBay and in the kitchens of antique collectors the world over. Fiestaware plates and bowls made prior to 1942, especially the orange ones, contain so much uranium in their glaze that they will set off Geiger counters from several feet away. Eating from them is a bad idea not so much because of the radioactivity, which is of the relatively harmless alpha type, but because uranium, like lead, is a heavy metal poison that can leach out of the glaze on contact with acidic foods.

Possession of up to fifteen pounds of natural uranium (or thorium, element 90) is perfectly legal for private individuals, so radioactive Fiestaware can be, and is, widely sold, collected, and used legally. True story: A colleague at my software company has a kitchen full of Fiestaware, and she eats off it every day. After borrowing my Geiger counter she now stores one particularly hot set of bowls slightly further away from the sink.

Leaving uranium, we say good-bye to the naturally occurring elements. From here on out, the elements exist on earth only at the pleasure of mankind—we have to make them in our nuclear reactors. The first of this new breed is neptunium.

◄ Pure uranium metal is perfectly legal to own (up to 15 pounds at any one time), and there are actually a few companies that sell it to element collectors. This 30-gram piece came from one such company.

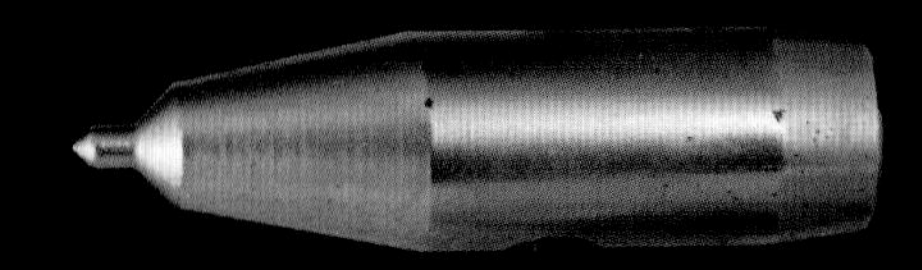

▲ A gold-colored titanium nitride coating protects the depleted uranium inside this small bullet from oxidation.

▲ An image of the Nagasaki atom bomb printed using uranium salted photo paper. The print is actually radioactive.

▲ The tiles behind the drinking fountains in the Art Deco grade school attended by the author and his children contain significant amounts of uranium in their glaze. The lighter colored ones read over 1000 counts per minute.

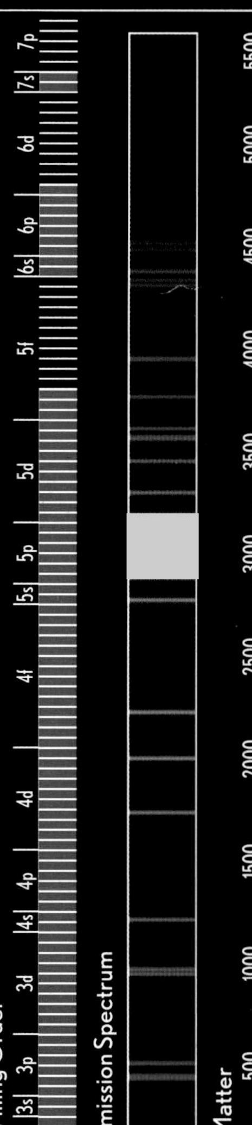

Uranium 92

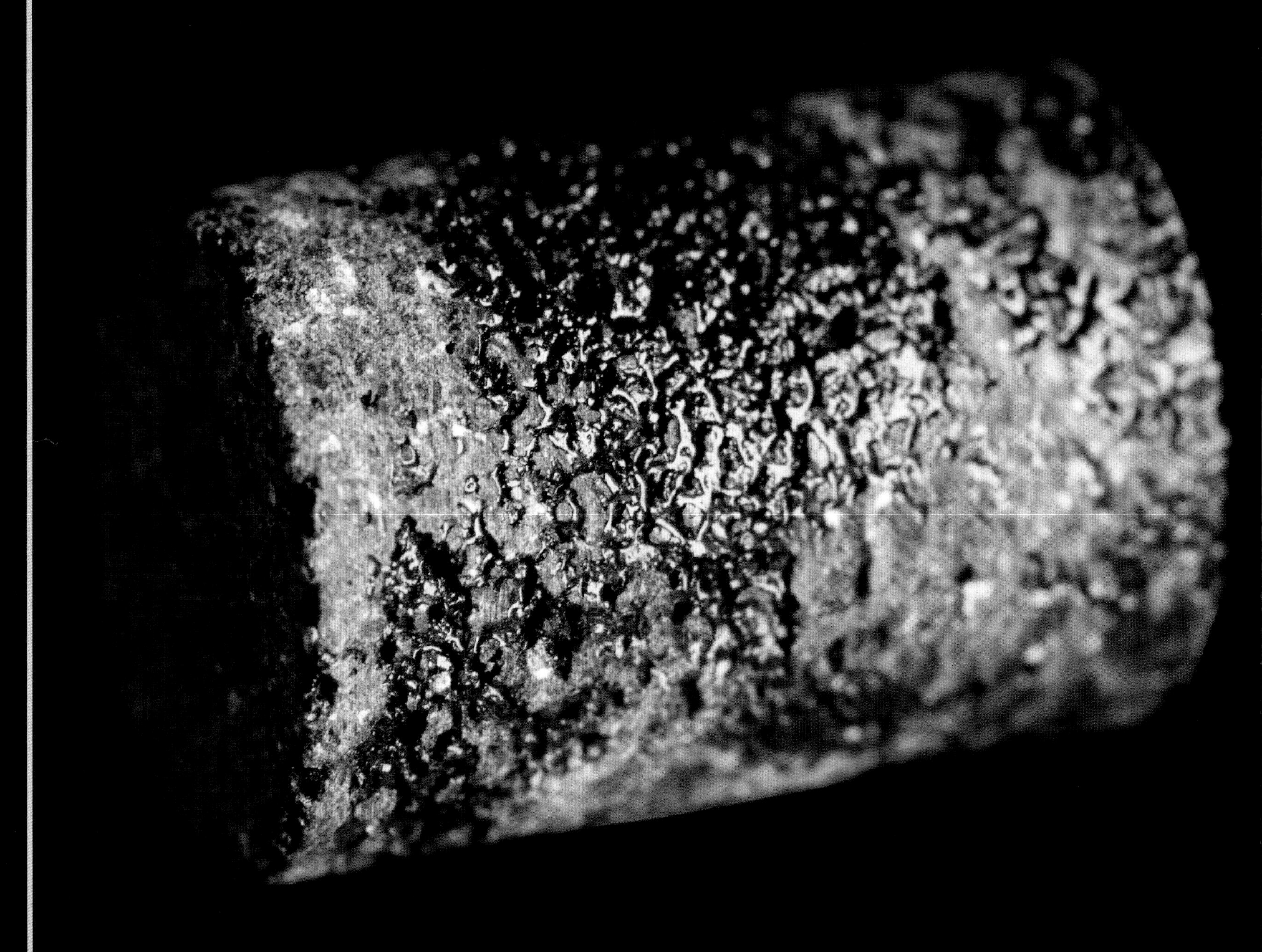

▲ Because this nuclear reactor fuel pellet contains uranium enriched in ^{235}U it is not legal to posses without a license.

Direct-reading radiation detector contains a luminous screen that begins to glow in the presence of dangerous levels of radiation.

Depleted uranium tank penetrator, with the uranium core visible inside the discarding sabot.

Green uranium "vaseline" glass is a popular collectable, and moderately radioactive. Combine vaseline glass and telephone insulators and you hit eBay gold.

A modern spinthariscope made with uranium ore so it is legal to sell.

Red Fiestaware made prior to 1942 is famously radioactive, but other colors and brands are as well.

Neptunium

Np

Neptunium

YOU MAY HAVE NOTICED a trend over the last nine elements: They are all radioactive, but the odd-numbered ones have very short half-lives while the even-numbered ones last much longer, in some cases billions of years. This trend continues through berkelium (97) and is due to the way the protons and neutrons pack themselves in the nucleus. Just as the noble gases are chemically stable because they have just the right number of electrons to form a complete outer shell, the nuclei of even-numbered elements in this range have the right number of protons and neutrons to form advantageous configurations.

Another, if shorter, trend is that elements 92, 93, and 94 are all named after planets. This trend started with uranium (92), named in 1789 after the planet Uranus, discovered eight years before the element. (The fact that uranium was discovered in 1789 is alarming when you consider that the phenomenon of radioactivity was not discovered until 1895, more than a hundred years later. During all that time, people had absolutely no idea that there was something very, very different about uranium—that unlike all other known elements, it could leap out of its container and bite you.)

Neptunium was the first transuranic (i.e., beyond uranium) element to be discovered (in 1940 at the University of California, Berkeley). By convention, uranium is considered the last naturally occurring element, but in fact very tiny amounts of neptunium should exist in uranium-bearing minerals due to nuclear side reactions triggered by uranium's decay.

There are no common applications for neptunium, but you almost certainly have some at home. The standard household smoke detector uses a tiny amount of americium (95) to generate alpha particles whose interaction with smoke particles can be detected. The isotope of americium used, ^{241}Am, has a half-life of 432 years, and its decay product is the ^{237}Np isotope of neptunium, with a much longer half-life of 2,145,500 years. The older your smoke detector is, the more neptunium you'll find has accumulated in it, to the point where after a few thousand years it'll pretty much all be neptunium (a few tens of millions of years after that, it will nearly all be stable thallium, element 81).

Continuing the trend of planet names (or at least used-to-be-planet names, depending on whether Pluto is defined as a planet this week or not), we arrive next at the most potent symbol of death and destruction in the modern age, the grim reaper of the elements, plutonium.

< Aeschynite, $(Y,Ca,Fe,Th)(Ti,Nb)_2(O,OH)_6$, from Molland in Iveland, Norway. Not really any neptunium in it, but it's radioactive and actual neptunium is unobtainable.

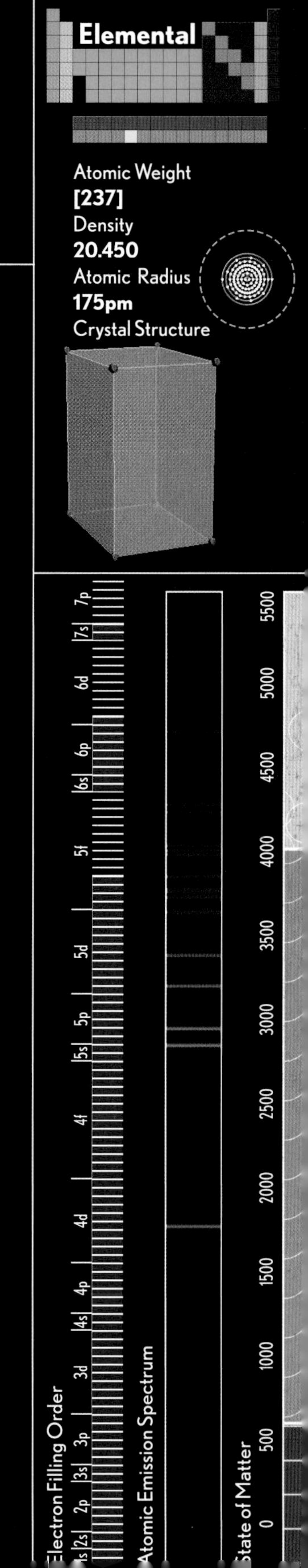

Plutonium

Pu

94

CAUTION
RADIOACTIVE PLUTONIUM -238
LESS THAN 3 CURIES 1973
DO NOT DISCARD. CONTACT
NUCLEAR BATTERY CORP.
COLUMBIA, MARYLAND
DATE OF MANUFACTURE 1973
SERIAL NO. AA-237-8

Plutonium

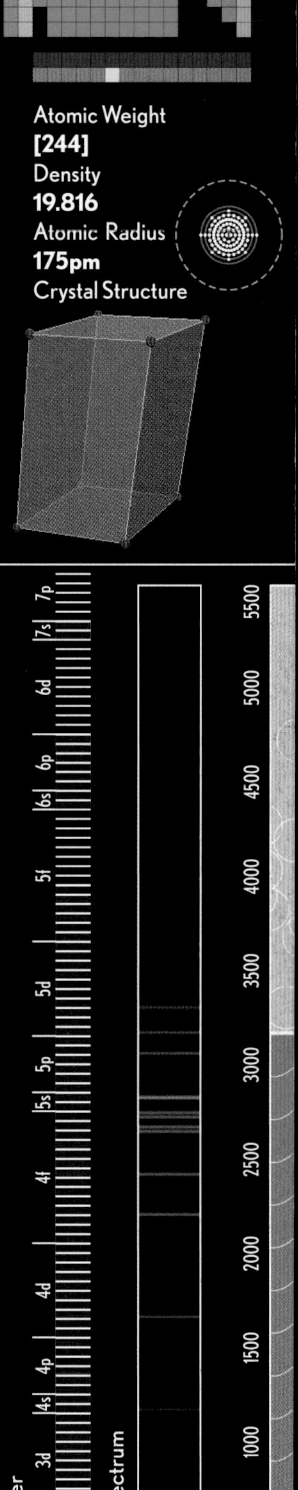

IT'S AN INCREDIBLY FORTUNATE thing that making atomic bombs is so very difficult—if it were any easier, more groups would surely have done it by now.

A uranium (92) atomic bomb is hard to make because the ^{235}U isotope you need is extremely expensive to separate from the much larger amount of useless ^{238}U found in nature. Expensive as in "if you're not the government of an advanced nation, you can't afford it." But if you can get your hands on a critical mass of ^{235}U, it's pretty easy to make a bomb: You essentially just build a cannon that shoots one subcritical lump of uranium into another—and boom.

Making a plutonium bomb might seem easier because it is not so hard to get enough plutonium. Sure, you need a nuclear reactor, but that's child's play compared to isotope separation. (In fact, at least one child has seriously tried: David Hahn constructed a breeder reactor as his Eagle Scout project in 1995 and got into *big* trouble when people realized it wasn't just a model. Some people think his miniature reactor might have actually worked.)

But while it's relatively easy to get plutonium, through an incredibly fortunate coincidence it's extremely difficult to turn it into a bomb. Plutonium splits so much more easily than ^{235}U that if you shoot two subcritical masses toward each other, they start reacting before they touch and blow themselves apart, before even a tiny fraction can fission. This is called a fizzle, and while it might spread radiation over a wide area, it won't melt the target city as intended.

To make a plutonium bomb work, you have to assemble the critical mass by imploding a sphere into itself using explosive "lenses," which have to be nearly perfect. Any asymmetry in the shock wave gives the plutonium a way to sneak out the side. Even today a plutonium fission bomb requires state-of-the-art metallurgy, pyrotechnics, and fabrication technology. Any amateur plutonium bomb would almost certainly fizzle.

Plutonium is often called the most poisonous element. The people at Los Alamos, where nearly all of America's plutonium is kept, were so hurt by this that they published a paper defending plutonium against what they consider its unfair reputation. Well, they would do that, wouldn't they?

What is undisputed is that private ownership of plutonium is absolutely forbidden, with one tiny exception. Pacemakers today use lithium batteries, but a few people, no one knows exactly how many, still have models powered by plutonium thermoelectric batteries. If you have one in you, you're allowed to keep it until you die. I have—and I swear this is true—gotten e-mails from undertakers asking me what to do with the radioactive pacemaker they found in one of their clients. As tempted as I might be to invite them to send me the plutonium for my collection, I have dutifully told each one that by law all plutonium must go home to Los Alamos, where it will be loved and cared for.

While plutonium is perhaps *the* most highly regulated and tracked of all the elements, the same is not true of all synthetic radioactive elements created in nuclear reactors. Take, for example, americium.

This plutonium pacemaker battery case is empty–fortunately. If it were full, possession of it anywhere outside a body would be a crime.

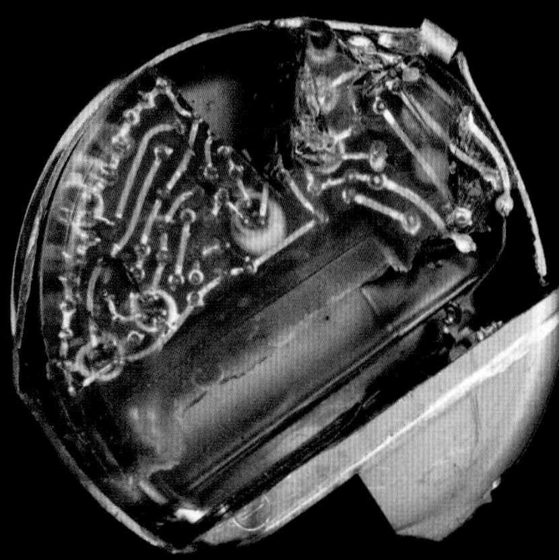

Outside and inside views of a pacemaker into which a plutonium thermoelectric battery (left) would fit.

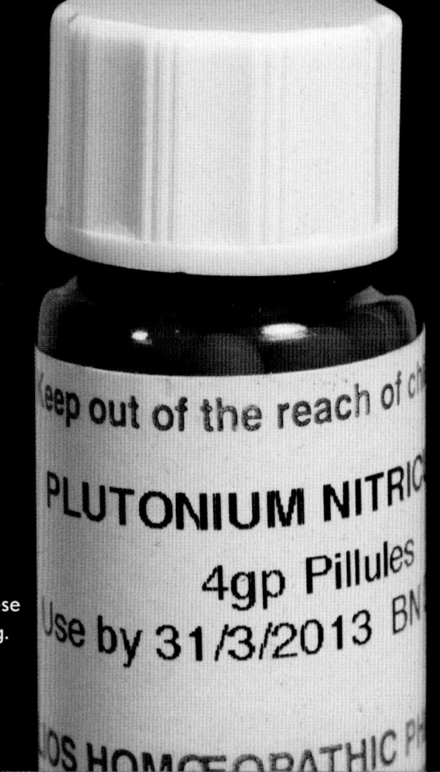

Homeopathic remedies are fraudulent products that contain none of their listed ingredients. In the case of these homeopathic plutonium pills that is a distinctly good thing.

Americium

Am

95

Americium

YOU MIGHT THINK that a synthetic radioactive element that follows plutonium (94)—and has a significantly shorter half-life—would be some kind of superbomb material, available only to scientists in secret laboratories. Perhaps a mad scientist is studying americium in a lair somewhere, but if you want some yourself you can simply walk into any neighborhood hardware store, supermarket, or Wal-Mart and buy some, no questions asked.

The reason is not that americium is fundamentally less dangerous than the elements around it. In fact, the commonly available isotope, ^{241}Am, is significantly *more* radioactive than weapons-grade plutonium, and at least as toxic. No, the difference is simply that there is a useful application for americium that requires only a very tiny amount, and for which a company was prepared to go through the effort required to carve out and get a regulatory exception.

Ionization-type smoke detectors contain a tiny foil button quite similar to the foils in the antistatic brushes discussed under polonium (84), only much smaller. The americium inside releases a steady stream of alpha particles, which travel through an air chamber and are picked up as an electric current on the other side. If even tiny numbers of smoke particles enter the chamber, they interfere with the stream of alpha particles, perturbing the current and triggering the alarm.

Should we worry about radioactive button sources in every home? Smoke detectors of this type are significantly faster than others in responding to common fires, and have no doubt saved many lives. And as with the polonium in antistatic brushes, the americium in smoke-detector buttons is well protected by a layer of gold (79). While not considered a good idea, people have swallowed such buttons with no ill effects: The gold, being a noble metal, withstands attack by stomach acid and allows the button to emerge intact and unscathed. Rejecting smoke detectors because of their americium would be very foolish.

With americium we reach the end of the line for element collectors. It is the very last element that is legal to own without expensive special licenses (which in general are granted only if you can demonstrate a legitimate reason for needing a given element).

Americium also starts a trend that continues all the way to the last of the currently named elements, and most likely will apply to all those yet to be named: The elements from americium on have been named for places or people.

Those honored so far have all been scientists of the highest order, starting with Marie and Pierre Curie.

The circuit board inside an ionization smoke detector. The vented metal can that normally encloses the ionization chamber has been removed so you can see the americium button in place inside.

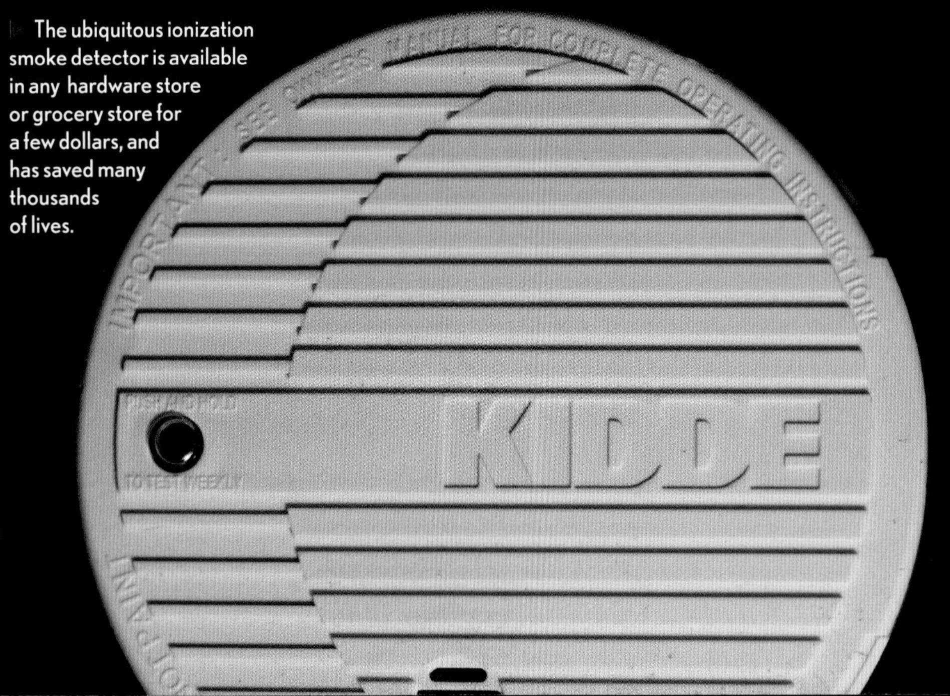

The ubiquitous ionization smoke detector is available in any hardware store or grocery store for a few dollars, and has saved many thousands of lives.

The radioactive americium button from inside a common ionization type smoke detector. Underneath the gold foil is 0.9 micro-Curies of 241Am.

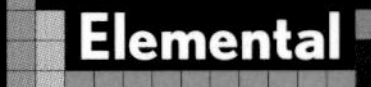

Elemental

Atomic Weight
[243]
Density
N/A
Atomic Radius
175pm

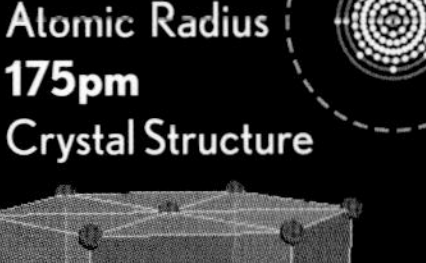

Crystal Structure

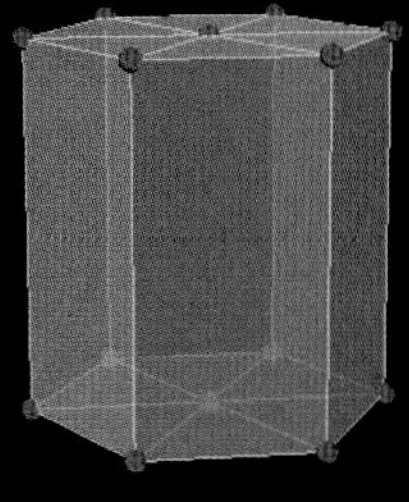

Electron Filling Order: 1s 2s 2p 3s 3p 4s 3d 4p 5s 4d 5p 6s 4f 5d 6p 7s 5f 6d 7p

Atomic Emission Spectrum

State of Matter: 0 500 1000 1500 2000 2500 3000 3500 4000 4500 5000 5500

Curium
Cm
96

Curium

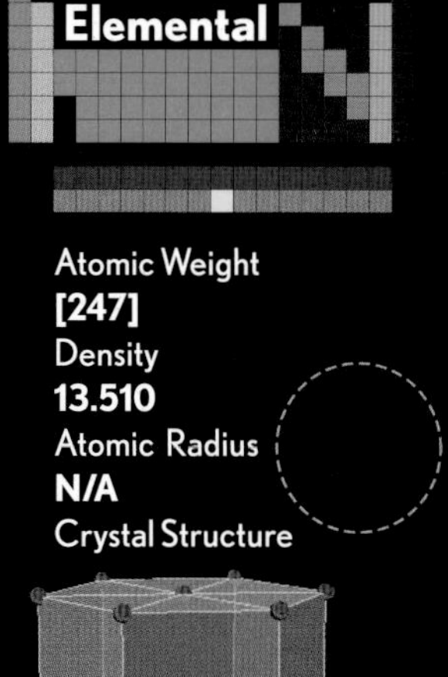

Atomic Weight
[247]
Density
13.510
Atomic Radius
N/A
Crystal Structure

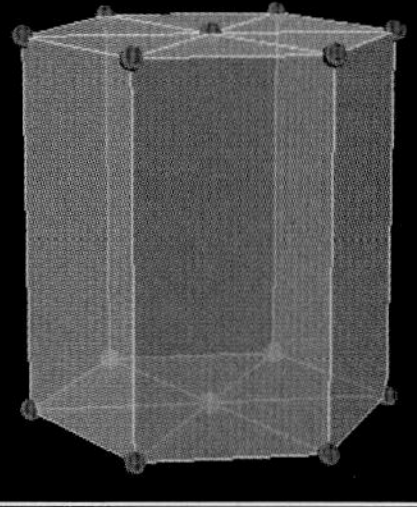

CURIUM, CURIOUSLY, was not discovered by the Curies. The dynamic duo of Marie and Pierre Curie discovered polonium (84) and radium (88), but not curium.

In fact, none of the elements named for people were discovered by those people, with the possible exception of seaborgium (106), depending on how you define "discovered by."

One reason is that it just wouldn't be cricket. Although scientists are just as likely to have very large egos as people in other professions, and some do everything they can to promote themselves, *appearing* to do so is just not something they can do in public. Donald Trump can put his name on his buildings, but any scientist caught trying to put his or her name on an element would be shamed out of the lab. (It would never work anyway: Element names have to be approved by a very tough committee of the International Union of Pure and Applied Chemistry.)

Moreover, the days are long gone when someone like Marie Curie could labor for months in her lab until she had concentrated an unknown substance (radium) down to the point where her beakers and funnels glowed in the dark (and contaminated her lab notebooks and even her cookbook to the point where they have been stored in lead-lined boxes to this day).

Since the advent of the era of "big science," in the wake of World War II's Manhattan Project, no element has been discovered by just one person. These have been joint discoveries, made by teams of dozens of researchers working cooperatively in a handful of large institutions. There would be no way to choose just one person to name the element after.

Curium was discovered by a large team led by Glenn T. Seaborg, Ralph A. James, and Albert Ghiorso, using the 60-inch cyclotron at the University of California, Berkeley. Its only applications relate to its extreme radioactivity: portable alpha-particle sources and so-called RTGs (radioisotope thermoelectric generators), which use the heat generated by radioactive decay to power instruments that must operate for long periods of time away from people and other power sources, such as those on space probes and the like.

If a new element is to be named for a person, the solution (again, with the exception of seaborgium) seems to be to choose important people who are safely dead, like the Curies. New elements have also sometimes been named for the place where they were discovered, which is a bit of a self-promotional loophole. If you're a leading nuclear scientist at the University of California, Berkeley, everyone knows that. If you can get an element to be named, say, berkelium (97), that's pretty much the next best thing to getting it named after yourself. Which is exactly what happened with, well, berkelium.

A medallion commemorates the 100th anniversary of the birth of Marie Curie.

Marie Curie, for whom curium is named.

Berkelium

Bk 97

Berkelium

THE LONGEST-LIVED isotope of berkelium, ^{247}Bk, has a half-life of 1,379 years. What this means is that if you had a one-pound block of berkelium and let it sit for 1,379 years, you would have only half a pound of berkelium left. If you let it sit for another 1,379 years you would have a quarter of a pound of berkelium, and so on.

The berkelium doesn't just vanish—it is transmuting, in place, into americium (95)—specifically the isotope ^{243}Am, which has a half-life of 7,388 years. After ten thousand years or so the block will be mostly americium, but that's just temporary. Even as it's building up, the ^{243}Am is decaying into ^{239}Np, which then very rapidly decays into ^{239}Pu with a half-life of 24,124 years.

After a couple hundred thousand years, most of the ^{239}Pu has decayed into ^{235}U, and there it sits for a very long time because ^{235}U has a half-life of 70 million years. But eventually, after several additional stages of decay, the end result is 5/6 of a pound of stable lead (82), ^{207}Pb.

Where did the other 1/6 pound go? Consider the first decay from ^{247}Bk to ^{243}Am. Americium has two fewer protons than berkelium, and its mass number is four less (243 vs. 247), meaning that two protons and two neutrons were lost. When the ^{247}Bk decayed, it shot two protons and two neutrons out in the form of an alpha particle, accounting for some of the loss of mass. (What physicists call an alpha particle is the nucleus of what chemists call a helium atom.)

Other stages in the decay—for example ^{239}Np to ^{239}Pu—change the element number (the number of protons) but not the mass number. Since the mass number doesn't change you might think that a ^{239}Pu atom weighs the same as a ^{239}Np atom, but this is not the case. In fact the ^{239}Pu is very slightly lighter—the extra mass in the ^{239}Np has been converted directly into energy according to Einstein's famous formula, $E=mc^2$ (in words, "energy equals mass times the speed of light squared"). The speed of light, c, is a very big number, meaning that a small amount of mass converts into a huge amount of energy.

So the answer is that the missing 1/6 pound has turned into a combination of helium (2) (from the emitted alpha particles) and pure energy. (And in practice, that energy means you'd never be able to keep an actual pound of berkelium on your desk, it would be far too dangerous.)

Virtually no practical applications have been found for berkelium. But, surprisingly for such a high-numbered element, there are actually a few real applications for californium.

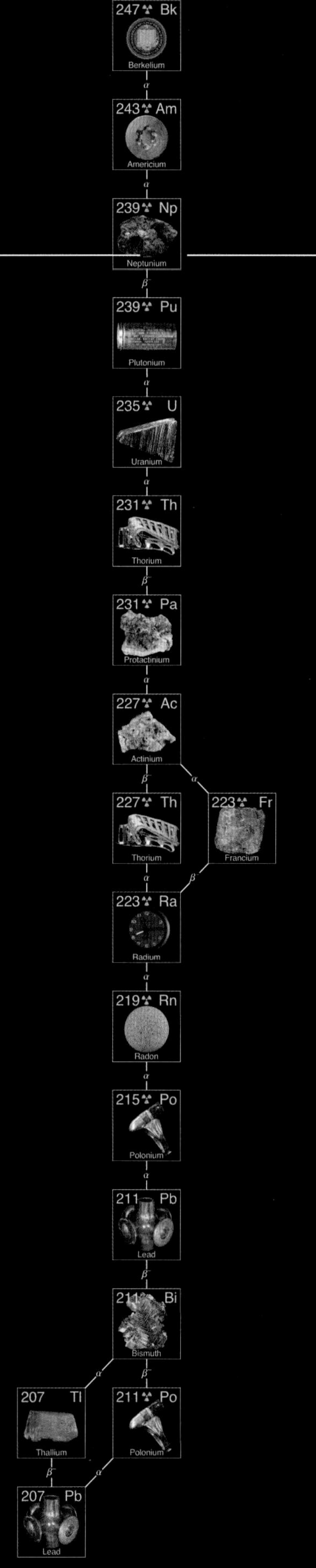

The decay chain of ^{247}Bk, described in detail in the text. In most cases a given isotope decays almost entirely into one new isotope, but sometimes there is more than one possible decay pathway. Shown here are those paths that occur at least 1% of the time. The chain stops when it reaches a stable element, in this case almost all the material ultimately ends up as the lead isotope ^{207}Pb. Yes, this is transmutation of the elements just like the alchemists dreamed of, only more expensive.

The great seal of the University of California at Berkeley, where Glen Seaborg discovered berkelium and many other elements.

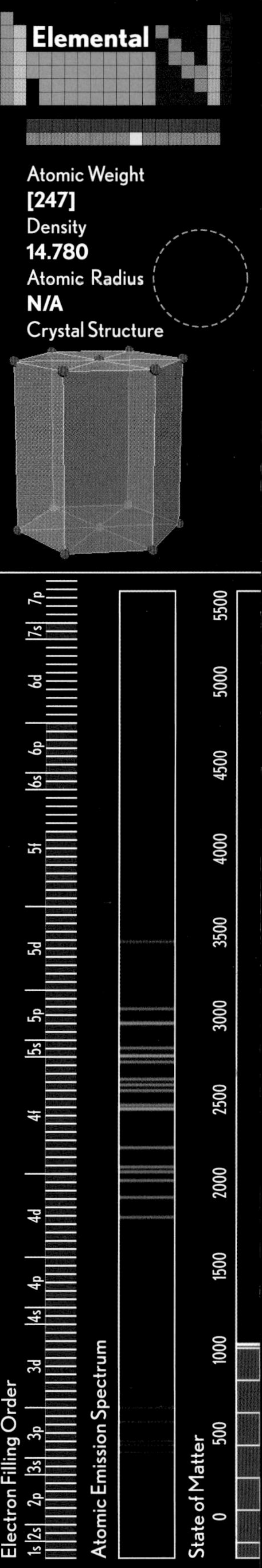

Californium

Cf 98

Californium

GLENN SEABORG is a name you run into a lot around this part of the periodic table. He is on the list of those credited with the discovery of californium, and also plutonium (94), americium (95), curium (96), berkelium (97), einsteinium (99), fermium (100), mendelevium (101), nobelium (102), and seaborgium (106).

The last one is of particular note because it is the only really unambiguous case of an element named not only for a person involved in its discovery, but also for a person who was not dead yet. This was so controversial that the 1997 agreement to allow it was reached only after a horse trade in which Seaborg's colleagues at the University of California, Berkeley, allowed their archrival, the Joint Institute for Nuclear Research in Dubna, Russia, to pick the name for element 105, which both groups claimed to have found first. And that is how it came to be that we now have both seaborgium and dubnium (105)—though it is said that, to this day, some people at Berkeley refuse to call 105 by its official name.

The naming of einsteinium and fermium might have been similarly controversial but for the fact that Cold War secrecy kept their discovery, and their proposed names, secret for so long that both Albert Einstein and Enrico Fermi had conveniently, if unfortunately, died by the time the existence of their elements could be revealed.

I promised application examples for californium, the last element that has any whatsoever. Californium is an extremely powerful neutron emitter, a property that makes it both extraordinarily dangerous and uniquely useful.

Of all the forms of radioactivity, the most dangerous is neutron emission.

◀ The great seal of the state of California, for which californium is named.

Because neutrons carry no charge, they are not repelled by either negatively charged electrons or positively charged protons. This allows them to pass through solid matter with relative ease. If they should happen to strike a nucleus, they are able to infiltrate and destabilize it. A beam of neutrons thus has the alarming property of converting perfectly ordinary matter into radioactive isotopes: Exposed to neutrons, you yourself become radioactive (with a half-life of 15 hours, primarily from the sodium, element 11, isotope ^{24}Na).

What makes neutron irradiation useful is that when a given element is turned radioactive and then decays, the type and energy level of the resulting radiation is highly characteristic of that particular element. For example, if you beam neutrons at a piece of rock and get back gamma rays of a particular energy, you can say for sure that there is gold (79) inside that rock.

This technique is called neutron activation analysis, and besides detecting gold it can identify oil in the bottom of a well, or explosives inside a shipping container or suitcase without having to open either. Neutrons can see straight through the solid steel hull of a ship. What californium does is provide a convenient, very small, very portable source of a whale of a lot of neutrons, one that can be deployed in portable inspection instruments such as those that are lowered down oil wells.

Now say good-bye to any form of usefulness. After californium, it is safe to say that the people and places for which the remaining elements are named are far more important, and more interesting, than the elements themselves. The most spectacular example of this principle comes in the case of einsteinium.

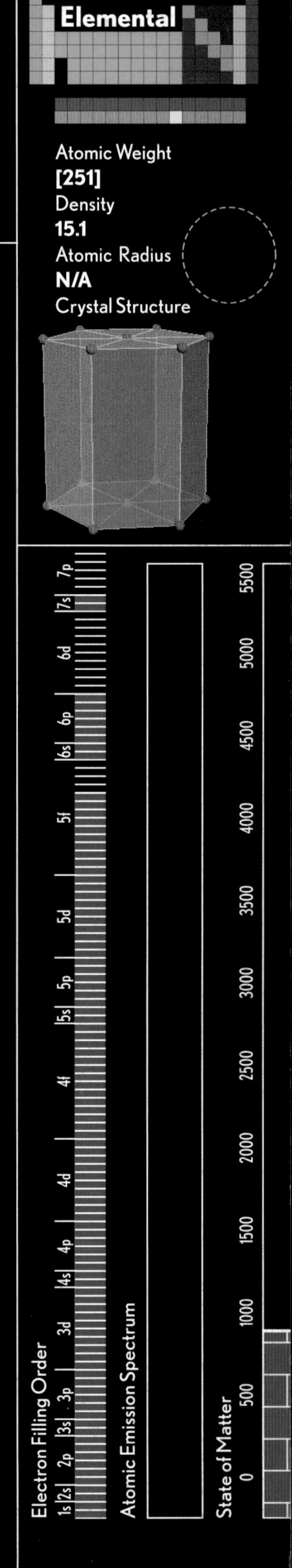

Einsteinium

Es 99

Einsteinium

GETTING AN ELEMENT named after yourself is not easy. Winning a Nobel Prize is trivial in comparison—there are more than 800 Nobel laureates, with more added every year, but just a handful of people are or can ever be honored with elements. Einstein, however, was a shoo-in. He was the most famous scientist in history while he was alive, and half a century after his death, he still has a Hollywood agent managing his image.

Everyone knows Einstein, but few know that he sent the single most-important letter of the 20th century, possibly the most important letter ever written. Even fewer know that this letter wasn't his idea, and he didn't actually write most of it. It was the letter that led to the atomic bomb.

Nuclear fission happens when a large atomic nucleus, say the nucleus of an atom of uranium (92), splits apart into two lighter nuclei. Sometimes this happens spontaneously, but if a neutron hits the right kind of nucleus, it can trigger the split immediately. When a nucleus fissions, a large amount of energy is released, but so is something else: one or more neutrons.

The "or more" part is what caused Leó Szilárd to have a sudden and profound vision of a dark future as he stepped off the curb onto Southampton Row in London on September 12, 1933. What he realized was that if somebody could build a device where one fissioning atom releases two neutrons that go on to strike and fission two more atoms, releasing four neutrons that fission four more atoms, then eight, then sixteen—mankind would have found a path straight to hell.

The simplest calculations showed that if you could actually initiate and sustain a nuclear chain reaction, the amount of energy released would be so much larger than anything yet experienced on earth that it would be difficult even to conceive of what could be done with it. Unfortunately, after the experience of World War I, Szilárd was pretty sure he wasn't going to like it.

Quite soon, Szilárd realized two things. First, that something very, very bad was brewing in German society, and second, that many of the best nuclear physicists were working in Germany. The only thing more horrifying to him than the thought of a nuclear chain reaction used in war was the thought that Nazi Germany could very possibly do it first.

He made a fateful decision to write a letter to President Roosevelt warning him that America needed to build whatever it was that could be built, and do it before the Germans. But who was he to write such a letter?

And so it came about that Albert Einstein put his name on a letter drafted for him by Leó Szilárd and arranged to have it personally handed to Franklin D. Roosevelt by a trusted friend. Five years, eleven months, and fourteen days later, the nuclear device named Trinity lit the sky over the deserts of Alamogordo.

The Germans never got anywhere close. For one thing, the scientists working in Germany so completely bungled their corresponding attempt to get the attention of top leaders that their bomb remained a low-budget university project. For another, the Nazi insistence on Aryan racial purity meant that people like Enrico Fermi simply went to work for the other guy.

◁ Albert Einstein, the most famous scientist of all time and thus a fitting person to name an element after.

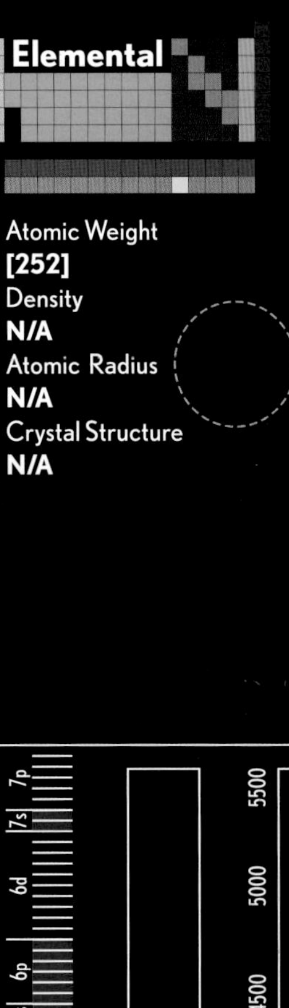

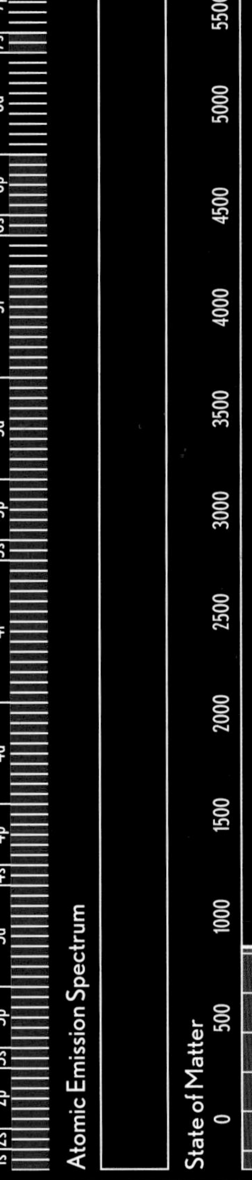

Fermium

Fm 100

Fermium

Elemental

Atomic Weight
[257]
Density
N/A
Atomic Radius
N/A
Crystal Structure
N/A

EVERY FIELD HAS its legends, told and retold until they take on an air of mythology. One is how Enrico Fermi created the first sustained nuclear chain reaction in a racquets court under Stagg Field at the University of Chicago. His Chicago Pile One, or CP-1, went critical on December 2, 1942, at 3:25 p.m.

As described under einsteinium (99), a nuclear chain reaction starts when a neutron hits and splits a heavy atom, releasing more neutrons, which go on to split more atoms, and so on. However, a number of factors stand between this simple arithmetic and an actual sustained chain reaction in a block of uranium (92).

Neutrons from uranium fission shoot out at a very high rate of speed, but uranium atoms are efficiently split only by neutrons traveling much more slowly. Furthermore, unless the block of uranium is very large, the neutrons are likely to exit it before they hit anything.

So while each uranium fission releases two or three neutrons, most of those neutrons do not lead to further fission, and the effective ratio of neutron production is much less than 1:1. In order to increase the ratio, you need to use either tons of uranium, or a particularly susceptible isotope, or you need to slow down the neutrons using what is known as a moderator. Or some combination thereof.

Fermi assembled a large rectangular pile (called a "pile") of several tons of uranium and uranium oxide alternated with blocks of high-purity graphite, an excellent neutron moderator. His careful calculations indicated that once the pile was finished, the neutron-production ratio would be larger than one, and the pile would be capable of an exponentially increasing chain reaction. Even if Fermi's experiment had not been located in the middle of a densely populated city, this potential needed to be kept under very careful control. The pile was designed with a set of "control rods" made of cadmium (48), which strongly absorbs neutrons. With the rods inserted into the pile, the cadmium captured enough neutrons to keep the neutron production ratio below one.

It was a tense few hours as Fermi's team slowly pulled the control rods out on that December day, carefully monitoring the neutron count coming out of the pile and testing again and again that the thing did indeed shut down when they pushed the rods back in. About the only thing they didn't actually test was the guy with an ax whose job it was to chop through the rope holding up the last-ditch set of emergency control rods.

The pile reached a ratio of 1.0006 at 3.25 p.m. and operated for 28 minutes, generating about half a watt of power. It wasn't much, but it was enough for Enrico Fermi's name to live forever in the legends of atomic power.

Of course, none of this has anything to do with the element fermium, which (just like the remaining eighteen elements) has no applications.

Enrico Fermi, for whom fermium is named.

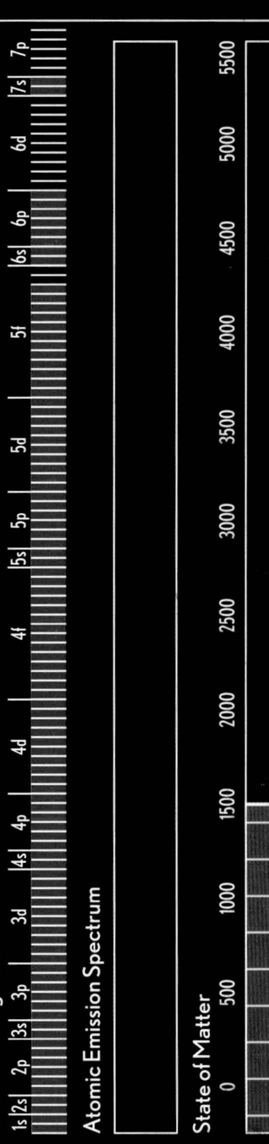

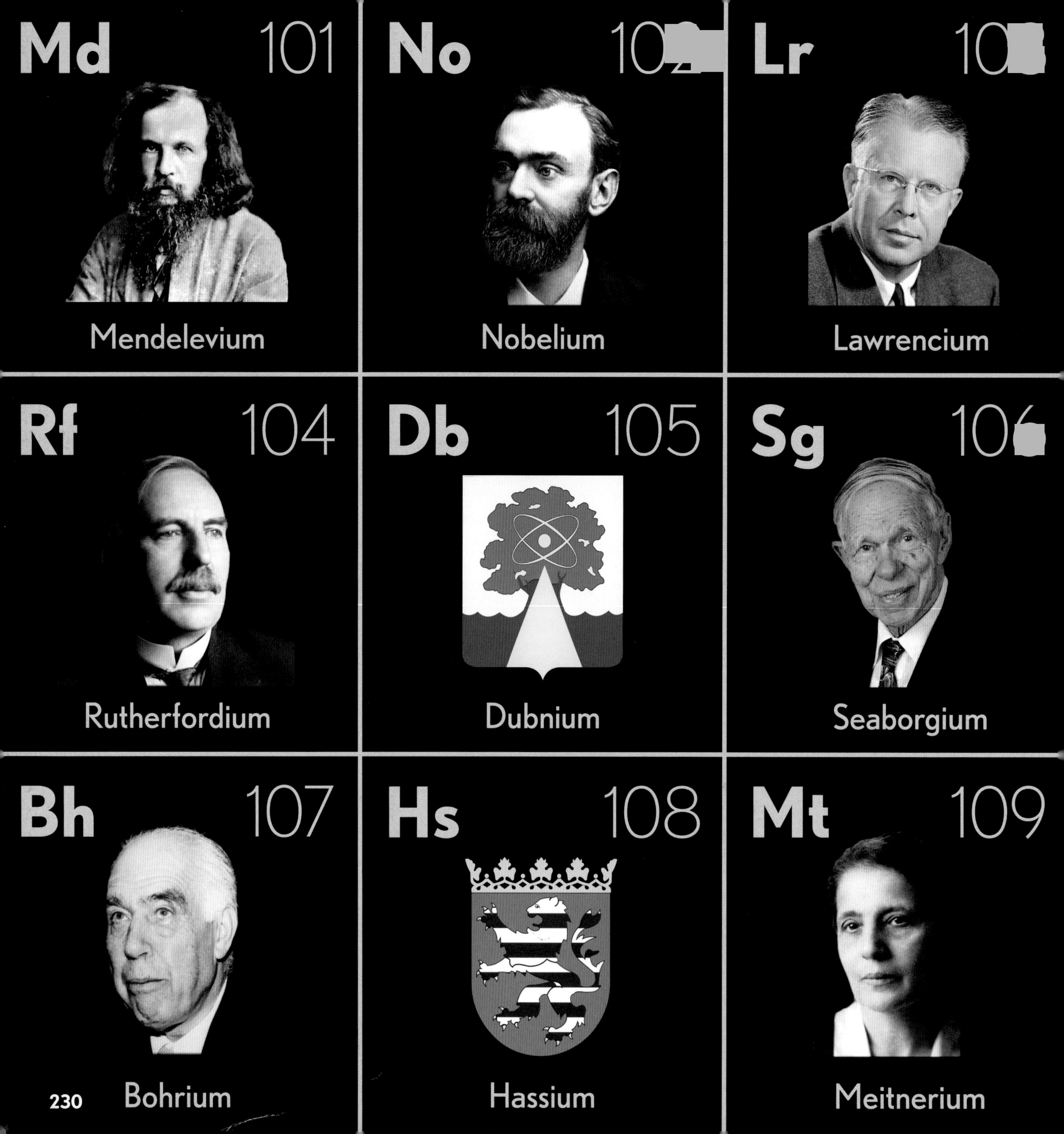
Md
101
Mendelevium
No
10
Nobelium
Lr
10
Lawrencium
Rf
104
Rutherfordium
Db
105
Dubnium
Sg
10
Seaborgium
Bh
107
Bohrium
Hs
108
Hassium
Mt
109
Meitnerium

ELEMENTS 101 THROUGH 109 are the range of elements over which things degenerate from "it has no applications but at least a visible amount has been created" to "you can list exactly how many atoms have been created and when."

By the time we reach meitnerium, we're talking about less than two dozen atoms, total. Nuclei in this part of the table are becoming too big and too unwieldy to hang together for more than a few hours. The longest-lived is mendelevium, with a half-life of 74 days, but the next longest is only 19 hours, for rutherfordium. The shortest is 43 minutes, for Lise Meitner's consolation element.

Most of the people honored up here in the transuranics have won Nobel Prizes, but not all of them. Dmitri Mendeleev did not win a Nobel Prize because they didn't yet exist when he invented the periodic table. Alfred Nobel didn't win a Nobel Prize because he invented Nobel Prizes. And Lise Meitner didn't win a Nobel Prize mainly because she was a woman.

But Meitner gets the last laugh. Many people believe she deserved to share, with Otto Hahn, the 1944 Nobel Prize in Physics for the discovery of nuclear fission, but a Nobel Prize is a cheap trinket in comparison to the distinction of having an element named in your honor. Because "hahnium" was once a serious contender for element 105's name, the rules say it can never again be used as an element name. Lise is in, but Otto is shut out for good.

Back in 1944 Lise Meitner wasn't hard to find, but when the Nobel Committee awarded the prize to Hahn alone, they had no idea where he was. They put out a plea that anyone who could find him should please inform the Nobel Committee so they could give him his prize. (What they didn't know is that Hahn had been captured by the Allies in the last days of the war in Europe and was being held in secret confinement along with a group of other top German nuclear scientists in a remote British house called Farm Hall. When a reporter got wind of the search for Hahn, he peeked over the perimeter wall and may have caught a glimpse of Werner Heisenberg exercising naked in the garden. Or maybe he didn't.)

As mentioned under californium (98), dubnium and seaborgium got their names after a lot of arguing. Lawrencium, on the other hand, was a natural choice since Ernest Lawrence built the first working cyclotrons, the machines used to discover many of the new elements in this range. Ernest Rutherford goes back even further: He first discovered that elements had nuclei. Niels Bohr in turn showed how the buildup of the periodic table could be understood in terms of electron orbitals.

That leaves only hassium to explain: Its name comes from the name of the German state of Hesse, where it was discovered, making it the German equivalent of californium. For the German equivalent of berkelium (97), we need look no further than darmstadtium.

Ds 110
Darmstadtium
Rg 111
Roentgenium
Cn 112
Copernicium
Uut 113
Ununtrium
Uuq 114
Ununquadium
Uup 115
Ununpentium
Uuh 116
Ununhexium
Uus 117
Ununseptium
Uuo 118
Ununoctium

WE HAVE NOW REACHED a range of elements about which you can safely say that, even though they have all been discovered, they don't actually exist. By which I mean: No atoms of any of them are known to exist on earth, unless as you are reading this someone happens to have their heavy-ion research accelerator switched on, trying to make some.

Darmstadtium is, of course, named for the German city of Darmstadt, home of the Gesellschaft für Schwerionenforschung (Institute for Heavy Ion Research).

Wilhelm Conrad Röntgen discovered x-rays, which makes it somewhat ironic that his element does not emit x-rays when it decays moments after being created.

Copernicium, discovered in 1996, but not officially named until 2010, has the distinction of being the only element other than Nobelium to be named for someone who didn't have much to do with chemistry or nuclear physics. Nicolaus Copernicus' main claim to the name, aside from being a great astronomer, seems to be that, like the scientists who discovered the element, Copernicus was German.

The names of all the remaining elements are, at the time of this writing, temporary placeholder names built from the digits in their atomic numbers, expressed via a mixture of Greek and Latin stubs for those digits.

Element 116, for example, is un-un-hex-ium, from the Latin *uno* for one and the *Greek hex* for six. While the mix of Greek and Latin in the word "television" can be attributed to sheer anti-intellectualism, in the case of element names it was done to arrange it so that none of the digits start with the same letter, allowing those elements' three-letter symbols to be constructed in a systematic way; for example, Uuh for *u*n*u*n*h*exium.

There is no fundamental reason for the elements to stop at 118. That's just the last one that fits into the standard arrangement of the periodic table. Since none higher has been discovered, there's no reason to add a whole new row. Yet.

Theoretical calculations indicate that there may be an "island of stability" around 120 (unbinilium) or 122 (unbibium). Those elements are not likely to be *stable*, but they might have significantly longer half-lives.

At the time of this writing, elements 113 through 118 have been discovered. Most recently, six atoms of Ununseptium (117) were reported in April 2010 by a joint Russian-American team working at a nuclear research facility at Dubna, Russia. Actually giving these elements names could take longer than it did to discover them: Each has competing claims of priority, and no one is going to agree to anything until every last argument has been heard by the naming committee.

And so it is that we come to the end of our journey through the periodic table not with a bang, but with a committee.

The Joy of Element Collecting

I STARTED COLLECTING elements in 2002 and thought that in 30 years I might have most of them. Thanks in large part to eBay—and my own insanity—by 2009, I had assembled nearly 2,300 objects representing every element, the possession of which is not forbidden by the laws of physics or the laws of man. You have seen many of these precious items in this book.

To quote ABBA, "What a joy, what a life, what a chance!" OK, maybe being an international pop star is more exciting than the life of an element collector, but it has its moments.

I particularly enjoy finding oddball elements in unexpected places. Who would have thought that you can find very pure niobium (41) in very impure piercing shops—the kind that make you feel like disinfecting yourself after you leave. Or that Wal-Mart sells simple rectangular blocks of pure magnesium (12) metal, about as plain as can be, with no other function than to be used to demonstrate the fact that magnesium is a flammable metal. (They sell them in the camping section: You shave off a bit with a hunting knife, then use the attached flint to light the shavings and thus your campfire.)

Some elements can be experienced in large quantities, like the 135-pound iron (26) ball I keep in my office for people to trip over. Others are best enjoyed in responsible moderation—keep too much uranium (92) in the office, and people start asking questions (keep over 15 pounds, and the Feds start asking questions).

Element collecting isn't a big hobby. Compared to the number of people who collect chemical compounds (minerals), polymers (Beanie Babies), or the same darn metals over and over again (coins), we element nuts are few and far between. Part of the reason is that even just storing your collection safely requires significant chemical knowledge. Sodium + damp basement = bang. But if you're willing to learn the ins and outs of each unique element, collecting them can be a tremendously rewarding experience.

See you all at periodictable.com, where I do my collecting and you can join the fun!

The author in his element(s). Shown on top of the world-famous wooden periodic table table are a small fraction of the author's nearly 2,300 samples of the elements and their applications.

New York: Oxford University Press, 2006.

Emsley, John.
Nature's Building Blocks: An A–Z Guide to the Elements.
New York: Oxford University Press, 2003.

Emsley, John.
The 13th Element: The Sordid Tale of Murder, Fire, and Phosphorus.
New York: John Wiley & Sons, 2000.

Eric Scerri.
The Periodic Table: Its Story and Its Significance.
New York: Oxford University Press, 2007.

Eric Scerri.
Selected Papers on the Periodic Table.
London: Imperial College Press, 2009.

Frame, Paul, and William M. Kolb.
Living with Radiation: The First Hundred Years.
Self-published, 1996.

Gray, Theodore W.
Theo Gray's Mad Science: Experiments You Can Do at Home—But Probably Shouldn't
New York: Black Dog & Leventhal Publishers, 2009.

Rhodes, Richard.
The Making of the Atomic Bomb.
New York: Simon & Schuster, 1995.

Sacks, Oliver.
Uncle Tungsten: Memories of a Chemical Boyhood.
New York: Vintage Books, 2002.

Silverstein, Ken.
The Radioactive Boy Scout: The True Story of a Boy and His Backyard Nuclear Reactor.
New York: Random House, 2004.

Sutcliff, W. G., et. al.
A Perspective on the Dangers of Plutonium.
Livermore, CA: Lawrence Livermore National Laboratory, 1995.

Acknowledgements

I ORIGINALLY HIRED Nick Mann as a lackey to sweep up gadolinium (64) dust around the shop, but he quickly rose to the position of sidekick, filming poison gases and automatic penny sanding machines for the Popular Science articles that went into my book Mad Science.

With The Elements, through hard work and tremendous skill in the photography of elements, and the fact that he agreed to have no social life for three months straight as the deadline approached, Nick has risen to the lofty position of coauthor.

The majority of the photographs in this book, while "from the studio of Theodore Gray," were in fact lit and shot by Nick. If not for his hard work, skill, and dedication, you would be reading this book next year as opposed to now.

These photos were then lovingly and expertly assembled by the book's designer, Matthew Cokeley, who together with my editor, Becky Koh, should be given an award for patience in dealing with authors who want to change everything even as the trucks are leaving the printing plant. Thanks to Hiroki Tada! for carefully cleaning up the well over 500 photographs in this book, so you can see only the true appearance of these lovely elements, not all the flaws and lies introduced by the camera. And, thanks to Nino Cutic for generating excellently scaled atomic emission spectra for all the elements for which data is available.

David Eisenman carefully edited every word in the book and is responsible for many factual and stylistic improvements. Being edited by David is a rigorous experience not unlike a tax audit, but in a good way.

Max Whitby has been my steady partner in the element business for many years. Together we have built a veritable empire of elements, of which this book is but one manifestation. His many valuable comments and suggestions on the manuscript are much appreciated.

Timothy Brumleve provided expert commentary on the more technical aspects of the chemistry of the elements, particularly the rare earths, of which his knowledge is unique.

Paul Frame provided advice and stories about radioactive quack medical products, and very kindly allowed us to photograph in his unique museum of radioactivity at the Oak Ridge National Laboratory. His coauthor, William Kolb, has also provided years of advice and guidance in all things radioactive.

John Emsley and Eric Scerri, two of the world's preeminent authorities on the elements, have been tremendously generous in providing advice and support for this project.

And now we come to the question of thanking contributors of individual facts about the elements. There are far too many to list, but here is a sample of the kind of issue that comes up when writing a book about the elements: In trying to figure out which mineral to use to represent technetium (43), Blaise Truesdell pointed out that technetium was found in African pitchblende in 1962, and could therefore be considered naturally occurring, even though it traditionally is not. Chris Kanter, on the other hand, tasted all the alkali metal chloride salts and lived to tell us which one tastes best (sodium chloride). Add another few hundred more of these and you've got a start at thanking everyone who's given input to the book, to all of whom I apologize for omitting them.

The samples in this book come from an incredibly diverse range of people, from random eBay sellers to distinguished professors to fellow element collectors, again far too many to list in print. Fortunately, that information is carefully cataloged in detail on my website, periodictable.com. Every photo in this book is there too, listed under the appropriate element and tagged with a link to the source. In quite a few cases, this would allow you to get one of your own, if you like.

Of course, I must thank my parents for my existence, but particularly also for a small crystal of rhodochrosite you will find listed under manganese (25), which was so coveted by the mineral dealer Simone Citon that, with the permission of my father, I traded that one crystal for most of the rest you see in this book. The remainder of the minerals come primarily from Sarah Kennedy at Jensan Scientifics. Both Sarah and Simone provided invaluable advice in deciding which mineral to use to represent a particular element.

And, finally, I must thank my family, Jane, Addie, Connor, and Emma (in order of height—though if Addie keeps growing, maybe not in order much longer), for putting up with the process of creating this book. I promise, no more books until next time.

Index